Folkert de Jong

Marine Eutrophication in Perspective

Folkert de Jong

Marine Eutrophication in Perspective

On the Relevance of Ecology for Environmental Policy

With 20 Figures

Springer

DR. FOLKERT DE JONG
International Wadden Sea Secretariat
CWSS
Virchowstr. 1
26382 Wilhelmshaven
Deutschland

e-mail:
dejong@waddensea-secretariat.org

Library of Congress Control Number: 2006923640

ISBN-10 3-540-33647-8 Springer Berlin Heidelberg New York
ISBN-13 978-3-540-33647-1 Springer Berlin Heidelberg New York

Springer is a part of Springer Science+Business Media
springer.com

Printed in The Netherlands

Cover design: Erich Kirchner
Typesetting: camera-ready by author
Production: Christine Jacobi
Printing: Krips bv, Meppel
Binding: Stürtz AG, Würzburg

Printed on acid-free paper 30/2133/cj 5 4 3 2 1 0

Preface and acknowledgements

The idea for this book, in which the results of a Ph.D.-study, carried out at the University of Groningen (NL), are presented, has its roots in my personal experiences with international marine pollution policies, in which I became professionally involved in the mid 1980s. At that time, marine eutrophication was high on the political agenda in Northwest Europe, and it was exciting to be part of the intensive political activities that happened between 1985 and 1990. After that, I became intrigued by the very rapidly decreasing political interest in the issue, while scientific and management activities continued at a high level of intensity. This confronted me with questions about the compatibility of the political and scientific processes and the relevance of ecology for environmental policy. These themes, as well as the fascinating history of marine eutrophication, are what this study is about.

I am much indebted to my supervisors Prof. Dr. Wim Wolff, Prof. Dr. Franciscus Colijn and Dr. Henny van der Windt, who provided me with valuable ideas and suggestions for both the contents and the structure of this book and, most of all, for their stimulating support all along the way. I am also grateful to Dr. Wolfgang Hickel for his extensive comments on first drafts of the manuscript.

I would furthermore like to thank the members of the review committee, Prof. Dr. Rutger Rosenberg, Prof. Dr. Victor de Jonge and Prof. Dr. Michiel Herweijer, for critically reading the final manuscript.

Special thanks go to Jens Brøgger Jensen, Dr. Gerhard Cadée, Prof. Dr. Victor de Jonge, Dr. Winfried Gieskes and Dr. Karel Essink, with whom I have held personal interviews.

I would also like to thank Dr. Justus van Beusekom, who has taught me a lot about the science of marine eutrophication, as well as Prof. Dr. Karsten Reise for providing useful advice at the start of the project.

What would a scientific study be without libraries? I am very grateful for the excellent support by the librarians of

Biological Centre, Haren;
University of Groningen;
University of Kiel;
Scripps Institute, La Jolla;
NIOZ, Texel;
ICES, Copenhagen;
Terramare, Wilhelmshaven;
Senckenberg Institute, Wilhelmshaven;

Bundesforschungsanstalt für Fischerei, Hamburg;
Bundesanstalt für Seeschifffart und Hydrography, Hamburg;
Environment Ministry, Oslo;
Secretariat Helsinki Commission, Helsinki;
English Nature, Peterborough.

Special thanks go furthermore to Roger Staves, Prof. Dr. Mike Elliott and Nicola Ricklefs for correcting my English, and to Gerold Lüerßen for his support in drawing maps and figures.

Finally, I thank my wife Tessy and my sons Daniel and Jonathan for their patience and their conviction that one day the work would really be finalised.

Acronyms

ACE	Advisory Committee on Ecosystems (ICES)
ACME	Advisory Committee on the Marine Enionment (ICES)
ACMP	Advisory Committee on Marine Pollution (ICES)
ASMO	Assessment and Monitoring Committee (OSPAR)
BAH	Helgoland Marine Station
BMP	Baltic Monitoring Programme (Helcom)
BOD	Biochemical Oxygen Demand
C	Carbon
CEC	Commission of the European Communities (European Commission)
CFC	ChloroFluoroCarbons
CPR	Continuous Plankton Recorder
DIN	Dissolved Inorganic Nitrogen
DIP	Dissolved Inorganic Phosphorous
DPEUT	Drafting Panel on Eutrophication (OSPAR)
DSP	Diarrhetic Shellfish Poisoning
EC	European Community
EcoQO	Ecological Quality Objective
EEC	European Economic Community
EEZ	Exclusive Economic Zone
EQO	Environmental Quality Objective
EU	European Union
EUC	Eutrophication Committee (OSPAR)
FAO	Food and Agricultural Organisation (UN)
GESAMP	Group of Experts on the Scientific Aspects of Marine Pollution (UN)
HAB	Harmful Algal Blooms
Helcom	Helsinki Commission
ICES	International Council for the Exploration of the Sea
IGO	Intergovernmental Organisation
IMM	Intermediate North Sea Ministerial Meeting
INSC-1	1st International North Sea Conference (Bremen, 1984)
INSC-2	2nd International North Sea Conference (London, 1987)
INSC-3	3rd International North Sea Conference (The Hague, 1990)
INSC-4	4th International North Sea Conference (Copenhagen, 1995)
INSC-5	5th International North Sea Conference (Bergen, 2002)
IOC	International Oceanographic Committee (UN)

IPCC	Intergovernmental Panel on Climate Change
JAMP	Joint Assessment and Monitoring Programme (OSPAR)
JMG	Joint Monitoring Group (OSPAR)
JMP	Joint Monitoring Programme (OSPAR)
MCWG	Marine Chemistry Working Group (ICES)
MMP	Monitoring Master Plan
MPB	Marine Pollution Bulletin
N	Nitrogen
NAO[I]	North Atlantic Oscillation [Index]
NAEP	National Agency for Environment Protection (DK)
NERC	National Environment Research Council (UK)
NERI	National Environment Research Institute (DK)
NEUT	Nutrients and Eutrophication Committee (OSPAR)
NIOZ	Nederlands Institute for Sea Research
NSTF	North Sea Task Force (OSPAR and ICES)
NUT	Nutrient Working Group (OSPAR)
OECD	Organisation for Economic Cooperation and Development
Oscom	Oslo Commission
OSPAR	concerning the Oslo and Paris Conventions (before 1992) or the OSPAR Convention (as of 1992)
Osparcom	Oslo and Paris Commissions (before 1992) or OSPAR Commission (as of 1992)
P	Phosphorus
p.e.	population equivalent
Parcom	Paris Commission
PCB	Polychlorinated Biphenyl
PRAM	Programmes and Measures Committee (OSPAR)
PSP	Paralytic Shellfish Poisoning
PWG	Policy Working Group (INSC-2) Preparatory Working Group (INSC-3)
QSR	Quality Status Report
RIZA	Governmental Institute for Water Purification (NL)
SACSA	Standing Advisory Committee on Scientific Affairs (Oscom)
Si	Silicon
SIME	Substances in the Marine Environment working group (OSPAR)
STWG	Scientific and Technical Working Group (INSC-2)
TWG	Technical Working Group (Parcom)
UES	Uniform Emission Standard
UK	United Kingdom
UN	United Nations
UNCHE	United Nations Conference on the Human Environment
US[A]	United States [of America]
USEPA	US Environmental Protection Agency
WFD	Water Framework Directive (EU)
WHO	World Health Oganisation
WRc	Water Research Centre (UK)

Contents

1 Introduction: Ecology for policy

"In recent years it has become impossible to talk about man's relation to nature without referring to 'ecology'. Such leading scientists as Rachel Carson, Barry Commoner, Eugene Odum, Paul Ehrlich and others have become our new Delphic voices [....] So influential has their branch of science become that our time might well be called the 'Age of Ecology'." (Worster 1994)

The above quote is from the preface to the second edition of Donald Worster's "Nature's Economy." The "Age of Ecology" refers to the period that started after World War II and that has lasted until today. Indeed, with the rise of environmental problems, biology, and more specifically ecology, has become indispensable in environmental policies. The role of ecologists in the post World War II society is best known for discovering environmental problems, with Rachel Carson's "Silent Spring" from 1962 as a classical example. Increasingly important, however, became the notion of ecologists as solvers of environmental problems. The latter role rapidly developed in the second half of the 1960s and the first half of the 1970s, as a result of strong ambitions on the side of ecologists and high expectations of authorities. In the first edition to Krebs' textbook on ecology (Krebs 1972) two reasons for studying ecology were given, namely to increase one's understanding of the world we live in and to provide a basis for practical action on environmental problems.

But not only practical action was considered necessary. A need was felt to use science as a basis for decision-making and management. According to Küppers et al. (1978), an increasing number of problems had an interdisciplinary character, and administrations could no longer rely on "accidentally applicable knowledge," but had a need for a systematic longer term knowledge, developed in accordance with defined problems. Nelkin (1987) attributed the vast growth of scientists employed by the administration to the increasing complexity of policy decisions, and the use of science as a source of authority by which consensus in public affairs could be reached: "Scientific standards have a universal appeal as an authoritative basis of rational decision making." According to Jasanoff (1990), regulatory agencies developed in the early 20th century and were mainly dealing with fact finding. The tasks of these agencies became increasingly complex. From the beginning of the 1970s new scientific duties emerged, such as sponsoring basic research, conducting inspections, performing risk assessments and developing analytical methodologies. At first the agencies could not cope with these tasks. For these reasons a scientific basis for public policy had to be developed.

The call for science as a basis for decision-making and management is still topical, as was illustrated by Gro Harlem Brundtland (1997) who stated:

"In ocean management, as in most other areas of human endeavour, close co-operation between scientists and politicians is the only way to move forward. Science must underpin our policies. If we compromise on scientific facts and evidence, repairing nature will be enormously costly, if possible at all."

It is clear from the above that there were, and still are, high expectations to science to support public policies. But what then is and what has so far been the impact of science? At first sight, the influence of ecology and ecologists on public policy is clearly visible in the changes that have occurred in environmental policies in the past decades. Environmental policies have become ecologised and are developing from sectoral (pollution, species protection) to so-called integrated ecosystem policies (see for example De Jong 1994). At the same time (ecological) science has become politicised, meaning that more scientists have become involved in decision-making, for example as civil servants (see among others Nelkin 1987; van der Windt 1992; De la Mothe and Dufour 1995), and scientific research has become more policy applied.

But has science indeed contributed to solving and managing environmental problems? Already in 1975 Nelkin described the rise of public expectations towards ecologists in view of increased awareness of environmental problems, and the corresponding problems ecologists were faced with when trying to live up to these expectations (Nelkin 1975). De la Mothe and Dufour (1995), in a more recent commentary in Nature, were very critical about the supposed capacities of science: "The scientific community [....] has for decades promised the public and politicians far more than it could deliver. The 'endless' frontier of science has not managed to translate itself into an 'endless solution'." The tension between science and politics is not only about the delivery of solutions, but also about the differences in attitude between scientists and politicians, as is illustrated by the following example: According to a member of the UK House of Lords' subcommittee on fisheries, "scientists should say clearly that there are certain visible trends, which, if allowed to continue, will lead to catastrophe. I feel if they had said this earlier, it would have had an impact on policy-makers" (Masood 1996). In contrast, a quote is presented by fisheries ecologist Niels Daan in the Dutch weekly magazine "Vrij Nederland": "I feel that it is a threat to scientific research that we are forced more and more to present hard statements. Also when it is not possible" (Van Wijnen 1995).

The central question of this book is whether ecology has indeed contributed to solving environmental problems and, if yes, to what extent and in which way. This question will be investigated in detail for the case of marine eutrophication in the North Sea and the Northeast Atlantic Ocean. Already in the 1950s marine ecologists had recognised marine eutrophication, the loading of the marine environment with phosphorus and nitrogen compounds, as a potential pollution problem. The issue started to achieve world-wide attention in the 1970s, be it predominantly from the side of marine ecology. The 1980s were the decade of political action to combat marine eutrophication in the North Sea, the Baltic Sea and the Northeast

Atlantic Ocean. Scientists connected serious oxygen depletion events in the Danish Belt Seas, the Kattegat and the German Bight with excess loads of nutrients from the mainland to the sea, caused by human activities. National and international political action followed, resulting in 1987, at the second International Conference on the Protection of the North Sea (London 1987), in the agreement between North Sea states to reduce by 1995 inputs of nitrogen and phosphorus compounds to the North Sea by 50%. However, this was required only for discharges of nutrients into areas "where these substances may cause pollution." Such areas would, as decided at the third North Sea Conference (The Hague 1990), have to be determined on the basis of scientific research. By doing so, politics had laid a heavy burden upon the scientific community and, consequently, marine eutrophication research had become substantially politicised.

1.1 Science for policy

Before embarking upon the analysis of the role of marine ecology in marine eutrophication policy, it is necessary to address in more detail why science is considered potentially beneficial for policy-making, and in which way science may contribute to the policy process. This will provide the theoretical framework for the analyses, and allow a more precise formulation of the main questions to be addressed. In the following sections two relevant aspects of science for policy will be addressed:

1. The necessity of science for decision-making;
2. The use of science in the different phases of the policy process.

1.1.1 Rational policy-making

What are the presumed capacities which make science so suitable to make a positive contribution to public policies? In the social studies of science the concept "rational policy-making" or "rational decision-making and management" is used to describe policies and management for which science is regarded a necessary condition (Brooks 1987; Nowotny 1987; Underdal 1990; Jasanoff 1990). Underdal (1989) concluded that there is general consensus among decision-makers and scientists that theoretical understanding of cause-effect relationships, as well as relevant descriptive information, are necessary conditions for rational management. Knowledge is both a tool for diagnosing an environmental problem and for prescribing remedial action. Underdal (loc.cit.) stated that science is not the only provider of knowledge, but "the more technical the measurement required, the more complex and less transparent the cause-effect relationships and the more stable the dynamics of the system studied, the greater seems to be the comparative advantage of systematic research over more impressionistic modes of generating knowledge." According to Underdal (loc.cit.), this has two implications:

1. Decision-makers will turn to science for advice on complex questions. Generally, science will not be able to give precise answers on short term (as usually required). From this it can be concluded that the claim of research to play a role in management is primarily based on what is described above, and not on its ability to provide instant answers to questions from policy makers.
2. Natural sciences seem to have an advantage over social sciences, although they are in principle not more relevant. Underdal stated: "the advantage of natural sciences and technology stems largely from their likely ability to make a greater marginal contribution in terms of advancing further beyond the informed judgement of the decision makers themselves or the immediate experience of the general public."

Young (1989) concluded that the scientific community is not only effective in the identification of unknown problems, but that it also "can play a significant role in the pursuit of compliance at the international level by operating as an organised interest group. This community is unusually, perhaps uniquely, transnational in character." Another advantage is that it is well organised. For these reasons "the scientific community is capable not only of transcending the parochial concerns of individual states, but also of bringing pressures to bear on national governments that exceed the pressures more localised groups can muster" (Young, loc.cit.).

1.1.2 The policy life-cycle

Different possible contributions of science to the policy process were mentioned above, but in order to analyse the role of science in the different phases of the policy process, it is first necessary to describe this process in more detail. A simple model of the policy process was developed by Winsemius (1986), on the basis of his experiences as Dutch Minister of Environment. He proposed the so-called policy life-cycle, an amended version of which is in Fig. 1.1.

The policy life-cycle consists of three phases: In the first phase, the discovery phase, the issue is recognised as a (potential) problem. In this phase there is high uncertainty and controversy regarding both the seriousness of the issue, and the need for and dimensions of abatement policies. In this phase the problem is not yet a political issue. In the political or decision-making phase the issue is placed on the political agenda, and negotiations about possible solutions to the problem will start, followed by political decisions. At the beginning of this phase there is still considerable controversy, which gradually decreases as more information becomes available. In phase three, the management phase, political decisions are implemented and management instruments developed and applied. This phase is, furthermore, characterised by a decrease in political interest and a further decrease in uncertainty.

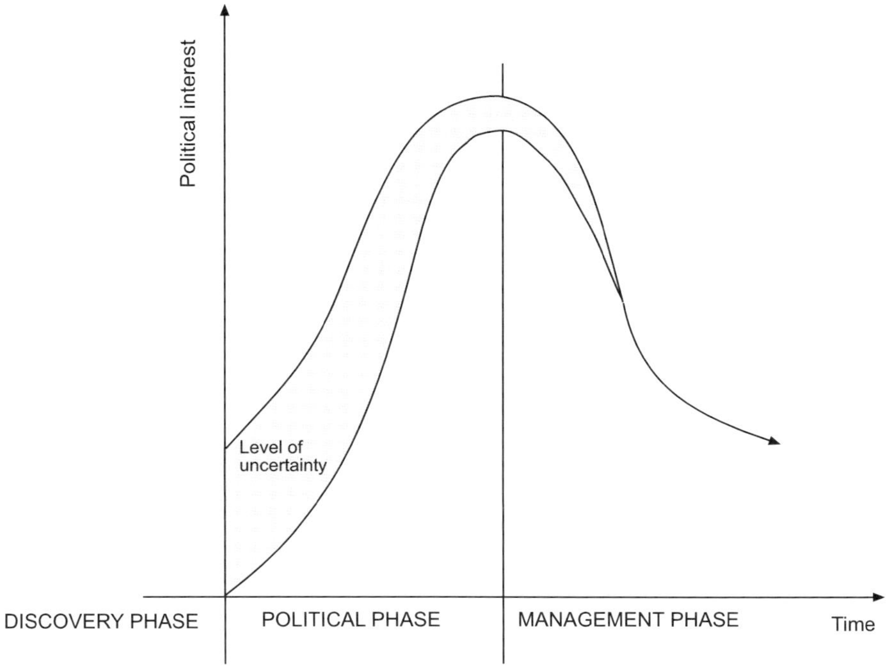

Fig. 1.1. The policy life-cycle. Modified from Winsemius (1986)

How can science best contribute to policy-making in the different phases of the policy life-cycle? Hannigan (1995), who regards environmental problems as social constructions, has identified three elements, the assembly, the presentation and the contesting of the problem. For the first element, which includes the discovery, the problem definition and the establishment of the main parameters, science is considered the central forum. The mass media play a dominant role in the presentation of the problem, and politics are mainly responsible for invoking action, mobilizing support and defending ownership (Hannigan, loc.cit.). As explained above (1.1.1), the rational policy-making model assumes an important contribution by science to political decision-making. Wettestad and Andresen (1990) have suggested that in the phase of political negotiations crude knowledge may be sufficient, but that in the following phase of implementation and compliance, i.e. the management phase, a further development of knowledge is necessary for the fine-tuning of policies. This is confirmed by the wish to gain more knowledge, which is part of most political agreements. Winsemius (1986) has pointed to the fact that in the decision-making phase measures usually have a crude character, and should first of all be effective. In the subsequent phase of solving the problem there will be increasing emphasis on efficiency.

In summary, the following five main goals may be served by science:

1. Assembly of the problem, i.e. the discovery, problem formulation and establishment of main parameters;
2. Providing the scientific basis for political decisions;
3. Reduction of uncertainty as a basis for justification of political decisions. An essential feature of the Winsemius model is that the uncertainty declines in the course of the process (figure 1.1). Lambright (1995) has underlined the importance of reducing uncertainty and concluded that an important factor in the success of the CFC case (the depletion of the ozone layer by chlorofluorocarbons [CFCs]) had been the speedy (2-3 years) narrowing of scientific uncertainty;
4. Specification (fine-tuning) of political decisions;
5. Development of management instruments necessary for the compliance with the new policies, i.e. monitoring, prediction and assessment.

1.2 Science for policy: a myth?

Although the above assessment of the role of science in policy, especially that of international science in international policies, was generally judged positively, other studies have come to much more negative conclusions. Several analysts contend that the impact of scientific knowledge on public policies is limited. According to Engelhardt and Caplan (1987), controversies over environmental issues often have the character of scientific controversies, although in many cases the "non-scientific" factors are the most important ones. They called these, scientific controversies with an overlay. This overlay can be ethical, cultural, economical or otherwise. Boehmer-Christiansen (1989) emphasised the importance of economic factors in decision-making regarding pollution control:

"the economic and technological consequences of proposed environmental regulations, as well as the capacity and willingness of political institutions to respond to their concerns, all contribute to the making of environmental policy at the national level and thus affect international decisions as well. Governments are rarely fully in control of the issues raised and domestic policies thus becomes a part of international negotiations."

When there are differences between countries regarding environmental issues, economic and technological consequences of pollution abatement become decisive factors in international negotiations. Boehmer-Christiansen (1987) concluded that in order to assess the role of science, first political priorities, ideological commitments, political stability (especially with regard to pressures from interest groups) and the connection with the "environmental learning process" of specific governments must be known, as well as the economic and technological consequences of specific abatement techniques.

However, apart from the above non-scientific factors, there are difficulties within the scientific community itself, as well as principal differences between the scientific and political processes, which hamper the use of science in policy. Because of these differences, several authors even distinguish between science carried out within the regular academic settings (academic science) and science for policy, i.e. scientific research with the purpose of generating information for the policy process (Brooks 1987; Nowotny 1987; Jasanoff 1990). Jasanoff (1990) has termed the latter type "Regulatory Science." The contextual factors which hamper the use of science for policy purposes, and which are covered in more detail below, relate to consensus building, dealing with uncertainty, dealing with values, as well as differences in time frames between the scientific and political processes.

1.2.1 Contextual factors

Consensus

Miles (1989) suggested that the most important factor in the application of knowledge is whether or not the knowledge is consensual. Young (1989) compared the CFC and acid rain cases with regard to the success of the scientific community in influencing the political agenda, and concluded that key factors were whether scientists could reach consensus amongst themselves – which is foreign to scientists, who "seek competition as a method of advancing knowledge" – and whether the scientific community could overcome "the natural tendency [....] to exhibit extreme caution in the interest of avoiding any appearance of overstating the inferences to be drawn from available evidence." Boehmer-Christiansen (1989) acknowledged that the scientific community has an "intrinsic power to encourage co-operation even among political opponents," but also concluded: "science itself, because of its own internal propensity for conflict, can only make a limited contribution to the resolution of international conflicts over public choices." Collingridge and Reeve (1986) underlined that modest critique is an important feature of the development of science. They argued that heavy critique is counterproductive to scientific progress and, in fact, not necessary because the so-called "error-costs" are low. In other words, not much harm will be done if "false" conjectures remain unchallenged. However, as soon as science is used in policy-making with high stakes, the level of critique will increase, and a scientific debate will start with a high level of controversy. Such scientific debates may continue for long periods of time, and will thus not be helpful in political decision-making.

Collingridge and Reeve (1986) identified a second factor responsible for increased scientific debate, namely the fact that political questions are generally not confined to one scientific discipline. Such was also found by Andresen (1989), according to whom the scientific community is characterised by uncertainty, disagreement, caution and difficulty of simple communication, in particular when the number of disciplines increases. This also includes the social sciences, especially in cases of resource management. Each discipline has its own basic values which do not necessarily coincide with those of other disciplines.

Complexity and uncertainty

According to Underdal (1989), uncertain knowledge in international negotiations has the disadvantage that those, in favour of new regulations, will have to prove their case. For opponents it is far easier to come with substantive critique than it is for the advocate to reduce or remove the uncertainty. This is especially so for science to be used in public policy ("regulatory science"), which often works "at the margins of existing knowledge, whereas academic science works within established paradigms" (Jasanoff 1990). Complexity and uncertainty are especially relevant for large-scale environmental issues. The apparent complexity of ecosystems has been a major obstacle in the development of general ecosystem theories and, consequently, understanding the system and predicting developments (compare Peters 1991; Edwards et al. 1994; Macgarvin 1995; Sagoff 2003).

Another problem with complexity of ecosystems is the management of information. Collingridge and Reeve (1986) questioned the ability of policy makers to manage and integrate the huge amount of often controversial expert information, generated by scientific research. One of the myths of rational decision-making is, according to these authors, the assumption that it should be based on the availability of full information. They used in this respect the term "synoptic rationality."

Dealing with values

Science mainly deals with facts, whereas politics mainly deals with value judgements and conflict solving. The question poses itself whether and to what extent value judgements and the resolution of conflicts can be improved by the use of science. Nowotny (1987) discussed the principal question whether scientific proof can be used to solve conflicts. She has worded this as follows:

> "It is worthwhile to recall the great appeal that the scientific method once commanded as a way of settling disputes, and the futile hope that was expressed again and again, in scientific and political utopias alike, that it would be possible to arrive at similar rational procedures for solving conflicts in the political realm."

Engelhardt and Caplan (1987) have compared different ways of closing disputes. They concluded that many controversies have important ethical or political aspects and cannot be solved by using "sound" arguments. In such cases negotiation is the only way of closing the dispute.

Differences in time frames

Time is a critical factor in regulatory science because decisions must often be taken before a consensus has been formed about the acceptability of evidence (Jasanoff 1990). Lambright (1995) used the term "readiness of knowledge" in the question whether or not science is prematurely introduced into policy deliberations. According to Lambright, the issue of readiness of knowledge is perhaps most visible in the environmental field. He concluded from an evaluation of the CFC case that the speeding up of the communication of available science had worked well, but "science by press conference [....] can also burn the provider and policy user if the information proves faulty." The latter was underlined by Jasanoff (1990), according to whom "ripeness of knowledge" is an important factor in reducing scientific controversy.

The ecologist Likens (1992) complained that sponsored studies must emerge at a specified time and usually in a specified format. According to Likens (loc.cit.), "these requirements mean that the process is fatally flawed from a scientific point of view. It is not possible to buy full and complete answers to complex environmental problems within a specified period of time." Also the customary quality review process in peer-reviewed journals is, for reasons of time, amounts and nature of the contents, "largely bypassed." "The result can lead to poor scientific communication as applied to complex environmental problems" (Likens, loc.cit.).

1.2.2 Science in the policy life-cycle

The next question is what will happen with science, once it has entered the policy process. The policy life-cycle model, proposed by Winsemius, is simple and straightforward, and assumes a regular course of events from discovery to solving the problem, making use of science to reduce uncertainty. But science may not always function in accordance with this model. Hischemöller et al. (1998) and Groenewegen et al. (1998) have questioned the linearity of the policy cycle. They concluded that environmental problems change in structure in the course of time, but that this change is not necessarily from unstructured to structured and from conflict to consensus, as assumed in the Winsemius model. Problems may also develop from structured to unstructured, depending, among others, on the emergence of new knowledge or changes in societal perception. According to Jasanoff (1990) scientific advice is not a "one-shot process". In complex decision-making, which takes often many years, there will be multiple rounds of consultation, whereby, as a result of the emergence of new knowledge (often as a result of purposeful scientific activity), a constant redefinition of the state of knowledge is necessary. Jasanoff (loc.cit.) stated that such changes must be taken into account in any comprehensive account of science policy. Central questions are which role parties play in the redefinition of the state of knowledge, and to what extent a change in the state of knowledge can be a reason for a change of policies. That new knowledge not necessarily contributes to the expectations of policy makers was underlined by Miles (1989), who contended that the outcome of scientific research may be a "wild card" in the policy process, in other words, an unknown variable. Likens (1992) argued that environmental problems are not necessarily solved by the delivery of appropriate science. Other scenarios are that a problem remains unsolved because it either does not become a political issue, or because the public and/or politics lose interest.

1.3 Matching science and policy

In the past decades, several solutions to overcome the above problems have been proposed and tested. In the literature dealing with the interaction between science and politics, an often heard solution for handling the incompatibility between science and politics is improved communication (Timberlake 1989; Andresen 1989; Underdal 1989; Wettestad and Andresen 1990; Porrit 1993). Communication problems are caused by the contextual factors listed above, as well as the "jargon" used by scientists, the "ivory tower" attitude of scientists and, most important, the inability of scientist to differentiate between policy-relevant and science-relevant facts. On the side of politics there is the inability to formulate proper scientific questions, and ignorance about the scientific methodology. For an effective communication between science and politics, including the translation of scientific knowledge into politically usable knowledge, specific mediators or mediating bodies are considered necessary (Andresen 1989; Timberlake 1989; Wettestad and Andresen 1990; Lambright 1995). The North Sea Task Force (NSTF), which will be addressed in detail in this study, and the Intergovernmental Panel on Climate Change (IPCC), are examples of mediating bodies established at the international level.

According to Lambright (1995), the principal actors in the interaction between science and politics are researchers, politicians and managers. Normally, that is under normal research conditions, these actors are distant. But in the case of policy-relevant science, that is when science is needed in the policy process, either advocated by scientists (science driven) or by politicians (policy pulled), the relationships intensify and the science-policy connectivity is enhanced. Increased science-policy connectivity is accompanied by the formation of institutions for the communication between science and policy. Hoogerwerf and Herweijer (2003) have defined a policy network as a group of actors, which develop sustained interaction and communication patterns, directed at solving certain policy problems. In this study I will use the term "science-policy network" for the network consisting of the scientific community, the political community and mediating bodies. The latter will be referred to as the "science-policy interface."

The most important functions of the science-policy interface are the translation of scientific knowledge into political language and vice-versa, and the closing of (scientific) controversies through negotiation. The persons working at the science-policy interface should, according to Timberlake (1989), be "as conversant with theory as the research scientist," but must also have a good understanding of the bureaucratic process. Generally, the participants in the science-policy interface are "civil servant scientists," i.e. civil servants with a scientific background. Jasanoff (1990) has analysed the Science Advisory Board (SAB) of the US Environmental Protection Agency (USEPA), the membership of which has developed into repeated assignments and informal interest balancing "into the special subculture of regulatory science rather than short rotation of 'top-flight research scientists'." This was, according to Jasanoff (loc.cit.), done under the authority of USEPA, which was aware of the fact that the position of "neutral" experts can generally be predicted in advance. It is, therefore, essential to have a balancing of different points of view from a diversity of backgrounds and interests. Board members are, generally, selected on broad scientific expertise, rather than specific knowledge. Especially in cases where there is insufficient knowledge, the members must be able to use subjective judgement, a quality which very specialised scientists usually do not have (Jasanoff, loc.cit.).

1.4 Aims and structure of the study

The aim of this study is twofold. First, to provide a comprehensive account of the history of marine eutrophication, comprising its discovery, the construction of the issue as a political problem, political decision-making and management. Second, to analyse the role of science in these different phases. The material provided in the following chapters is intended to serve both purposes, but is should be noted that in some cases the descriptive material is more elaborate than strictly necessary for the second aim.

The analysis of the role of science in decision-making and management with regard to marine eutrophication focuses on three aspects, relevant for the interaction between science and policy, the normative, the structural and the temporal aspect.

The normative aspect

The starting point is the model of rational decision-making and management. According to this model science is a necessary basis for decision-making and management. In the foregoing, several problems with the application of this model have been addressed, which are expected to be relevant for marine eutrophication as well. Marine eutrophication is a highly complicated issue. This complexity is related to the size, the openness and the dynamics of the marine ecosystem. Moreover, nutrients are an essential natural feature of the marine ecosystem, and it is therefore hard to separate man-induced from natural causes. For these reasons, marine eutrophication is a very scientific environmental problem, which may explain the strong demand from politics to science to help structure and manage the problem. Because marine eutrophication is an international issue, it is also possible to investigate the role of the international scientific community in policy-making. Finally, policies to combat marine eutrophication are about reducing nutrient inputs, demanding large investments in sewage treatment and agricultural practices. This requires justification, both in terms of management efficiency and expected improvements in the marine environment.

The central issue addressed in the analysis is whether and how science has contributed to decision-making and management. This query is the starting point for a more detailed analysis, which will focus on the following questions:

A. How has the notion of rational decision-making with regard to marine pollution and marine eutrophication developed over time?

B. What has been the impact of science on the policy process? This question can be further specified as follows:

B.1 What has been the role of ecology in the construction of the marine eutrophication problem?

B.2 Has ecology been used as a basis for political decision-making?

B.3 Has ecology been used as a basis for justifying decisions?

B.4 To what extent has new knowledge, i.e. knowledge that has become available after decisions have been taken, influenced the political status quo.

B.5 Has ecology contributed to the fine-tuning of decisions and the elaboration of management instruments, i.e. monitoring, prediction, assessment and validation?

C. Have contextual factors influenced the impact of science? These are:
C.1 Complexity and uncertainty;
C.2 Consensus within the scientific community;
C.3 Different time frames;
C.4 Dealing with values.

The structural aspect

The interaction between science and politics takes place within certain structures, in this study referred to as the science-policy network. Of particular relevance is the so-called science-policy interface, in which the communication between science and politics is facilitated. The main structural question for the marine eutrophication case is:

D. What has been the role of the science-policy network for the use of science in the policy process? More in particular, the following questions will be addressed:
D.1 Which structures have been developed for the interaction between science and policy?
D.2 Have there been changes in these structures and if yes why and how?
D.3 How have these structures, in particular the science-policy interface, functioned with regard to the use of science in policy making?

The temporal aspect

The policy life-cycle (figure 1.1) will be used as the temporal framework for the analysis. The main questions addressed with regard to the temporal aspect are:

E. Have developments with regard to marine eutrophication in the North Sea and the North East Atlantic Ocean followed a pattern of discovery, decision-making and management, and what has been the role of science in these different phases?

Outline

The general structure of this book reflects the temporal aspect, in accordance with the three phases of the policy life-cycle: the discovery phase, the political or decision-making phase and the management phase. The discovery of marine eutrophication is covered in Chap. 3, in which an overview of the development of marine eutrophication during the period 1950-1980 is given, including the main scientific issues at stake, the scientific relevance of marine eutrophication and the assessment of the severity, as well as the political awareness of this particular problem. Chapter 3 is relevant for the analysis of questions A, B1, C1, C2

Chapter 4, "The politics of marine eutrophication," covers the period 1980 to 1990 and focuses on marine eutrophication in the North Sea. The analysis in Chap. 4 focuses on the role of ecology in the agenda-setting and political decision-making with regard to marine eutrophication. The information presented and the analyses carried out in this chapter relate to questions A, B, C, D and E, with particular emphasis on A, B1 and B2.

In Chap. 5 the international management of marine eutrophication is described. The focus is on the North Sea and the Northeast Atlantic Ocean. The period covered is 1990 – 2005. The analysis in this chapter is about the use of ecology for the fine-tuning of political decisions and the development of management instruments. Questions addressed are B, C, D and E with an emphasis on B3, B4, B5 and D.

Because marine eutrophication is a special field within marine pollution, its political and management developments are rooted in those of marine pollution in general. This is the reason why in Chap. 2 a general introduction to the development of marine pollution is given, together with a description of the emergence of a marine pollution science-policy network. Chapter 2 focuses on questions A and D.

In the final Chap. 6 the main findings from Chaps. 2 to 5 are summarized and discussed from the perspective of the whole policy life-cycle, and compared with other international cases of environmental policy. This chapter also presents and analyses alternatives to the rational policy-making model.

Material

The material used in this study consists of scientific literature and reports of official meetings, supplemented with a small number of interviews with key persons. The scientific sources have, as much as possible, been limited to proceedings of major international conferences and key scientific review articles. This is for two reasons. The first is a practical one. In the past 50 years the scientific literature related to marine pollution and marine eutrophication has grown exponentially. A description of some 50 years of scientific developments is only possible through the selective use of aggregated scientific material. Secondly, and very relevant for this study, is the question why international scientific conferences on specific topics were organised. This is particularly interesting if such a conference is the very first one in its field. Several of these "first-time" conferences were organised in the first decade of marine pollution research, i.e. the period 1959–1970, and are analysed in this study. In particular, the prefaces to the proceedings of these conferences, as well as the recommendations formulated by the participants, provide useful material for the analysis of historical and scientific developments.

2 The art of discharging

"The oceans are great 'holes' in the ground, and their enormous volume together with the relatively rapid stirring of the ocean water, do allow the possibility of safe disposal of very large amounts of waste, if due care, based on adequate scientific knowledge, is exercised and if a generous respect for other people's use of the oceans is maintained." (Føyn 1965)

Marine eutrophication is a special field within marine pollution, and its history, as well as its scientific and political features, are closely connected with those of marine pollution in general. Therefore, a description of relevant developments of marine pollution is a prerequisite for a proper evaluation of marine eutrophication. Before embarking upon marine eutrophication as such, this chapter addresses some basic developments in the field of marine pollution. It concerns, in particular, the role of marine ecology and the development of a science-policy network for the management of marine waste pollution.

Today, marine pollution is a well-known environmental problem and the subject of much marine ecological research. A comprehensive national and international legal regime is in place to prevent and control discharges of polluting substances to the marine environment. But when was marine pollution first seen as a societal problem, and what were the main causes of the problem? What was the role of science, in particular marine ecology, in the discovery of marine pollution, and what was expected of this branch of science to solving the problem? When and how were the legal instruments that are now in place developed? Addressing these questions is a must for an historical account of marine pollution. It is, moreover, very relevant for the analysis of the development of the role of science in solving societal problems, and the concept of rational policy-making.

The developments in the field of marine pollution, described and analysed in this chapter, cover the period 1950 to 1980 and focus on substances other than oil and nuclear wastes. For both oil and nuclear wastes specific regimes have developed, which are less relevant for the analysis of the marine eutrophication case.

2.1 The sewage problem

2.1.1 Waste disposal in the marine environment

Already in the 19th century the discharge of sewage to surface waters caused problems in the big cities, and led to the construction of sewers and purification systems. The sewer system was a British invention and was first introduced to the continent in Marseilles in 1891 (Koch 1960). The first International Conference on Waste Disposal in the Marine Environment almost exclusively dealt with the question how to discharge sewage in such a way that it would not cause nuisance or harm. The conference, which was held in 1959 in Berkeley, California, had been initiated by the California State Water Pollution Control Board with the objective "to bring together scientists and engineers concerned about waste water disposal, to encourage an exchange of knowledge and experience on an international basis, and to stimulate research in this broad subject area" (Pearson 1960).

In 1956 the Board had started a co-ordinated programme for planning and co-ordinating research relating to marine waste disposal. It had, however, become obvious that there was a lack of scientific data on the subject, a shortage of trained scientists and inadequate communication between workers in the field (Pearson, loc.cit.). In a keynote address to the conference, the chairman of the Board, Rawn, underlined the advantages of sewage discharge to the marine environment by the statement "To be able to relegate the entire job of secondary sewage treatment to a few holes in the end of a submarine pipe [....] presents a picture of such great allure as to capture the imagination of the dullest" (Rawn 1960). He, however, placed these words into historical perspective by adding

"if this meeting had been held in 1930, or a little before, the marine environment wouldn't have been given much, if any, thought. Discussion would have centred on much narrower objectives, principal among which would probably have been the anticipated size and extent of a sewage field over an operating ocean outfall and how to prevent the sewage field from reaching nearby shore waters."

What had changed, according to Rawn, was an increasing population in the coastal area and an increase in recreational activities, also offshore. Moreover,

"Those engaged in the study and protection of marine life are becoming gravely concerned lest the effects of wastes, concentrated in the sea at or near seaboard communities, seriously deplete marine fauna by despoiling breeding areas, killing fish food, or destroying mature fish by repeated dosing of water with substances toxic to fish life" (Rawn, loc.cit.).

Also Koch, Director of Water Supply and Sanitation of the city of Paris, underlined the changes that had occurred:

“discharge of urban wastes into the sea is no new problem, [....] in the early days of modern drainage it was even regarded as an ideal solution. The vast quantity of water in the sea, the fact that rivers inevitably discharge into it everything they receive on their way, the ceaseless movement of waves and swell, and the nature of the coastal area as a whole, provided a series of arguments which at the time appeared pertinent and decisive. [....] Nowadays, however, the actual character of the effluent discharged into the sea is called into question because of the massive building-up of many coastal areas, the increasing popularity of watering-places, and the consumption of increasingly large quantities of shellfish, especially in France. All these facts call for the strictest precautions because of their bearing on public health and even on the national economy" (Koch 1960).

Of the 26 contributions presented at the Symposium, five dealt with effects of wastes on marine life, two addressed public health aspects and the remainder technical aspects of discharging waste, such as ocean dispersion and the development of monitoring techniques. The emphasis was on cases from the United States, mainly from California, but some very interesting contributions from Europe were presented as well. The biological studies clearly showed the infant state of coastal marine biology and marine pollution biology. For the first time a comprehensive survey of the benthic fauna along the Californian coast had been carried out (Hartman 1960). Clendenning and North (1960) demonstrated effects of various polluting substances (diluted sewage sludge, heavy metals, oil and synthetic detergents) on the giant kelp (*Macrocystis pyrifera*). The giant kelp is the largest plant that grows in the sea, and it forms beds in the coastal sea of southern California. The plant is harvested for use in pharmaceutical products, and the beds are an important habitat for other marine species. Although the beds constantly changed as a result of natural factors, they had almost disappeared from the waters off Los Angeles and San Diego. The above studies had been commissioned by the Californian Water Pollution Control Board, in the framework of the earlier mentioned research programme.

2.1.2 The WHO inventory

At the conference the results of an enquiry, initiated by the World Health Organisation (WHO), into the situation in Europe regarding the treatment of sewage, were presented (Koch 1960). Sixteen European countries had provided information. The results of the enquiry showed that in The Netherlands more than 200 treatment plants were in operation and another 100 planned. The cases in which sewage was discharged directly into the sea were becoming scarcer with increasing human use of the coastal area. As a specific problem case the sewage of The Hague was mentioned. Through an outlet of 400 m length the sewage was discharged untreated and as a result

"secondary pollution, due to proliferation of aquatic flora, which thrive on sewage, can be seen nearly all along the Dutch coast, but it is particularly bad at Scheveningen and has evoked keen public indignation, even if it is not particularly dangerous to human health. [....] The Hague Public Works Department has therefore had to prepare plans for a large biological purification plant" (Koch, loc.cit.).

For the United Kingdom it was reported that all coastal settlements were fully provided with sewers and that, generally, the sewage was discharged into the sea either raw or after primary treatment. The report went on to say:

"Nowadays the point of discharge is chosen very carefully, and placed as a rule beneath the low water mark at a spot from which the prevailing currents will carry the sewage out to the sea or at any rate keep it away from frequented beaches. [....] The sanitary risk seems, however, to be only potential. Under the present conditions there is very little danger of contracting disease on a British beach; the discharge of sewage can be and often is satisfactory. Some improvements are being carried out, others are probably needed; but funds are limited and in many cases it has been more necessary to spend money on abolishing pollution inland than at the seaside" (Koch, loc.cit.).

The report from the Federal Republic of Germany provided details about the sewage situation in the German coastal zone. Of the daily generated 240,000 m^3 household sewage, some 100,000 m^3 was treated. Based upon the results of research carried out so far, it was concluded that wastes should be discharged sufficiently far out to the sea "so that its dilution will exclude all danger to the public health. Under certain circumstances it might also be necessary to carry out biological treatment and disinfection of waste water" (Koch, loc.cit.).

For Norway, the Oslo Water Purification Board had submitted a report, according to which the only serious problem in Norway was in the Oslo Fjord. The sewage of 600,000 inhabitants and the wastes of a quarter of Norwegian industry were discharged untreated into the fjord, which has a very limited water exchange with the open sea. The situation in the fjord was so bad that steps had to be taken to remedy the situation. Details about the nature of such steps were not provided. It was finally made clear that the marked pollution had attracted the attention of the Institute of Marine Biology of the University of Oslo.

BOX 1: ORGANIC MATERIAL, MINERALIZATION AND BOD

Organisms possessing the availability to photosynthesise, the so-called primary producers (see Box 3), take up carbon dioxide and convert it into organic compounds. Mineralization is the process in which carbon dioxide is released again from the organic material, thereby completing the carbon cycle. Also other minerals are released, such as phosphorus and nitrogen compounds. It is mainly the bacteria which are able to break down organic material, but also fungi play an important role.

In the aerobic mineralization process oxygen is used, which, in aquatic environments, is extracted from the water. The mineralization of large amounts of organic material may lead to low oxygen concentrations, which is detrimental to water organisms, such as fish and crustaceans. The level of organic pollution is, generally, expressed in terms of the Biochemical Oxygen Demand (BOD). The BOD is determined by measuring the decrease of the oxygen concentration in a water sample within a specified time (usually five days: BOD_5) under controlled conditions.

BOX 2: TREATMENT OF SEWAGE: PRIMARY AND SECONDARY

The simplest method of sewage treatment is the so-called primary treatment. Sewage is collected in a reservoir in which most of the coarse, non soluble, parts are removed mechanically, i.e. by settling and sieving. This type of treatment is also called mechanical treatment. Roughly 30% of the BOD is removed in this process.

Secondary treatment is much more complicated. It is also called biological treatment because it makes use of bacteria to mineralise organic material. Through biological treatment the organic load of the sewage, expressed as BOD, is reduced by some 90%. However, only some 25% of the phosphorus and 10–20% of the nitrogen compounds are removed with the sewage sludge.

The pollution of the Rhine was addressed in a contribution by the director of public works of the city of The Hague, Bolomey. He informed the meeting that the Rhine was not only the main source of drinking water for The Netherlands, but also heavily polluted by the 40 million inhabitants and the industrial centres in its drainage area. Although many purification plants had already been built or were under construction, and industries would have to clean their effluent for salts, phenols and other toxic substances, the problem had become so formidable, Bolomey said, that an international commission had been set up to improve the situation (Bolomey 1960). Bolomey specifically pointed to the problems of pollution of the beaches of the Dutch North Sea coast, caused by the Rhine plume. The Rhine water was transported to the north along the Dutch coast and carried sewage to the beach within six to eight hours after entering the sea, a period much too short for biological degradation of the organic material. Bolomey therefore made a plea for biological purification along the Rhine:

"Nevertheless, although it may be added that the situation is really not alarming, and is indeed even much better than on many beaches in the world, in my opinion it must also be clear that no further pollution of the estuary can be tolerated and that existing outfalls along the Rhine must in due course be purified biologically, especially in the interest of our well known beach of Scheveningen."

2.2 Water pollution research

Whereas in the above conference the emphasis was almost exclusively on sewage as a (potential) polluting substance, already three years later, at the First International Conference on Water Pollution Research, London, 1962, an increasing interest in other polluting substances, in particular heavy metals, became visible. Although most of the presentations dealt with the technical aspects of mixing and dilution of sewage discharged into coastal waters, there were also a number of contributions addressing the possible effects of wastes on the marine environment, covering a wider spectrum than sewage.

North (1964) presented the results of extensive studies into the possible causes of the decline of the giant kelp (compare 2.1). In another US contribution an extensive overview was presented about fate and effects of oil in the marine environment (Zobell 1964). The results of toxicity tests for copper and chromium were presented in a UK contribution by Raymond and Shield (1964). In a contribution, entitled "Water Pollution and Minamata Disease," for the first time a paper was presented at an international symposium about what was many years later to be considered as the most classical example of marine pollution, the Minamata case. The information was presented by Raisaku Kioura of the Tokyo Institute of Technology who was, during the symposium, heavily criticised for his stance that it was not mercury that was responsible for the poisoning of some 90 inhabitants of the Japanese city of Minamata, but some other "casual substance" (Kiyoura 1964). Moore (1964) opened the discussion by pointing to the problems related to the "Minamata disease," which "span a wider range of disciplines than any one person can hope to master, in various fields of medicine, chemical engineering and marine biology." The 88 cases of Minamata disease that had been diagnosed between 1953 and 1961 had, according to Moore (loc.cit.), been caused by organic mercury poisoning, "although many features of the pathogenesis of the disease still await clarification." What makes this discussion so interesting is the fact that a clear indication had been given of the complexity of the impacts of polluting substances in the marine environment. Dealing with marine pollution was apparently much more than finding the proper technical solutions for diluting the substances.

2.3 The contribution of ecology

At the two international conferences described above, there was a strong emphasis on the technical aspects of discharging wastes. However, also the need for more knowledge about the biological impacts of polluting substances had been underlined. In 1966 for the first time a scientific conference exclusively dealing with the ecological aspects of marine pollution was held. The conference "Pollution and Marine Ecology" was organised by the Environmental Sciences and Engineering Study Section of the US National Institute of Health. The objectives were to "delineate the status of knowledge and areas of critical research needs related to pollution and the marine environment and to focus attention of existing and potential researchers on these areas of need" (Olson and Burgess 1967). The conference had a strong North American colour. Of the 29 contributors only two were non-US experts, namely from Norway and The Netherlands. The timing of the conference coincided with important proposed amendments to water quality policies in the US, among which an effluent tax and a required permit for effluent discharges (Hall 1967). The scope of the conference was very clearly illustrated by the contribution by Royce (1967), who referred to the new national strategy in the battle against pollution, and the essential role that water quality standards were to play: "we can hope that the standards will be better if the ecologists can forecast the effects of alternative uses of the waters." He immediately expressed his doubt, however, about ecologists' capacity to do so by saying that "in view of the lack of good ecological information, there is little doubt that industries and cities with direct, immediate and powerful interests, will use the water as they wish without regard to biota." In order to increase the role of ecologists in decision-making, Royce (loc.cit.) proposed a strategy, "which will confront the diverse groups of ecologists with the plans and decisions with regard to marine ecology; this will stimulate the collection of more adequate data which will foster basic theoretical studies and will help train people in the field."

In his keynote address Ketchum defined pollution as "any substance added to the environment as a result of man's activities which has a measurable and generally detrimental effect upon the environment." He underlined the importance and the capacity of the marine environment to dilute and disperse the wastes of "an active, vigorous and affluent society." At the same time he stressed that there were limits to this capacity, which had in many cases already been exceeded. "This is," Ketchum continued to say, "not necessarily the fault of the engineer or the industrialist. It is, in part at least, the fault of the ecologist. Rapid technological developments have far surpassed the necessary scientific understanding of these problems" (Ketchum 1967). He presented a flow scheme of the main processes that determine fate and effects of pollutants in the sea. The scheme clearly showed that not only dilution and transport of wastes, but also concentration as a result of biological, chemical and physical processes occurred. The details of the latter were hardly known (Ketchum, loc.cit.).

2.3.1 Computers and modelling

One session of the conference was dedicated to computers and modelling. Paulik (1967) expressed the view that "Digital simulation holds the greatest promise for ecologists as a thinking tool. It clearly extends Man's ability to describe and manipulate complex systems." He considered it inevitable that in the future, computer simulation models would form a basic tool for the planning of research programmes and the development of management policy. North (1967) concluded that his model had already contributed to the understanding of the responses of the giant kelp community to human impacts. It was, however, not yet accurate enough to be used for predictions because of "mysterious localised parameters," which might alter the consequences of impact considerably.

The main critique on the use of models focussed on the quality of data and the variability of ecosystems in real life. Thomann (1967) admitted that Paulik's paper had indicated that "a new level of sophistication is now present in the field of ecology." He also warned "not to let the model building aspects run several laps ahead of our ability to observe real data." Baalsrud (1967a) stressed that the answers of computers could not make the results better than the data in themselves were. To this he added: "We all appreciate that so many in this audience are concerned about the use of computers, but we must not forget that we first of all have to improve the actual data."

2.3.2 Process control

The by far most comprehensive contribution to the conference was given by H.T. Odum. In his paper, entitled "Biological Circuits and the Marine Systems of Texas," several ecosystems along the coast of Texas were described, and their energy flows analysed according to sophisticated circuit analogons (Odum 1967). In an assessment of human impact on such systems Odum went into detail about eutrophication, and concluded with regard to the changes it may cause: "This is bad if clear water is desired; however, it is good if food production is desired. The control of any particular situation therefore depends on man's handling of those loop connections." With the latter, the feedback loops in his energy circuits were meant, through which, as Odum believed, any system could be regulated in such a way that it would suit man's purposes: "Man's problem is, through loop control, to design and help the self-design of systems of complexity which will be compatible with his needs." As an example of such a self-design system Odum described a closed bay system, which was seeded with populations of organisms from all over the world. Next waste was added and, according to Odum (loc.cit.):

"the various vast diversities of creatures preadapted in the world's environment may provide us with the network parts from which self-design can give us an entirely new ecosystem. Hopefully this will be a competitive and effective treatment for Man's present waste and a system which will give him yields in terms of sport fishing, aesthetic values, and even clear water."

2.3.3 Homeostasis

Pomeroy (1967) stressed the inherent importance of the ecosystem concept in the presentations dealing with energy transfer. To this he added the concept of homeostasis as another running theme of the conference. Natural ecosystems had evolved in such a way that most of them possessed inherent stability. Accordingly, solutions to pollution problems should take into consideration "the need to maintain reasonably homeostatic systems" (Pomeroy, loc.cit.). He also pleaded for the recovery of bacterial protein from sewage. In contrast to highly toxic and relatively dilute materials, recovery of which was economically not feasible and which should be diluted as rapidly and completely as possible, domestic sewage constituted "abundant and concentrated sources of high grade chemical-bond energy."

2.3.4 Interactions with the physical-chemical environment

In the session about interactions between biota and the physical-chemical environment, Postma (1967) presented examples of how pollutants could be concentrated instead of diluted after entering the sea. Dissolved material would, generally, be transported from regions with high to regions with low concentrations, but with suspended material the opposite might happen, which was, according to

Postma (loc.cit.), often neglected in publications about waste disposal. He illustrated the behaviour of suspended material for three cases, namely beaches, estuaries and tidal areas. In all these cases the material was trapped and concentrated in nearshore areas, together with the attached wastes. Postma pointed to a secondary effect of the accumulation of suspended material. The organic fraction would be decomposed and act as a source for minerals in nearshore areas. This could be beneficial for primary production, although high amounts of inorganic matter might hamper light penetration and thus primary production. Postma concluded by giving the "ideal" composition of the least harmful waste effluent. It would be such that all harmful organic and inorganic substances would be in solution and thus transported offshore, whereas the harmless inorganic waste would be in particulate form and be available for primary production in the coastal zone. He admitted that it would not be simple to regulate the composition of waste products in such a way.

2.3.5 Biological indicators

The heart of the conference was the session about parameters of marine pollution. In introducing the session, McKee (1967) first described a parameter as a measuring device or a yardstick, and, consequently, as something that had to be quantitative. A second prerequisite for this session was to be clear about the definition of marine pollution. At the start of the conference Ketchum had already given a definition, but McKee preferred the one used by the California State Water Quality Board, which had defined water pollution as "any impairment of its quality that adversely and unreasonably affects the subsequent beneficial uses of such water." This definition contains two basic adjectives, namely "adverse" and "unreasonable." There had to be evidence that changes in parameter values as a result of human influence were adverse to one or more beneficial uses and if so, that such an adverse impact was unreasonable, meaning that it was more than trivial or superficial. In principle, every substance was a potential pollutant, but "There is a threshold value below which the concentration of each potential pollutant cannot be measured, or below which no adverse effects are discernible. It is one of the responsibilities of ecologists to determine such threshold values" (McKee, loc.cit.). According to Oglesby (1967), ecologists had already at the beginning of the century realised that it would be desirable to have certain organisms, indicative of the overall biological effects of pollution. So far, the search for such organisms had not been successful in terms of general applicability. The dilemma of the ecologist was that industrialisation and population growth had developed faster than ecological understanding of polluted ecosystems. Therefore, Oglesby concluded, "if the ecologist is to play any decision-making role in the massive national effort of pollution cleanup and prevention which is now beginning, he must do so with opinions based on something a great deal less than a complete understanding of the ecological principles involved." Wass (1967) stressed that it was hardly possible to determine proper indicator organisms without an understanding of the biota under natural conditions, but that the knowledge of such natural systems was limited. Bartsch (1967) underlined the importance of biological studies of estuaries and coastal waters, if only

"to mark for the future where we stood on this score in the late 1960s. Such knowledge was not available in time to use as reference point in assessing the ravages of pollution of the nation's rivers. In many areas there is still time to do this in coastal waters and to be a jump ahead of blossoming coastal cities and expanding coastal industries."

He referred to the many studies in estuaries and coastal waters presently going on, and pointed to a possible advantage of the fact that marine pollution had been neglected until fairly recently. It would allow for an approach, which differed from the saprobic system applied to freshwater pollution, especially in Europe, and which, according to Bartsch, was not very suitable for the marine environment.

Benthic species and communities were, generally, considered best suited for assessing possible effects of pollution because of their constant presence, relatively long lives and sedentary habits (Wass 1967; Stein and Denison 1967). Copeland (1967) favoured the use of community metabolism as "a simplified summation of the total conditions of all the cycles for circuits of the entire community." The overall condition of the community was indicated by the species diversity, a principle developed in the early 20th century, Copeland said. Lowered species diversity might be a good indicator of stress from pollution, although natural stress factors, such as salinity changes, might have the same effect.

In a paper, dealing with the limitations of indicator organisms, Stein and Denison (1967) again addressed the definition of pollution because this determined to a large part the tools that had to be used in the evaluation of pollution. They differentiated between the "preservationists" concept and the "multi-discipline approach." Within the framework of the first, nothing would be allowed to be discharged because the natural condition would be altered. The authors were clearly not in favour of this concept because "this starry-eyed philosophy would prevent any further use of our water resources and reverse the economic development of the nation." In contrast, the multiple-use concept allowed for the use of the resource in any manner, provided that other beneficial uses were not damaged. Most of the work on pollution problems done so far had been carried out by sanitary engineers and concerned the treatment of sewage. The emphasis was on preventing human diseases and, consequently, biologists involved were mainly bacteriologists. The growing recognition that pollution also affected aquatic flora and fauna had led to an increased involvement of aquatic biologists and biological oceanographers. According to Stein and Denison, this recognition had caused a demand for improved evaluation techniques and increased attention for indicator organisms. The authors were convinced that, generally, indicator organisms were better suited for assessing water quality than chemical analysis. They warned, however, for the use of single organisms, which might lead to arbitrary conclusions, and recommended that for investigating a pollution problem "as large a segment of the biological community as possible" had to be selected. The parameter to be measured should be species diversity. According to Stein and Denison (loc.cit.) "The central idea is that in non-polluted environments there is a diversity in the qualitative and quantitative structure of the community. In polluted environments there is less diversity."

2.4 Emerging limits

From the foregoing it may be concluded that in the first half of the 1960s scientists were generally positive about the possibilities for discharging waste in such a way that harm to the marine environment could be kept within safe limits. A remarkably different picture emerged during the International Symposium "Biological and hydrographical problems of water pollution in the North Sea and adjacent waters," which was held 19–22 September, 1967, on the island of Helgoland (Germany). In his opening address to the Symposium, chairman Otto Kinne stressed that the release of wastes into the marine environment was much more complicated than generally thought. He particularly addressed the North Sea, which is a shallow sea with a slow water exchange with the Atlantic, and which is an important fishery and recreational area. Outside the three mile territorial zone there existed no legal regime regulating dumping and discharges of wastes (Kinne 1968a). However, various industries had voluntarily requested scientific advice about intended dumpings. Such requests were handled by the German Hydrographic Institute, the Federal Institute for Fisheries Research and the Federal Institute for Limnology. Moreover, in view of the intended dumping of acid wastes from the titanium dioxide industry, the German Research Society had founded a Working Group "Effects of wastes in the coastal region," with the aim of investigating possible effects (Kinne, loc.cit.).

2.4.1 Pollution cases

The developments described by Kinne had, without doubt, played an important role in the preparation of the Helgoland Symposium. But certainly also the increasing evidence of mercury poisoning in the Japanese Minamata (section 2.2), and examples of pollution from the neighbouring Netherlands coastal area (see below) will have contributed to the atmosphere of warning, which emanated from the gathering. Last but not least, half a year before the Symposium the oil tanker Torrey Canyon had grounded off the southwest coast of England, causing massive public worry. Two reports about the wrecking of the Torrey Canyon were presented, one during the actual symposium and one at a meeting of a working group on oil pollution and abatement, directly following the symposium (Cooper 1968). In his symposium presentation, Cooper stated that industry should principally bear the costs of waste disposal and accidents. Cooper furthermore stressed the need for developing international legislation to combat marine pollution, and worded the role of scientists in this process as follows: "Sound law must be based on sound observations of a kind only scientists can make." He finally proposed the formation of an international organisation that could act as a "fire brigade" in cases of accidents such as the Torrey Canyon.

Koeman et al. (1968) reported about pollution of sandwich terns with chlorinated hydrocarbon insecticides. Since the mid 1950s the population size of the sandwich tern (*Sterna sandvicensis*) colony on the uninhabited island Griend in the Dutch Wadden Sea (figure 3.2) had declined from 20,000 to 1000 pairs. In 1964 residues of insecticides were discovered in birds, dying with neurotoxic symptoms. A detailed study in the years 1965 and 1966 made clear that the birds had been poisoned by telodrin. Since this insecticide was not being used in Europe, the most probably source was a manufacturer located at the mouth of the Rhine River. Close contacts with this industry resulted in strict precautionary measures to prevent the substance from entering the waste stream, and in the following years telodrin concentrations in mussels showed a strong decline (Koeman et al., loc.cit.).

The above case was a clear indication of the fact that Rhine water is transported along the coast in a northerly direction and is only slowly diluted with seawater. Korringa (1968) gave another example, illustrating this fact. In 1965 a clandestine dumping of copper sulphate had occurred on a Dutch beach. After two weeks the polluted water, which was transported in a small band along the coast, had reached the Wadden Sea. The copper sulphate had only been diluted by a factor of five and now threatened the mussel cultures in the western Wadden Sea. Fortunately, increasing winds caused a more rapid dilution to relatively harmless concentrations (Korringa, loc.cit.). He concluded: "This simple, well-documented case could teach us that a thorough knowledge of zonation and current patterns is a prerequisite if one plans to dispose waste in a given coastal section of the sea."

2.4.2 Waste classification

That marine pollution was considered a global problem, became clear from the results of a United Nations' questionnaire into marine pollution, which had been circulated to 66 nations in 1966. 36 cases of acute danger to marine organisms and 20 of danger to human health had been reported (Tomczak 1968). 25 countries proposed the establishment of an international convention for the prevention of pollution. One of the aims of the questionnaire was to evaluate a proposal for the classification of wastes, drafted by the UN organisations IOC (Intergovernmental Oceanographic Committee) and FAO (Food and Agriculture Organisation). The classification system of the UN was one according to five impact categories, namely damage to marine life, danger for human health, interference with fisheries, interference with other marine activities and reduction of amenity. Tomczak criticised the proposed system because it did not take into consideration differences in hydrographic conditions. A more sophisticated classification scheme was presented by Weichart (1968). It consisted of four depth zones and five danger classes. The scheme had been developed upon request of the German Ministry of Traffic, with the aim of harmonising the dumping of industrial wastes. In the introduction to his paper Weichart remarked that the introduction of wastes into the sea could, generally, not be promoted or rejected because this was dependent upon the physical, chemical and biological nature of the wastes and the quantities discharged. The first and foremost condition for permitting a discharge was that negative effects on marine plants and animals, shipping, fisheries, sports and tourism and, not in the least, human health, would possibly have to be kept small (Weichart, loc.cit.). The model of Weichart was criticised most for its generalisation of the various depth classes. Class two, for example, ranged from 20 to 1000 m depth and covered all of the North Sea, without taking into consideration differences in hydrology and biology. Weichart stressed that each dumping would have to be evaluated for the local and regional conditions of the site of discharge. Biologists would have to play an important role in this assessment (Weichart, loc.cit.).

2.4.3 Recommendations

In his closing address, Otto Kinne stated that marine pollution had rapidly become a problem "of grave concern in the North Sea" (Kinne 1968b). He continued to say:

> "Papers and discussions presented at this Symposium have conveyed a sense of urgency in regard to dealing with this problem; they have also made clear that marine sciences will have to go a long way to provide solutions [....] They have demonstrated that pollution research is a rather new domain for many of us – all too sudden confronted with its dangerous acuteness – and that it is characterised by an unexpected high degree of complexity."

Kinne considered the symposium as "a first critical assessment of the body of knowledge available on pollution in the North Sea." It had become clear that marine pollution had to be dealt with on a regional level, so as to be able to account for differences in physical, chemical and biological conditions. He saw as the primary goal in water pollution research "a sound scientific controlling and forecasting capacity."

During the symposium an ad-hoc committee had prepared a series of recommendations, one of which was directed to the International Council for Exploration of the Sea (ICES), which was asked to establish

> "a committee to be concerned with the pollution effects of the northwest European seaboard and, particularly, the North Sea. The committee should stimulate and coordinate the production of such scientific information as is required by the responsible agencies to counteract and control pollution in the North Sea, including the preparation of a list of such substances which are known to be particularly harmful to marine life and human activities and which, therefore, should be totally excluded from dumping into the shallow parts of the North Sea. It is hoped that this committee will succeed in providing member governments with the scientific basis to take immediate measures against particularly harmful or potentially irreversible pollution effects in the North Sea, even before the urgently required formal international conventions are established."

2.5 Marine pollution and IGOs

The increasing engagement of scientists in policy and management aspects of marine pollution also became visible in the activities of international governmental organisations (IGOs), such as ICES, NATO and UN organisations. The activities within these bodies were predominantly of a scientific nature, but the emphasis differed from the scientific symposia described above. Not only were basic data on inputs and human pressures collected, but, on the basis of scientific knowledge, also explicit proposals for policy and management were formulated.

2.5.1 The ICES working group on pollution of the North Sea

The extent of the problem of marine pollution was hardly known in quantitative terms. For the North Sea ICES had, in 1968, on the basis of national submissions, carried out an inventory of types and sizes of discharges. The data submitted were very incomplete, and hardly any information was available about the contents of the discharges. The International Council for the Exploration of the Sea (ICES) was established in 1902, following a decision of an international conference in Stockholm in 1899. The main reason for establishing ICES had been to stimulate and co-ordinate research into changes in fish stocks, particularly in the North Sea and the Baltic Sea (Hempel 1978a). Until the second half of the 1960s ICES had exclusively been dealing with matters of fisheries science and oceanography.

As described above (2.4.3), the international symposium "Biological and hydrographical problems of water pollution in the North Sea and adjacent waters" (Helgoland 1967), had adopted a recommendation directed to ICES, requesting this organisation to establish a working group to deal with marine pollution matters. At the ICES Council meeting of that same year, it was decided to install a working group under the Fisheries Improvement Committee "for the purpose of assembling factual data regarding substances harmful or potentially harmful to fisheries being discharged or likely to be discharged into the North Sea and adjacent areas" (ICES 1969). One year later, in 1968, the ICES Council decided to establish a working group for the Baltic Sea with the same remit.

The ICES "Working Group on Pollution of the North Sea" consisted of an "expert in the field of pollution," each from Belgium, Denmark, the Federal Republic of Germany, The Netherlands, Sweden, Norway and the United Kingdom, assisted by two members of the ICES Hydrography Committee (ICES 1969). The group, chaired by Cole from the fisheries laboratory in Lowestoft (UK), held two meetings, the first in London in February 1968, the second in IJmuiden, The Netherlands, in June of that same year. During this period a report was drafted on the basis of information provided by the members of the group. The following types of information were collected and evaluated:

1. Legislation regarding marine pollution;
2. Sewage pollution;
3. Pollution by industrial wastes;

4. Pollution by pesticides;
5. Pollution by oil, including oil-removing chemicals;
6. Toxicity studies, methods and results;
7. Dispersion studies.

The report was completed with a comprehensive bibliography, containing almost 300 entries on marine pollution and marine pollution legislation issues (ICES, loc.cit.). Radioactive waste was not addressed because the group considered bodies, such as the European Nuclear Energy Agency, better qualified to deal with this issue. With regard to pollution legislation, the general picture emerged that, by the end of 1968, all North Sea countries had legislative powers to control pollution of inland and coastal waters up till the three nautical mile border of the territorial waters, although there were large differences between the countries as to the nature of the pollution control. From the description of the national situations it may be concluded that there were no fixed standards for evaluating waste discharges. In Denmark, for example, waste might not be discharged when "considerable pollution" would arise. The German law controlling pollution of waterways, which was amended in 1967 so as to include coastal waters, stated that wastes might not be discharged without permission of the authorities. A similar situation existed in the United Kingdom. The dumping of wastes outside the territorial waters was in principle not subject to legislative control, although, generally, those discharging sought advice from authorities. The Netherlands were in the course of adopting a new law controlling water pollution. This law would also exercise control over sea dumping of wastes outside territorial waters, provided the wastes had been carried through Dutch ports.

2.5.2 The United Nations and marine pollution

GESAMP

In 1967 the Intergovernmental Oceanographic Committee (IOC) of UNESCO discussed "the urgent problem of marine pollution," and recommended that the UN organisations explore the possibility of forming a joint group of experts "to ensure that the necessary scientific information is put at the disposal of those agencies responsible for conservation of resources, pollution control and abatement" (International Marine Science V(4) 1967). Following this recommendation, the United Nations Joint Group of Experts on the Scientific Aspects of Marine Pollution (GESAMP) was established, which held its first meeting in March 1969. The remit of GESAMP was to provide scientific support to the Inter-Governmental Maritime Consultative Organisation (IMCO), the Food and Agricultural Organisation (FAO), the United Nations Educational, Scientific and Cultural Organisation (UNESCO) and the World Meteorological Organisation (WMO). In later years the sponsorship also included the World Health Organisation (WHO), the International Atomic Energy Agency (IAEA), the United Nations Environment programme (UNEP) and the United Nations (UN) (Windom 1991). The first two meetings of GESAMP (1969 and 1970) dealt, among others, with the identification and dangers of specific categories of pollutants. One of the first tasks carried out by GESAMP was the preparation of the FAO technical conference on marine pollution (see below).

Marine pollution and sea life

In 1970 FAO organised a major conference on the effects of marine pollution on living resources in the marine environment, called "Marine pollution and sea life." In the preface to the conference proceedings it was stated that "The Conference provided the first world forum where experts from all backgrounds and disciplines concerned, examined the problems of pollution of the sea in relation to sea life and fisheries" (Popper 1972). The main reason for organising the conference was to analyse the experiences of all sectors, so as to be able to formulate strategies to pollution problems. Moreover, the timing of the conference would ensure a contribution to the preparations of the United Nations Conference of the Human Environment, which was to be held in Stockholm in 1972 (Section 2.6), as well as to the United Nations Conference on the Law of the Sea (Popper, loc.cit.).

According to Ruivo (1972), critical reviews had, to date, only been undertaken for the North Sea, the Baltic and the Mediterranean. For other parts of the world the necessary machinery to carry out such studies was not yet available, but they were considered essential for the development of regional monitoring schemes. In addition, the need for ecological baseline studies was underlined, although the difficulties in establishing these, "in view of the inherent variability of ecological systems," were clearly recognised (Ruivo, loc.cit.). Such difficulties were clearly illustrated in the contributions of Lewis (1972) and Glover et al. (1972). Lewis (1972), in a study of rocky shore populations of several species (including *Balanus balanoides* and *Mytilus edulis*), found the faunal variation within groups of unpolluted and polluted sites to be so high that it would be hard to find significant differences. With regard to chronic pollution he concluded:

> "without a massive expansion of ecological and reproductive data by simultaneous multidisciplinary studies not only will we be unable to detect significant long-term changes, but we will even remain unaware of the most suitable or important species and methods to build into a monitoring program."

Lewis acknowledged the efforts that would be needed, but concluded: "They must, however, be faced, for short-term, superficial surveys that ignore community dynamics may have immediate public relation value but will contribute little else."

Glover et al. (1972) presented the results of 22 years of sampling with the Continuous Plankton Recorder (CPR), with the aim of discussing monitoring and research implications of long term variability. The CPR is a device that is towed behind ocean-going ships at a depth of 10 m, at monthly intervals, along twenty standard routes in the North Sea and the Northeast Atlantic Ocean. It had been in operation since 1948. The authors introduced their contribution by underlining the possible long-term effects of chronic pollution and the problems of detecting such changes in view of the natural variability of ecosystems. They stated: "it has been argued that there is an urgent need to advance our understanding of variability and its causes in natural ecosystems. In particular, it will be essential to establish ecological 'base-lines' so that the effect of pollutants can be identified against natural variation." The analysis of the results of 22 years of plankton monitoring revealed a dramatic and highly significant decline in total number of copepods per sample for both the North Sea and the Northeast Atlantic, a significant decline of total zooplankton biomass in both seas, and a significant shift of the North Sea phytoplankton spring bloom from early March in 1949 to late April in 1969. In attempting to relate these trends to physical changes, Glover et al. (loc.cit.) discussed the possible influence of changes in global radiation and temperature. For both parameters, available data were scarce. Time-series of solar radiation from the Soviet Union showed a strong decrease since the end of the 1940s, which could, according to Glover et al., "contribute towards patterns of variation in the plankton of the kind described in this paper." It was, however, unclear to what extent the atmospheric changes were part of a natural cycle or caused by human activities. "The same inconclusive comments," Glover et al. continued, "could be made, with varying force, with regard to all other aspects of the marine environment which we have not considered here, for example [....] observed variations of sea temperature." They referred to the work of Rodewald, published in 1967, who had demonstrated major trends in sea surface temperature, which he had partly attributed to changes in the atmospheric pressure over the North Atlantic. For parameters not included in weather monitoring programmes the situation was even worse. It would, according to Glover et al. (loc.cit.),

> "be quite impossible, for example, to compile a record of fluctuations of nutrients in time and space, during the past twenty years in the North Atlantic. [....] the absence of such field records has the effect of sterilising knowledge gained from laboratory experiments which have shown, for example, that light, temperature and nutrients are, indeed, critical to fundamental biological processes of marine organisms."

They concluded that with the current state of knowledge and monitoring, it would not be possible to separate natural variation from pollution effects, and that it would be necessary "to implement field monitoring programmes designed to provide the basic data for environmental research as a whole; the problems of pollution cannot be considered in isolation from those of ecology in general."

In the general discussion it was underlined that baseline studies would be particularly important in the light of chronic effects of long-term exposure to persistent pollutants. It was proposed to use ecologically defined management regions as administrative concepts in pollution control. Baseline studies, including physiological criteria, would be an essential component of such an approach (Ruivo 1972). With regard to the value of biological monitoring, it was stressed that "irrespective of how sophisticated chemical monitoring techniques become, they will not provide information on the status of the biological components of an ecosystem" (Ruivo, loc.cit.).

2.5.3 The NATO North Sea science conference

Under the auspices of the NATO Science Committee, a conference was held in Aviemore, Scotland, in November 1970, with the aim of providing a scientific basis for the management of the North Sea. A second aim of the conference was "to provide a pattern for the integration of knowledge in regional marine areas, especially those where the impingement of man threatens their resources" (Goldberg 1973a). It was in this respect the first time that an attempt had been made to draft a resource map of the North Sea. In doing this it had become clear that there was a lack of scholars to consider all relevant issues (Goldberg, loc.cit.).

In his preface to the conference proceedings, the chairman, Goldberg (loc.cit.), declared:

"the mood of the participants appeared to me to differ quite substantially from that of some of our colleagues dedicated to the concept that the North Sea is in a terribly unhealthy state. The North Sea has suffered some insults due to the activities of man, primarily along its coasts. The open North Sea appears healthy. The increasing inputs of man must be closely monitored to ascertain just what levels of materials and energy this zone can accommodate and still retain its viability."

In the introductory chapter, Goldberg continued with underlining the importance of the environmental sciences for predicting the consequences of man's activities. According to Goldberg "sufficient information should be obtained to allow rational planning decisions to be made. In making a decision, a balance must be struck between unrealistic preservation and careless exploitation" (Goldberg 1973b).

Although the first goal of the conference was to provide a basis for resource management, most of the contributions can best be classified as reviews of the current state of knowledge in the various disciplines, presented in the categories "Physical Oceanography," "Geology," "Meteorology," "Biology," "Chemistry," "Living resources" and "Non-living resources." For each of these categories the participants had drafted long series of recommendations, focussing on gaps in knowledge, rather than concrete management proposals. But, Goldberg stated, "The implementation of these recommendations can provide a springboard for effective management of the North Sea."

Physical Oceanography

Three fields of priority research were recommended for the category Physical Oceanography, namely the North Sea water budget, long-term investigations of currents and vertical variation in currents. With regard to the water budget it was remarked that estimates of the main inflow via the Straits of Dover varied with a factor of four. In this category also the present stage of mathematical modelling was discussed. Water circulation models were "fairly well understood." For passive dispersion models, those that aimed at predicting the fate of non-interacting substances, displacement equations were reasonably well known, but improvements and better understanding were needed for eddy diffusion and atmospheric influences on diffusion. For the active dispersion models, in which it was tried to incorporate hydrological, non-interacting and interacting substances and organisms, it was concluded that for very few of the relevant parameters numerical values could be given, and that in many cases their roles were not understood (Goldberg 1973b).

Biology

Goldberg (1973b) mentioned three possible causes of the changes observed in parts of the North Sea ecosystem, namely changes in pattern and intensity of fishing, increasing waste disposal and natural changes in the physical environment. In order to determine the relative importance of these factors, it would be necessary not only to understand the dynamics of the food chain, but also factors determining the long-term stability of the ecosystem. Large gaps in knowledge still existed with regard to primary production, the role of micro-organisms in the food web, bottom communities and the place of commercial fish stocks in the food web.

In a review of zooplankton, Fraser (1973) stated that the North Sea was one of the most studied areas for plankton. After having given comprehensive descriptions of types and validity of sampling and hydrographical factors, Fraser briefly discussed the effects of pollution. He wondered whether "with the increase in attention to pollution in recent years [....] the zooplankton in the North Sea can be expected to reveal any changes." He first addressed sewage, which contains nutrient salts and bacteria. The first might increase plant production, the latter were a food source for filter feeders, so, generally, an enriched plant production together with an increased number of filter feeders might be expected. According to Fraser local increases of nutrients from sewage, drainage from agricultural land and upwelling might stimulate red tides. With regard to other types of pollutants he concluded that toxic effects would not be easily determined in plankton because it was generally at the bottom of the food pyramid.

Chemistry

The areas where, in the context of pollution and pollution control, further research was considered highly necessary, were estuarine chemistry, bioaccumulation of trace substances, airborne material, nutrients, oceanic dispersion of substances and toxicology. For the category toxicology it was stated:

"even if pathways of chemical substances through the system were adequately understood, any conclusion on the capacity of the North Sea to accept wastes ultimately depends on the standards which are set as acceptable levels of substances in human food or at critical stages in the food web. We feel that, with the exception of radioactive materials, knowledge of the toxicological significance of many of these substances is almost totally lacking" (Goldberg 1973b).

2.6 Pollution control

By the end of the 1960s pollution became a public and political issue. Growing problems in dealing with wastes of all kinds, a number of serious incidents, such as the stranding of the Torrey Canyon in 1967, publications like "Silent Spring" (1962) and a changing social climate by the end of the decade, had caused a rapidly increasing awareness of environmental problems, both by scientists, administrators, the general public and politicians (Cramer 1987; Leroy 1989). Several quotes of scientists, presented in the foregoing, illustrate this fact. Also the publication of journals, specifically dealing with pollution, underlined that pollution had become an issue of public concern. In 1970 the first issue of the Marine Pollution Bulletin was published. In the editorial of the first number it was stated that publicity generated by pollution incidents "does not help create a climate of opinion in which effective legislation is possible and the vital scientific activity in research and monitoring is encouraged" (Clark 1970). The aim of the journal was to help the co-ordination of research by disseminating information about pollution on a world-wide scale.

The journal "Ambio was first published in 1972. According to the editorial of the first issue, the aim of the journal was to document and communicate scientific information on environmental issues, in the framework of "the need for more effective communications between science and society."

2.6.1 The Stockholm Conference on the Human Environment

Political answers to public concerns became visible in the first half of the 1970s, which was marked by the development of a large number of national and international instruments and structures, intended to better understand and protect the environment, in particular the freshwater and marine environment. The United Nations Conference on the Human Environment (UNCHE, Stockholm, 1972) was, at least from the environmental perspective, the culmination of this increased awareness. In addition, as will be shown below, the conference, was a major impetus for the development of a scientific-political network for the control of marine pollution. Maurice Strong, Secretary-General of UNCHE, characterised it as a conference "where science and politics meet" (Strong 1972). The draft proposals to the conference reflected, according to Strong (loc.cit.), "the best available advice and input from governmental and non-governmental sources and experts, and from the scientific, academic, intellectual and industrial communities." The emphasis of the conference was on political issues, but this, Strong continued to say, "in no way diminishes or alters the need for science to provide many of the key answers to today's environmental problems."

Problems with pollution of the marine environment received considerable attention at the conference. A working group, which had been installed to prepare a convention on marine dumping, had not succeeded in finalising its work before the Stockholm Conference. To stimulate further action, leading to the adoption of such a convention, the conference declared that "States shall undertake all possible steps to prevent pollution of the seas by substances, of which can be expected that they pose danger to human health, living resources and marine life, destroy natural beauty or interfere with other, legitimate uses of the sea" (UNCHE, §7). With regard to marine pollution, the Action Plan stated that governments would have to adopt and support general principles regarding the quality of seawater and the control of pollution. Emphasis would have to be put on the most important pollutants, i.e. heavy metals, sewage effluents and oil. Of the 25 recommendations in this part of the Action Plan, nine were about marine pollution.

2.6.2 Marine pollution science-policy networks

Stimulated by the Stockholm Conference, several international conventions for preventing and combating marine pollution were established. For this study, the Oslo and Paris Conventions, covering the Northeast Atlantic Ocean and the Helsinki Convention, covering the Baltic Sea, are most relevant. The Oslo Convention, in full "The Oslo Convention for the Prevention of Marine Pollution by Dumping from Ships and Aircraft," was signed February 15, 1972. It entered into force in April 1974. The Oslo Convention, which covers the waters of the Northeast Atlantic Ocean, including the North Sea, but not including the Baltic Sea, principally applies a system of licensing of substances to be discharged or dumped. A differentiation is made between so-called black- and grey-listed substances. Black-list substances may in principle not be dumped or discharged. The black list contains organohalogen compounds, mercury, cadmium, persistent oils and plastics. Grey list substances are, among others, organic phosphorus and tin compounds, the heavy metals copper, chromium, zinc and lead, as well as non-persistent plastics, and may only be dumped or discharged with a license. The Convention also provides conditions for dumping. Contracting parties must establish complementary or joint scientific and technical research programmes and institute complementary or joint monitoring programmes, to monitor the distribution and effects of pollutants. For the implementation of the Convention, a commission was set up (Oslo Commission: Oscom), which held its first meeting in October 1974. To facilitate the work of Oscom, the Standing Advisory Committee for Scientific Advice (SACSA) and a secretariat, based in London, were established.

The Paris Convention aims at the prevention of pollution of the marine environment from land-based sources. The area under consideration is the same as covered by the Oslo Convention, i.e. the Northeast Atlantic Ocean. The Paris Convention entered into force in 1978. It also applies the methodology of black and grey lists. Pollution of the marine environment by substances, which are on the black list, must be eliminated, if necessary by stages. Pollution of substances from the grey list should be strictly limited. The system of black and grey-list substances reflected the general feeling within the scientific community that the sea could be used as a medium to receive wastes, under certain scientific premises. As is the case with the Oslo Convention, the implementation of the Paris Convention is a responsibility of a commission, the Paris Commission (Parcom), which is assisted by a Technical Working Group (TWG) and by a secretariat. The latter is the same as for the Oslo Commission. One more body is shared by the two Conventions. It is the Joint Monitoring Group (JMG), which must supply scientific advice about monitoring carried out in the Convention area. In 1978 the Commissions decided to establish a Joint Monitoring Programme (JMP). The International Council for Exploration of the Sea (ICES) had played an important role in the preparation of the programme, since it had, upon request of the Commissions, carried out a baseline study for various pollutants.

A more elaborate description of the development and contents of the Oslo and Paris Conventions, as well as other international regulations relevant for marine pollution prevention, which emerged in the first half of the 1970s, is in Annex 1[1]. With the exception of the Helsinki Convention, none of these international regulations addressed marine eutrophication. The emphasis was on hazardous substances, most notably PCBs, organic pesticides and heavy metals. In the Rhine Convention and the European Economic Community (EEC) Hazardous Substances Directive, (see Annex 1), phosphates, nitrites and ammonia were part of the grey list. This can be explained by the fact that these regulations also dealt with the freshwater environment, for which eutrophication was a political issue.

The international conventions, including their technical working groups, together with scientific advisory bodies, can be regarded as science-policy networks. For the North Sea and the Northeast Atlantic Ocean such a network consisted of the Oslo and Paris Commissions (Osparcom) and their standing groups SACSA, TWG and JMG, together with ICES.

2.7 Summary and conclusions

In this chapter some basic features of the development of marine pollution have been addressed, including the causes of the pollution problem, the increasing awareness of scientists, the identification of the main problems, the definition of marine pollution, the role of science in the control of marine pollution, the development of societal awareness and the development of legal instruments.

2.7.1 The causes of marine pollution

Problems in the marine environment with wastes other than oil started after World War II, with increasing industrialisation and urbanisation in general, and of the coastal zone in particular. First, lakes and rivers had been used as recipient of domestic and industrial waste, but in the course of the 1960s direct discharges into the sea became regular practice. Marine pollution is clearly rooted in the discharge of sewage. The increase of the population generated increasing amounts of sew-

---

[1] Both in Sect.2.6.2 and in Annex 1 the descriptions relate to the situation as at the start of the Conventions. These have undergone considerable changes in the past decades, both with regard to contents and structure. In 1992 the Oslo and Paris Conventions were merged into the OSPAR Convention (see further 5.3.2)

age, which were discharged into the sea through sewer systems, either directly or via rivers. However, as a result of increasing industrialisation, the sewage problem was soon overtaken by pollution problems with heavy metals and organochlorinated compounds.

2.7.2 The role of science

Until the end of the 1950s mainly "discharge" engineers and authorities responsible for public health were involved in mastering possible negative effects of discharges. As was shown in the description of two major international conferences, held in 1959 and 1962 (2.1; 2.2), the emphasis was on the technical aspects of outfall construction in relation to dilution characteristics of the receiving water. In the second half of the 1950s monitoring and research programmes, accompanying discharges, became operational in California. These included, although still limited, investigations of the biology of the coastal area, and were carried out in co-operation with university researchers.

In the course of the 1960s ecologist were increasingly called upon to assist in mastering the growing problems with discharged wastes. From the proceedings of the conference "Marine Pollution and Ecology," held in 1966 (2.3), it may be concluded that there was generally an optimistic attitude as to the possibilities of ecology to assist in solving marine pollution problems. H.T. Odum, for example, envisaged the control of marine pollution as finding the proper valves in the relevant energy loops. Furthermore, the use of digital computers was considered very promising for ecology. There was a general understanding about the use of biodiversity as an indicator of marine pollution. Problems identified were a lack of knowledge about non-polluted reference areas and a lack of proper indicators of the quality status of ecosystems.

Marine pollution research in the first half of the 1960s predominantly focussed on oil and domestic wastes, and those involved in its study and control generally considered the problem as a necessary and unavoidable element of economic growth, to which solutions could be provided by science and technology. In the second half of the 1960s this picture changed in two respects. First, the emphasis slowly shifted from domestic waste problems to discharges from industrial sources. This was, without doubt, related to a number of serious pollution cases, which became widely known to the scientific community and the general public. The second aspect relates to the general attitude regarding waste discharge. So far, this attitude was a positive one: society could make use of the enormous diluting capacity of the oceans, be it with the necessary precautions. But at a scientific symposium at Helgoland (1967) the limits to this dilution capacity were underlined (2.4). It was proposed to regulate dumping and discharges on the basis of the danger of the concerned waste and the particulars of the receiving water. Regulation could also mean to forbid a substance from being discharged. Marine pollution research would have to play a major role in controlling and forecasting pollution. A major problem, identified at the Helgoland Symposium, was the absence of a legal regime outside the three mile territorial waters, by which discharges could be controlled. It was, therefore, recommended that the International Council for Exploration of the Sea (ICES) would start investigating and collecting the necessary scientific information, needed for the establishment of such a legal regime.

2.7.3 Public awareness

The rapidly changing social climate and the increasing public awareness of pollution, which evolved in the second half of the 1960s, started to exert its influence on the scientific community. Pollution, which had so far been an exclusively scientific issue, now also became a matter of public concern. Both from outside and from within the scientific community appeals were directed at scientists to take their responsibility in finding solutions to pollution problems. In an essay in the third issue of Ambio this was worded as follows (Hollander 1972):

> "Thus, many scientists have come to see that we have an obligation to participate, not only as citizens but as professionals, in the great work of improving and maintaining a viable environment on the earth, to use our knowledge to help right the ecological wrongs that have been and are being committed in the name of scientific and technological progress."

Public worries and pressures, however, also raised critique in scientific circles. Roskam (1970) stated that the increased attention for environmental problems could not only be explained by a factual increase of pollution but also in the attitude towards pollution. As an example he mentioned the fact that, as a result of the installation of a new sewage outfall for the city of The Hague, a considerable improvement in the sanitary quality of the bathing beach of Scheveningen had occurred. Public protests, however, were much stronger now than in the past, although the factual pollution situation had improved (Roskam, loc.cit.).

2.7.4 Science and marine pollution management

The increased engagement of scientists in policy and management aspects of marine pollution became visible in the activities of international governmental organisations IGOs), such as ICES, NATO and UN organisations. The activities within these bodies were predominantly of a scientific nature, but also explicit proposals for policy and management were formulated. However, the comprehensive inventories and analyses, carried out by intergovernmental organisations in 1968–1971, clearly revealed the tremendous gap that existed between political needs and scientific possibilities. The extent of the problem of marine pollution was hardly known in quantitative terms. In 1968 ICES had carried out an inventory of types and sizes of discharges into the North Sea, on the basis of national submissions. The data submitted were very incomplete and hardly any information was available about the contents of the discharges. Moreover, there was no scientific basis to estimate fate and effects of potentially toxic substances. Toxicological knowledge was limited to the effects of a few specific substances or groups of substances, most notably hydrocarbons, radionuclides and methyl mercury, on individual species. The ecological relevance of bio-accumulation and persistence was generally acknowledged, but long-term effects on marine ecosystems could only be estimated in vague terms. A principal factor hampering the analysis of man-induced changes was the variability of marine ecosystems, both spatial and temporal. The lack of baseline studies was in this respect identified as a major problem. Generally, it was recommended to increase research efforts and to set up monitoring programmes and baseline studies.

2.7.5 Marine pollution science-policy networks

The scientific work, carried out by IGOs, partly paralleled and partly preceded international negotiations about the establishment of pollution control instruments. During the period 1971–1976 a series of international regulations to control dumping and discharges of hazardous substances to the marine environment was agreed upon. The structure of all these regulations was designed to determine which substances should be forbidden to be dumped or discharged, and which substances should be allowed but regulated. This clearly reflected the general feeling within the scientific community that the sea could be used as a medium to receive wastes, under certain scientific premises.

Marine eutrophication was hardly addressed in the international regulations that were initiated in the first half of the 1970s. The emphasis was clearly on hazardous substances, most notably PCBs, organic pesticides and heavy metals. As a result of these international and corresponding national developments, a scientific-political network, dealing with pollution issues, started to grow in the first half of the 1970s. The science-policy network for the North Sea and the Northeast Atlantic Ocean consisted of the Oslo and Paris Commissions and their standing groups SACSA, TWG and JMG, as well as ICES.

3 The discovery of marine eutrophication

"The qualification belonging to the word 'eu' (=good) has gradually lost its meaning." (Van Bennekom et al. 1975)

What is marine eutrophication and when and by whom was it discovered? What are the causes and effects of marine eutrophication and what were the expectations towards marine ecology in supporting marine eutrophication policies? These are some of the main questions addressed in this chapter, in which a comprehensive picture of the development of marine eutrophication during the period 1950–1980 will be given. Answers to these questions are a necessary basis for the analysis of the political aspects of the marine eutrophication case, covered in Chaps. 4 and 5. The issues addressed in this chapter can be divided into four categories:

1. The nature of marine eutrophication;
2. The relative importance of marine eutrophication research;
3. The assessment of marine eutrophication and its relevance as a policy problem;
4. The role of ecology in marine eutrophication policy and management.

In Sect. 3.2, which constitutes the major part of the current chapter, several cases of marine eutrophication in Europe and the USA are described, providing answers to questions from categories 1 and 3, in particular with regard to the causes and effects of marine eutrophication, the main topics in the study of marine eutrophication, the relevance of marine eutrophication research for the prevention and remedying of eutrophication problems, the main controversies about how to manage marine eutrophication, as well as how scientists involved in marine eutrophication research valued the issue. Section 3.2 is preceded by a brief overview of freshwater eutrophication (3.1), so as to illustrate the main differences and similarities between marine and freshwater eutrophication.

In Chap. 2 the rapid development in the 1970s of marine pollution research and marine pollution control instruments was described. In Sect. 3.3 of the current chapter the importance of marine eutrophication research, relative to marine pollution research in general, will be investigated. It is expected that this type of information will provide answers to questions about the contents of international regulations for the management of marine pollution, and the political awareness of marine eutrophication as a pollution problem.

The latter question is addressed in more detail in Sect. 3.4. Was there awareness of marine eutrophication as a topical or potential pollution problem outside the academic scientific community, and had the issue entered the science-policy networks described in Chap. 2?

In the final Sect. 3.5 the findings of this chapter are summarised and main conclusions are given with regard to the nature of marine eutrophication, the relative importance of marine eutrophication research and the assessment of marine eutrophication, both by the scientific community and official bodies. Finally, on the basis of these conclusions, the potentials and limitations of the rational management model will be discussed, with specific emphasis on the role of marine ecology in the management of marine eutrophication.

BOX 3: NUTRIENTS, EUTROPHICATION, PHYTOPLANKTON AND PRIMARY PRODUCTION

Eutrophic literally means, "well nourished." In ecology a differentiation is made between oligotrophic, mesotrophic, eutrophic and hypertrophic water systems, depending on the relative amounts of the nutrients phosphorus, nitrogen and silicon compounds available.

Oligotrophic systems are relatively poor in nutrients, whereas hypertrophic systems have an excess. Nutrients, light availability and temperature are the main factors determining primary production, which is the growth of phytoplankton and other water plants. Plankton is the collective name for all organisms living in fresh and marine waters in a suspended state. Of the plankton, the phytoplankton uses light, carbon dioxide and plant nutrients to grow. Phytoplankton itself is the main food for so-called secondary producers, zooplankton or animal-plankton.

3.1 Freshwater Eutrophication

In the 19th century the discharge of sewage to freshwater systems caused problems in the big cities, and led to the construction of sewers and purification systems (chapter 2). Sewage contains a high percentage of plant nutrients. As a result of sewage discharges, most freshwater systems – lakes, rivers, canals and ditches – became more eutrophic. Already in 1947 it was reported that during the 19th century 47 lakes in Europe and the United States had changed from oligotrophic to eutrophic (Milway 1968).

There are some important connections between eutrophication of fresh and marine waters. First, growing problems with freshwater eutrophication have increasingly led to the diversion of discharges to rivers and the sea. Second, as will be shown in this chapter, the study of freshwater eutrophication has substantially influenced ecological research into marine eutrophication problems. That is the reason why the first section of this chapter starts with a brief overview of international developments regarding freshwater eutrophication. Three major events will be described: the Vollenweider study into freshwater eutrophication, commissioned by the OECD, the 1967 international conference "Eutrophication, causes, consequences, correctives" and the 1970 international conference "The limiting nutrient controversy." These three events provide details about the eutrophication situation in Europe and the United States and about the central research and management issues concerning freshwater eutrophication by the end of the 1960s.

3.1.1 The OECD survey

In 1965 the problem of eutrophication was raised in the Organisation for Economic Co-operation and Development (OECD), and it was decided to carry out a survey of available scientific information. The survey was co-ordinated by Vollenweider and the results put forward in the report "Scientific fundamentals of the eutrophication of lakes and flowing waters with particular reference to nitrogen and phosphorus as factors in eutrophication" (Vollenweider 1970). Vollenweider's main conclusion was that nitrogen and phosphorus were probably the most important nutrients responsible for eutrophication, with phosphorus usually the initiating factor. He presented concentration levels for assimilable phosphorus and inorganic nitrogen compounds, above which "a body of water is in danger with regard to its trophic level." Vollenweider warned that these yardsticks had no universal value because morphometric, hydrological, optical and climatic factors were important as well. He stated:

"The relation between these factors and the actual production, the biocoenotic factors and other effects on the metabolism of waters is still not properly understood, and cannot as yet be reduced to a quantitative formula which is sufficiently meaningful to describe the majority of cases in a satisfactory manner; however, the general approach to be followed in this domain has already become clear" (Vollenweider, loc.cit.).

Vollenweider had also analysed the main sources and the temporal development of nutrient loading of lakes. He made a distinction between point sources and diffuse sources. The first consisted of sewage systems and industrial effluents, and could be estimated "with any degree of accuracy." Detergents made an increasingly important contribution to the phosphorus loads. With regard to diffuse sources, Vollenweider concluded that their relative importance had changed considerably as a result of new farming methods and air pollution:

"Artificial fertilisers and natural manure being washed out of the soil in growing quantities or led directly into the waters constitute a particularly serious threat. In many places and depending on local conditions, the nutrient supply from such sources may now be equivalent to the quantities received from point sources."

Vollenweider furthermore concluded that in Europe diffuse sources accounted for more than 50% of nitrogen loading, whereas most of the phosphorus derived from point sources. As a consequence, Vollenweider continued, "it is thus reasonable to assume that the present state of eutrophication of many waters could be improved by near-complete elimination of all point sources at least." Because of a lack of experimental data it was, however, not possible to predict the time required for recovery. Vollenweider finally addressed techniques for the removal of phosphorus and nitrogen from sewage, and concluded that these had now achieved a fairly advanced stage of development. Phosphorus elimination would be most preferable from the point of view of state of technology, efficiency and costs. But he also emphasised that attention be given to controlling diffuse sources.

The survey had also made clear that there was a need for targeted research to support the management of large lakes, and to this end a symposium was held in Uppsala in 1968. Vollenweider's report was first presented at this Symposium. The proceedings of the symposium revealed why the OECD had such a particular interest in eutrophication:

"The symptoms and manifestations of eutrophication of a lake show rapidly with a very great increase in algae and water plants which become widespread. Modifications occur in the quality and quantity of the fauna and flora, very rapid increases occur, often in the form of blooms of blue-green algae, and chemical changes are considerable. Increased biological activity uses up dissolved oxygen in the water and this in turn allows an increase in suspended organic matter that can no longer be oxidised. Reduced transparency of the water further reduces photosynthesis and the evolution of oxygen. As a result there is likely to be deterioration in the facilities for fishing, bathing, tourism, navigation, water supply and other lake usages" (Milway 1968).

Although it was recognised that "the phenomenon of eutrophication has an extremely complex mechanism and is far from fully understood," there was no doubt about the fact that nitrogen and phosphorus played a predominant role (Milway, loc.cit.).

3.1.2 Eutrophication, causes, consequences, correctives

Also in the United States there was growing concern about eutrophication of lakes and other watercourses. In 1965 the National Academy of Sciences decided to organise an international symposium, with the aim of discussing the state of knowledge and developing recommendations for the effective management of problems and for the course of future research. The symposium was held in 1967 and the proceedings published in 1969 under the title "Eutrophication, causes, consequences, correctives" (Rohlich 1969). Extensive overviews were presented of the situation in European, Asian and North American lakes. The development of eutrophication was best illustrated by Thomas (1969) for Swiss lakes, for which time-series of phosphate concentrations of up to 25 years were available. In the Bodensee annual average phosphate concentrations had increased from zero in 1940 to 50 mg/m^3 in 1964. In the Zürichsee annual average phosphate concentrations had increased from 69.9 mg/m^3 in 1946 to 234.8 mg/m^3 in 1966. Thomas informed that in 1966 Switzerland had set a standard of 2 mg/l phosphate for the effluent water from purification plants. In 1967 the state of Zürich had requested all communities in the catchment areas of the Zürichsee, the Greifensee and the Pfäffikersee to eliminate phosphates from sewage, a request that was now being implemented. Thomas stated: "Elimination of nitrates or other ions is out of the question."

For the North American Great Lakes no such time-series were available, but, according to an account of Beeton (1969), “Many of the changes that have taken place in Lakes Erie, Michigan and Ontario indicate accelerated eutrophication.” In the summary of the symposium it was stated: “the most ecologically sound approach to the problem is to prevent the introduction of nutrients resulting from man’s activities” (Rohlich 1969). Reduction of inputs was considered more effective than the eradication of nuisance organisms because these might be replaced by other types.

Several methods for reducing the nutrient loading of lakes were discussed. The removal of phosphorus from sewage was considered to be of substantial aid in controlling eutrophication. Also the use of non-phosphate detergents was recommended. In the light of the increasing use of commercial fertilisers, it would be necessary to modify agricultural processes, so as to reduce enrichment of surface waters. It was recommended not to spread manure on frozen soils. The Symposium also considered it necessary to use waste to increase the fertility of the land: “Forests and agricultural soils, if not eroded, have a remarkable capacity for accumulating and retaining mineral ions, especially phosphorus.” However, the first method to be applied to reduce nutrient loading of lakes was the diversion of effluent to rivers:

“Diversion from lakes to rivers can be justified in part by the fact that rivers have a greater ability to mix and aerate pollutants, and thus can handle more effectively an effluent with a high biological oxygen demand and nutrient content. Rivers, unlike lakes, are more readily rescued from a state of over fertilization. Estuaries, because of tidal flushing, have a greater potential than rivers for disposal of organic waste” (Rohlich, loc.cit.).

3.1.3 The limiting nutrient controversy

In 1970, at its annual meeting, the American Society of Limnology and Oceanography decided that it should take “a more active leadership role in public affairs relating to aquatic resources.” As a first step, conferences would be organised to “provide an open forum for the exchange of basic information and to foster communication between academe, state and federal agencies, and industry” (Likens 1972). Because of the current political and economic interests, it was decided to organise a conference on the importance of the various nutrients regulating or limiting eutrophication of aquatic ecosystems. The interest was caused by the controversy whether phosphorus or carbon would limit primary production in freshwater ecosystems. According to Likens (loc.cit.)

“the controversy is now emotionally charged following legislative proposals to remove phosphorus from detergent formulations as a step towards controlling eutrophication. The proceedings of the conference could provide the public and politicians with “some useful guidelines with regard to this problem.”

At the conference, entitled "Nutrients and eutrophication: the limiting nutrient controversy," mainly papers dealing with freshwater eutrophication were presented. Two presentations dealt with eutrophication in estuaries. Throughout the conference two main streams of discussion were discernible. The first was a scientific dispute about limiting factors and the consequences for management. The second mainly focussed on ways to remove phosphorus from either sewage, detergents or both. The results of the many research projects, presented at the symposium, made it very clear that there was not one single limiting factor. Depending on time, place and historic developments, it was shown that either phosphorus, nitrogen, light or a combination was the key factor in algal growth. In most cases, however, phosphorus had been found to be the most important regulating factor. Interestingly, carbon, the trigger of the symposium, was generally regarded not to be the limiting factor. Several warnings were expressed regarding the choice of one limiting nutrient. Goldman (1972), for example, stated: "The limiting-factor concept is useful but is complicated in application by the reality of multiple limiting effects within species as well as within the phytoplankton community as a whole." For large ecosystems, problems would be even greater, as expressed by Fuhs et al. (1972):

> "In the analysis of large ecosystems, however, the aggregate of these difficulties mounts rapidly with the number of subsystems involved. This and the uncertainty regarding the nature of the most important limiting factor that operates at any given time and place may produce residual errors that are large and very difficult to analyse."

Derr (1972) even considered the limitation of algal growth through the limitation of one nutrient as "fraught with dangers." He feared: "We might simply set the stage for growth of a dormant species that can survive very low inputs of the controlled nutrient." He made a plea for good sewage treatment in general, which was required for health and aesthetical reasons and which, with limited additional costs, could be extended with nutrient removal facilities. In this respect, however, he only mentioned phosphorus removal. In the panel discussion, the "scientific caution," expressed during the symposium, was counteracted by several advocates of a pragmatic approach. Bartsch (1972) stated: "People are not standing still while the scientist tries to come up with all the precise answers. There is not time for this; we must proceed with the knowledge we now have." He referred to regulations in the state of Michigan to control phosphorus in municipal sewage effluent. Panel member Winter informed that the Council on Environmental Quality, after an exchange of views with several experts, had come to the conclusion that phosphorus was the most important nutrient to control (Winter 1972). Panellist Vallentyne (1972) compared the attitudes of the scientist and the pollution control man: "the university man tends to become preoccupied with details and complexity [....] The pollution control man, on the other hand, tends to abstract those parts of the relationship that pertain to control." Vallentyne also referred to a member of the house of representatives, who believed that "limnologists will still be arguing 100 years from now about whether this or that nutrient is more growth-controlling in different lakes" and that "regardless of whether or not phosphorus is growth-limiting in a lake, it can be made to be growth-limiting by removing it from man-made sources that enter the lake."

3.2 Marine Eutrophication

Freshwater eutrophication problems had already in an early stage led to a shift of inputs from rivers and lakes to estuaries and the sea. Føyn (1965) described the developments in London, where in 1864 a system had been put into practice for the collection of sewage, which was discharged in the Thames estuary at ebb tide. Freshwater flow from land to the sea is one of the main sources of nutrient supply of coastal waters. Much more than temperature or light is the availability of nutrients important for primary production. Ocean areas with a high primary production are confined mainly to coastal waters because these receive nutrients in the form of organic and inorganic substances from land. The central parts of the world oceans can, in terms of primary production, be regarded as marine deserts (Tardent 1993). The discharge of organic material and nutrients to coastal waters was therefore considered potentially beneficial. It would stimulate primary production and, hence, the growth of fish and shellfish. Ketchum (1969), in a contribution to the Symposium "Eutrophication, Causes, Consequences, Correctives" (see above), stated that moderate eutrophication of estuaries could be of benefit to mankind. In excess, it would lead to undesirable developments, but "Even while we are fertilising natural waters to excess and creating undesirable and obnoxious conditions in our rivers and estuaries, we are concerned about available food supplies for the rapidly expanding populations of the world."

In this section "early" cases of marine eutrophication and marine eutrophication research are presented. The period covered is 1954–1980 and the cases analysed are from the US and Europe because for these most documentation is available. In Sects. 3.2.1 to 3.2.3 coastal eutrophication cases from the US are covered. Sections 3.2.4 to 3.2.6 deal with European cases, Sect. 3.2.7 with estuaries.

3.2.1 The US West Coast

Possible effects of nutrients on the marine ecosystem were investigated in the framework of the monitoring of discharge sites, of which there existed 135 in California in 1961, two of which with a volume of more than 200 million gallons per day (Ludwig and Onodera 1964). The monitoring of plankton was not mandatory in any of the programmes, but several measurements of the standing crop had been carried out by dischargers and regulatory agencies. On the basis of investigations of the Allan Hancock Foundation (University of Southern California), it was concluded: "The data indicate some correlation between high concentration and proximity to outfalls, but no particular species were found to exhibit aversion or affinity to the environment surrounding an outfall" (Ludwig and Onodera, loc.cit.). It was also reported that nutrient concentrations in the southern San Francisco Bay were "extremely high," possibly as a result of sewage discharges, and that the high nutrient concentrations coincided with average plankton concentrations of 2 million cells per litre. In coastal areas such values were only found in plankton blooms (Ludwig and Onodera, loc.cit.). It was, furthermore, concluded that monitoring the standing crop of phytoplankton was of little use for assessing the impact of wastewater discharge. They suggested that measurements of biological productivity of the affected zone would be more valuable but "due to the complexity of the subject, a major research programme will be required to develop practicable parameters." They furthermore pointed to the paucity of specific scientific knowledge about the effects of wastes on marine life of economic value. Only in the immediate vicinity of the outfall was it possible to determine effects. Ludwig and Onodera (loc.cit.) stated:

"while it seems certain that the wastes do materially affect the marine ecology over considerable areas beyond the vicinity of the discharge, it is not possible to relate ecological situations and values to particular waste discharges. While tending to inhibit certain species, waste effluent stimulates other species, and the overall biological productivity of the affected waters may be increased."

Tibby and Barnard (1964) had found that the increase in phytoplankton standing crop in coastal waters was not as high as could be expected on the basis of the amounts of nutrients discharged. The phytoplankton increase was a factor of five, compared to adjacent unaffected areas, but this was considered small, since the normal range was usually in the order of a factor two to three. They put these observations into perspective, by pointing to the importance of information about the turnover rate of organic matter for the monitoring, for which they suggested some possible methods, such as following the water mass by tagging it and measuring the productivity with carbon-14. One of the purposes of the survey of Tibby and Barnard was to establish background concentrations of phosphorus, silicon and nitrogen compounds. They measured these substances along the coast all year round at different depths.

3.2.2 The N/P ratio

One of the oldest, well-documented cases of coastal eutrophication is from Great South Bay and Moriches Bay, small, shallow bays, situated along Long Island, with a limited water exchange with the open sea. In 1954 an article appeared in the Biological Bulletin in which the following event was described:

"In recent years Great South Bay and Moriches Bay have supported an extremely heavy growth of phytoplankton which characteristically appears early in the spring and persist throughout the summer and fall. [....] These dense growths of algae have greatly reduced the value of the surrounding region as a recreational area, and are also considered to be the principle cause of the failure of what was formerly a prosperous oyster industry in Great South Bay. Correlated with and suspected as a cause of the algal blooms is the existence of a large duck industry which now consists of over 40 individual farms centred along the tributary streams and coves of Moriches Bay. These farms are so situated that their waste products eventually enter the bays, greatly enriching the water and presumably creating conditions conducive to the development of the plankton blooms" (Ryther 1954).

In 1971 Ryther and Dunstan published an article in Science about their investigations into the limiting factor for primary production in Moriches Bay and Great South Bay (Ryther and Dunstan 1971). They first discussed in general terms the relevance of the fact that the average concentrations of dissolved phosphorus and nitrogen in seawater are nearly the same as the average concentrations of these substances in plankton. This phenomenon had been comprehensively described by Redfield et al. (1963), according to whom the atomic ratio of total nitrogen and total phosphorus in plankton (phytoplankton + zooplankton) is about 16 to 1. According to Redfield et al. (1963) changes in the composition of seawater as a result of biological activity of the plankton (primary production, decomposition) reflect this ratio. They stressed, however, that it concerns a statistical value, which is only approached in large water masses. The latter issue was the central point in the discussion by Ryther and Dunstan (1971), who stated that the observed relationship "may be important in regulating the level or balance of nutrients in the ocean as a whole and over geological time. It is certainly not effective locally or in the short run." They, therefore, concluded: "it has become increasingly clear that the concept of a fixed nitrogen to phosphorus (N/P) ratio of approximately 15 to 1, either in the plankton or in the water in which it has grown, has little if any validity." According to Ryther and Dunstan, a "normal" nitrogen to phosphorus ratio in algae did not exist, and values between 5 and 15 were most commonly encountered. They considered a ratio of 10 as a "reasonable working value." Depending upon the kind of algae and the availability of both nutrients, values of less than 3 and more than 30 might occur (Ryther and Dunstan, loc.cit.). For seawater, they argued that the 15 to 1 atomic ratio might be typical for the ocean as a whole, but not for the euphotic zone, which represents 2% of the oceans volume. Only where upwelling of deep-water masses occurred, or where there was mixing with water from deeper layers, might the 15 to 1 ratio be approached. Generally, in surface waters, "the two elements appear to bear no constancy in their relationship" (Ryther and Dunstan, loc.cit.). On the basis of available literature, Ryther and Dunstan concluded that in surface waters nitrogen compounds were generally depleted faster and more completely than phosphate. As a result, the nitrogen to phosphorus ratio might vary between 15 and practically zero, the latter in circumstances where all detectable nitrogen had been assimilated. The results of their own investigations in Moriches Bay and Great South Bay, and also in the New York Bight, supported these literature findings. In both cases phytoplankton growth coincided with depletion of nitrogen compounds and not of phosphate.

The conclusion that phytoplankton growth was limited by nitrogen compounds, was experimentally confirmed by adding different nutrients to water samples from the investigated areas. Ammonium enrichment generally stimulated phytoplankton growth, whereas phosphate enrichment did not. Ryther and Dunstan also observed that phytoplankton density and phosphate concentration roughly coincided along the investigated trajectories, making phosphate concentration a suitable index of organic pollution. It was also a convenient index because phosphate analysis was reliable and relatively easy. This was the reason why, according to Ryther and Dunstan (loc.cit.), "it is a short and easy step to the conclusion that phosphate is the causative agent of algal growth, eutrophication, and the other adverse effects associated with organic pollution." They warned, however, that "In the sea, such is far from true."

Ryther and Dunstan had calculated that the atomic nitrogen to phosphorus ratio in domestic sewage was about 5. They argued that, even in the (unlikely) case of complete elimination of phosphorus, the ratio of nitrogen to phosphorus in the effluent would still be 10. With an assumed average cellular ratio of 10, no reduction of algal growth could be expected. Moreover, if phosphate in detergents would be replaced with nitrilotriacetic acid (NTA), the eutrophication process might be enhanced because the degradation of NTA yields nitrogen compounds that may be used as a nitrogen source by some algal species. Ryther and Dunstan finished by stating: "To replace a portion of the phosphate in this sewage with a nitrogenous compound, and to then discharge it into an environment in which eutrophication is nitrogen limited, may be simply adding fuel to the fire."

Ryther and Dunstan's paper was discussed in a news item in the April 1971 issue of the Marine Pollution Bulletin under the provocative title "Futility of Phosphate Detergent Ban" (Clark 1971). Here it was stated that the move in the United States towards banning the use of phosphate containing detergents as a means of controlling eutrophication, could turn out to be futile because it was nitrogen that was limiting algal growth. According to Clark (loc.cit.), "authorities have chosen to control the discharge of phosphorus rather than nitrogen because its source is the more distinct, being almost entirely the detergent in the sewage, and because it can be removed at the treatment plants relatively easily and cheaply."

3.2.3 Red tides

Already in 1957 Brongersma-Sanders (1957) had pointed to the possible stimulating effect of fertilised river water on the development of red tides in the coastal sea. According to Ludwig and Onodera (1964), there was no conclusive evidence that high nutrient concentrations would necessarily cause plankton blooms. "Red tides," which were common off the Californian coast, had sometimes been attributed to high nutrient concentrations, but such blooms also occurred in areas where there were no discharges.

On September 14, 1972, a toxic dinoflagellate bloom was identified in the coastal waters off Massachusetts. It was the first "red tide" recorded in Massachusetts history and, in the words of Bicknell and Chapman Walsh (1975),

> "The sudden, almost accidental discovery precipitated a public health emergency in the state. Faced with a potentially lethal problem, and with frustratingly limited information, the Massachusetts Department of Public Health had to assume the worst."

The occurrence of the Massachusetts "red tide" was reason enough for the Massachusetts Institute of Technology (MIT) to organise the "First International Conference on Toxic Dinoflagellate Blooms," November 4–6, 1974, in Boston (Mass., USA). Additional reasons for organising the Boston Symposium were outbreaks and increases in intensity of Paralytic Shellfish Poisoning (PSP) world-wide, and a spreading to new areas (Prakash 1975). The objectives of the Symposium were to evaluate the state of the art, to identify gaps in knowledge and research needs, and "to design effective control and management schemes which will protect the public health, the environment, the fisheries resources, and the coastal economics from negative effects of such blooms." According to Prakash "in present day context, research and management are interdependent; research information is vital for development of regulations, guidelines and general management policy." Prakash stated that there was "sufficient indirect evidence," pointing to discharges from land as a prerequisite for dinoflagellate blooms. He mentioned in this respect humic substances and organic material from sewage discharges. Prakash quoted as examples Tokyo Bay and the Oslo Fjord. Hartwell (1975), however, concluded that coastal upwelling had played a key role in the three blooms of *Gymnodinium tamarensis* that had occurred in the coastal waters of Maine in 1972–1974. According to Joyce and Roberts (1975), who had investigated red tide occurrences along the Florida west coast, these events were recurring, natural phenomena. In the summary of the session on oceanographic conditions, the duration and intensity of upwelling of nutrient rich water were considered the major factors in the development of blooms. Human activities, such as dredging and waste discharge, had aggravated the problem (LoCicero 1975).

BOX 4: RED TIDES AND OTHER PLANKTON BLOOMS

Phytoplankton cells of the species *Gonyaulax* and *Gymnodinium* sometimes occur in such high concentrations in seawater that the surface is coloured red. This phenomenon, called "Red Tide," occurs regularly in semi-tropical areas and is associated with high nutrient concentrations, especially of phosphates. The algae can produce a neurotoxin, which is dangerous for men and fish. Consumption of shellfish, contaminated with algal toxin, may cause paralytic shellfish poisoning (PSP). Also in Europe cases of PSP are known, all connected with mussels from very eutrophic harbour areas.

The general term for a high concentration of phytoplankton, which has developed in a relatively short period of time (e.g. days or weeks), is "bloom." Several other toxic and nuisance plankton blooms are known. An example of a toxic bloom is that of *Chrysochromulina polylepis* which killed large amounts of fish from fish farms along the Norwegian coast in 1988. Blooms of the species *Phaeocystis* can generate large amounts of foam that may wash ashore and cause nuisance on beaches (Korringa 1968; Tardent 1993).

The second International Conference on Toxic Dinoflagellate Blooms was held in 1978, in Key Biscane, Florida. In the introductory session, Provasoli (1979) provided an overview of the current scientific state of affairs. He first referred to the fact that there was now much evidence that new blooms started from resting cysts of dinoflagellates. These cysts were formed in periods not favourable for growth, and excystment was triggered by temperature change. The viability of the plankton cells after excystment was determined by light and the availability of trace elements. The latter was also the reason why red tides were largely coastal phenomena. Coastal waters had high concentrations of nutrients, organic matter and vitamins. Provasoli (loc.cit.) stressed the importance of metals, which are toxic to dinoflagellates, and chelating substances, to which metals can be bound. He also discussed the so-called "preconditioning" of water, caused by algal excretions. There were cases in which these excretions from previous growth stimulated current development, but also cases of inhibition were known. Provasoli (loc.cit.) furthermore addressed competition between plankton species as a crucial factor in the development of a red tide. He stated that nutrient competition was an important, and perhaps even the principal factor, determining species composition and population dynamics of phytoplankton communities. In addition, factors such as motility, growth rate and sinking rate, played a role in which species would eventually form a bloom. To this, Provasoli (loc.cit.) added that, unfortunately, essential data on specific growth rates, nutrient uptake rates and limiting nutrients were lacking.

In several of the contributions to the symposium, in which cases from all over the world were presented, the factors, put forward by Provasoli, were discussed in more detail. Interestingly, only in two cases increased anthropogenic nutrient inputs were mentioned as a causative factor of increases in frequency and intensity of red tides. In the Rumanian Black Sea a positive correlation between dinoflagellate cell numbers and concentrations of both nitrate and phosphate was found for the period 1960–1977 (Mihnea 1979). The author therefore concluded that the "strong eutrophication" observed in the past twenty years was, next to physical factors such as light, an explanation for the amplitude of the phenomenon. Tangen (1979) discussed dinoflagellate blooms in Norwegian coastal waters. Of the 31 blooms recorded since 1935, 24 had occurred in the Oslo Fjord. According to Tangen (loc.cit.), the large summer-autumn populations in the Oslo Fjord were conditioned by a large supply of nutrients to the surface layers, originating from sewage discharges. In most contributions, however, hydrographical factors and trace elements were considered essential for the development of dinoflagellate blooms. Margalef et al. (1979), for example, stated that, although an abundant nutrient supply also favoured red tides, the accumulation of dinoflagellate cells depended mainly on hydrographical factors. Anderson and Morel (1979) had investigated the causes of red tides in the Cape Cod area, where the 1972 bloom, described in the introduction to this section, had occurred. They presented a model in which both dormant overwintering cysts and variations in trace metals were the main factors determining the distribution of *Gonyaulax* sp. populations. Since metals were toxic to the dinoflagellates, stimulating them to form cysts, the presence of chelating substances, originating from land drainage, could reduce this toxicity by binding the metals and, thus, increase the viability of the population. The issue of metal toxicity was also the subject of a special workshop held during the symposium.

A world-wide increase in frequency and intensity of dinoflagellate blooms was not substantiated during the symposium. Only in some of the contributions such was mentioned, but in several others dinoflagellate blooms were, on the basis of available historical records, described as regularly occurring phenomena.

3.2.4 The Oslo Fjord

The earliest well-documented case of coastal eutrophication in Europe is from the Oslo Fjord. Already in the 1950s problems with sewage discharges into the Oslo Fjord had drawn the attention of the Oslo University Marine Institute (Koch 1960). At the first International Conference on Waste Disposal in the Marine Environment (see also 2.1.1), Føyn (1960) elucidated the activities of this institute. According to Føyn (loc.cit.) there was a clear relationship between sewage pollution and marine eutrophication. He stated that in the last decades sewage disposal had caused heavy pollution, and that some years ago the situation had become so serious that the Institute had started to register changes in the fjord by means of regular surveys, and "to try and find means to stop the eutrophization process." Føyn explained in detail the development of a process for the removal of phosphorus from sludge prior to discharging. The primary effect of sewage disposal was the distribution of organic substances, but, Foyn reasoned, there was an important secondary effect, caused by phosphates and nitrates "liberated from sewage and sludge in the seawater, which induce a heavy pollution of plankton algae." In order to reduce this effect, Føyn stated, "it should be sufficient to eliminate one of the fertilising compounds. For instance the phosphates."

In 1962, continuing problems with pollution of the Oslo Fjord as a result of the discharge of untreated sewage, had led the city of Oslo and adjacent communities to sponsor an extensive research programme, the results of which should provide a basis for technical measures to control the situation (Baalsrud 1967b). To further test Føyn's hypothesis that the cause of the problem was the degradation of the organic material and, consequently, the fertilisation of the water with plant nutrients, scientists of the Water Research Institute had carried out experiments with water from the fjord, enriched with different quantities of sewage and inoculated with algae. After a week of growth, the oxygen demand of the inoculated samples was much higher than of the control samples, in which no algal production had occurred. Baalsrud therefore questioned the efficiency of biological treatment plants, which would remove only 30–40% of the phosphorus and nitrogen compounds, a value that was comparable to that of mechanical treatment. He furthermore concluded: "while BOD is a valuable parameter for biological purification of sewage, it is of limited importance for evaluation of the polluting material on a recipient water." The Institute had, therefore, developed a bioassay for testing the growth potential for algae of water polluted with sewage. The assay had confirmed the research finding that the fertility increased significantly from the outer to the inner fjord. Baalsrud (loc.cit.) was very optimistic about the potential of the bioassay to predict future developments:

> "In the case of the bioassay method, we can take a sample of the recipient water as it is today and add to it the wastes which it may receive in the future. When we attain complete confidence in the method, we shall be able to say that in order to keep conditions in the recipient water below a certain level of nuisance the type of purification should be 'such and such' and the efficiency of the purification required could be suggested."

The uptake of nutrients by phytoplankton was a second important issue of research, about which Baalsrud reported. The traditional view, according to Baalsrud, was that nitrate and phosphate were the limiting factor for phytoplankton growth and that, hence, the control of these two elements would result in the control of algal growth. He referred to recent investigations, which had made clear that the situation was more complicated. First, cells may take up phosphate in greater amounts than needed, which allows for continued growth after depletion of the phosphate in the growth medium. Second, algae are also able to economise with nitrogen compounds. One of the implications, Baalsrud said, was that it was difficult to relate in a quantitative way concentrations measured in the seawater with developments in phytoplankton stocks. He concluded that it was not yet possible to provide the necessary basic information for designing a purification plant because "First, we do not know which nutrient or nutrients we should be concerned about and, second, we do not know the relationship between nutrient removal in the effluent and effect on the recipient."

Also at the 1967 Helgoland Symposium (see 2.4) the Oslo Fjord case was prominent. Almost a complete session of the Symposium was devoted to the Oslo Fjord. In six presentations the results of the studies by the Norwegian Institute for Water Research, carried out in 1962–1965, were presented. It could be established that discharged sewage was responsible for some 75% of the nitrogen and phosphorus loading of the fjord (Føyn 1968). In a discussion concerning the relative importance of the various plant nutrients, Føyn concluded that silicon was of less importance than the nitrogen compounds. For nitrate nitrogen, high surface water concentrations in winter and low levels in the productive season indicated its importance for phytoplankton growth (Føyn, loc.cit.). For ortho-phosphate, a close negative correlation was found with the number of plant cells in the surface layer. To Føyn this demonstrated that at the present state of pollution, phosphate was probably the most important factor, and ways should be found to remove it prior to discharging the sewage (Føyn, loc.cit.). On the basis of a long time-series of fish catch statistics for the Oslo Fjord, Ruud (1968) showed a gradual, but irregular increase in cod landings from the end of the last century to 1930, after which a sudden collapse occurred. This collapse could not be attributed to a decrease in fishing effort. Ruud (loc.cit.) concluded that until the early thirties the effects of sewage fertilisation had been beneficial, but then the negative effects of over-fertilisation became apparent, and now the water of the Fjord was so polluted that it hampered the hatching of fish eggs and the development and growth of fish larvae.

3.2.5 The North Sea

In the cases described above, problems observed in the environment, such as excessive algal growth, had led to scientific research into the causes and possible remedies. For the North Sea, no cases of adverse eutrophication phenomena had been reported. Only increased levels of nitrogen and phosphorus compounds had been documented in the 1960s. These increased levels, together with the fact that further increases were to be expected on the basis of high inputs, mainly from the Rhine River, stimulated scientists to start searching for possible biological effects, caused by the enhanced nutrient levels.

The distribution of nutrients

A comprehensive account of the distribution of nutrients in the North Sea was given by Johnston at the NATO North Sea Science Conference (see also 2.5.3). For nutrients there were, according to Johnston (1973), contrary to heavy metals, "fairly ample data on which to test any theories." The problems with regard to nutrients were, Johnston continued to say, "multidimensional" because their concentrations varied with time, position and depth. The situation was simplest in winter when the activities of phytoplankton and zooplankton were at a low level and nutrient concentrations highest. He presented overviews of winter values of nitrate, phosphate and silicate, taken from the Serial Atlas of the marine Environment (Johnston and Jones 1965), showing relatively high nitrate and phosphate levels in the inflowing oceanic water round the Shetland Isles and along the northern European coast. Variations in fertility were only to a minor degree determined by these winter values. For primary production, factors such as vertical mixing, solar irradiation and turbidity were very important. Johnston referred to Steele, who had investigated these parameters in the 1950s. The assessments of plant production by Steele relied on changes in phosphate, a parameter which had, Johnston stated, "the inherent advantage that the available form of phosphorus occurs mainly as free phosphate which is easy to estimate." Nitrogen was quantitatively more important as a nutrient element, but its many forms (nitrate, ammonium, nitrite and organic compounds) posed a much bigger analytical problem. At the end of the 1950s no reliable methods for the analysis of nitrate and ammonium were available (Johnston 1973).

Johnston also discussed the possible impact of anthropogenic inputs of nutrients to the North Sea. This impact could be traced as a band of nutrient-rich water along the east coast of the United Kingdom and the west coast of the European mainland. Johnston considered the enrichments as "reasonably well balanced" because they were largely from plant and animal origin. Industrial effluents might, however, add one nutrient only. He mentioned silicate which was available in much higher quantities than necessary for marine life, especially in the German Bight, and which was derived from industrial inputs.

In 1975, Folkard and Jones of the United Kingdom Fisheries Laboratory of the Ministry of Agriculture, Fisheries and Food, published the results of a comparison of nutrient levels from January to March 1974 in Dutch North Sea coastal waters, with measurements carried out in 1961 and 1962 (Folkard and Jones 1975). According to the study 1974 phosphate values were significantly higher than those from 1962. In the area off Hoek van Holland they were increased threefold. There were no differences in nitrate concentrations, but nitrite levels were also a factor of two to three higher. Since offshore values had not changed, Folkard and Jones (loc.cit.) concluded that the increase was the result of terrestrial inputs, possibly sewage. They wondered, however, why only phosphate had increased and not also nitrate and offered, as an alternative explanation, that the increase might be due to detergents. Folkard and Jones (loc.cit.) warned: "If these trends continue and eutrophic conditions become established, harmful side effects on both the biological balance of the region and the recreational amenities of the coastal resorts may become more evident." They referred in this respect to a meeting of the "ICES Working Group on the International Study of the Pollution of the North Sea and its Effects on Living Resources and their Exploitation," which had recommended that nutrients be regularly monitored (see also 2.5.1 and 3.4.1).

Also scientists from the Netherlands Institute for Sea Research (NIOZ) reported on the eutrophication of Dutch coastal waters. In 1970, at the NATO North Sea Science Symposium (see 2.5.3), Postma (1973) presented a rough calculation of the budget of organic matter of the North Sea. His aim was to evaluate the relative importance of the increased land-based inputs of nutrients for the overall North Sea primary production. Postma compared the inputs of nutrients from the Atlantic with those from the main rivers and concluded that both sources were of the same order of magnitude. For P he estimated inputs of 70,000 tonnes from the Atlantic and 70,000 from rivers. For N his estimates were 550,000 and 700,000 tonnes for the Atlantic and the rivers respectively. Of the latter, the Rhine River accounted for 28,000 tonnes of P and 280,000 tonnes of N. On the basis of rough estimates of primary production, Postma furthermore concluded that the average annual production of the southern North Sea had increased from 250 gC/m^2 to 350 gC/m^2, as a result of increased nutrient inputs. Postma envisaged an increasing trend in nutrient inputs from land. He stated that in the early thirties the phosphate load carried by the Rhine had been one tenth of that in 1970, caused mainly by the increase of population and livestock, better sanitation and greater use of fertilisers. He specifically referred to the increased contribution of nutrients by the Rhine River (Postma, loc.cit.).

Van Bennekom et al. (1975) made a more precise calculation of the effects of Rhine inputs on nutrient concentrations in Dutch coastal waters. In the introduction to their publication the authors stated: "The qualification belonging to the word 'eu' (=good) has gradually lost its meaning." They referred to American reports of massive blooms of blue-green algae in lakes and estuaries, such as the publication by Jaworski et al. (1972) at the symposium "Nutrients and Eutrophication" (see 3.1.3). On the basis of findings by Nitta (1972) and Postma (1973), van Bennekom et al. (1975) remarked: "the impression exists that increasing nutrient loadings of rivers and other waste streams have caused higher phytoplankton standing stocks, affecting for example, the yearly amplitude in oxygen saturation values." They stated that on the basis of an analysis of nutrient changes in the southern North Sea, more exact conclusions about the nutrient regime could be developed. At the same time, however, they acknowledged: "Conclusions about the influence of these nutrient regimes on plankton populations are much more speculative because interrelations of plankton cycles and physico-chemical environmental factors are not known in sufficient quantitative detail." Van Bennekom et al. (loc.cit.) in particular addressed the influence of Rhine water and Rhine nutrient loading on Dutch coastal waters because this river was by far the most important source of freshwater discharge into the southern North Sea. They calculated that in Dutch coastal waters the influence of Rhine water was relatively high compared to the influence of water entering the North Sea through the Strait of Dover, and they therefore concluded: "the Southern Bight of the North Sea seems particularly suited for studies on the effect of cultural eutrophication." On the basis of monitoring data from the Rhine at the Dutch-German border, carried out in the framework of the activities of the Rhine Commission (see 2.6.2), they showed that the Rhine loads of dissolved nitrogen and phosphorus had increased considerably in the period 1955 to 1970 (compare Figure 3.1). Phosphate loads had increased more rapidly because of the introduction of phosphorus containing detergents at the beginning of the 1960s. Van Bennekom et al. (loc.cit.) found that, as a result of the increase in N and P concentrations in the coastal waters, silicon was now even the first depleted nutrient, whereas 40 years earlier it used to be abundant relative to N and P. They stated that in a narrow stretch along the Dutch coast phosphate was "not fully consumed [....] Only *Phaeocystis poucheti* seems able to consume all P [....] during a short period in spring. Dissolved nitrogen compounds are never fully consumed." They compared this with the results of research carried out by Ryther and Dunstan (1971) (see 3.2.1) who had found that nitrogen was the first depleted nutrient. Van Bennekom et al. (1975) also referred to Schott and Ehrhardt (1969) who had carried out extensive nutrient measurements in the central North Sea in 1968 and on the basis of these had concluded that nitrogen was the limiting nutrient for phytoplankton growth. Van Bennekom et al. (1975) explained the difference as being caused by the high N/P ratio of Rhine water. Even though the phosphate loads of the Rhine had increased considerably (compare Figure 3.1), they considered it "entirely reasonable" that phosphate could still be the limiting factor in spring in coastal waters because of the average N/P ratio in algae of 15. Phosphate depletion would limit a further expansion of the spring bloom of the *Phaeocystis* dominated phytoplankton community.

Lucht and Gilbricht (1978) published results of their investigations into changes of the Elbe nutrient loads and the effects on nutrient levels in the German Bight. On the basis of flow data and nutrient concentration data, monitored since the beginning of the 1960s, they found a negative correlation between flow and phosphate concentrations in river water and a positive correlation between flow and nitrogen concentrations. They also established a significant correlation between phosphate winter concentrations in the Elbe and at Helgoland, a fact which, according to the authors, "underlines the assumption that man-made eutrophication influences the southern North Sea" (Lucht and Gilbricht, loc.cit.). For nitrogen compounds there was no such correlation. Lucht and Gilbricht (loc.cit.) also made calculations of the yearly cycle of nutrient concentrations and ratios in the Elbe and at Helgoland. The atomic N/P ratio in the Elbe was, according to Lucht and Gilbricht (loc.cit.), high compared with the biological demand of 16, but had decreased as a result of increasing P concentrations. Their calculations also showed decreasing values for the N/P ratio at Helgoland over the period 1968–1974, and the authors therefore stated: "This result indicates that P will lose its possibly limiting influence upon phytoplankton development if its concentration increases further with respect to N." Hagmeyer (1978) confirmed the calculations by Lucht and Gillbricht (1978). On the basis of phosphate measurements in the seawater at Helgoland, he established a significant linear increase over the period 1962–1974.

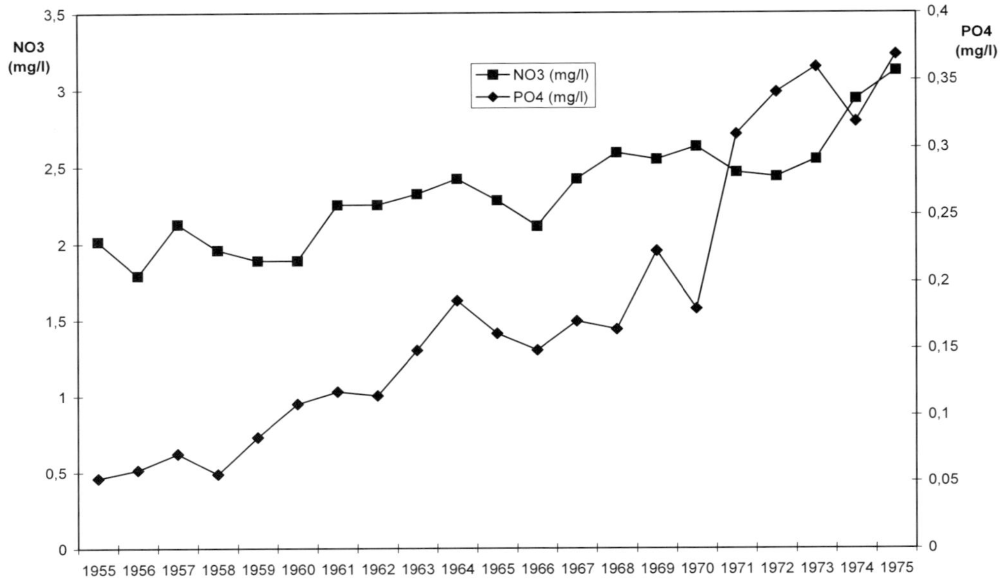

Fig. 3.1. Rhine phosphate and nitrate concentrations 1955–1975 at measuring station Lobith (German-Dutch border)

Nutrients and plankton growth

In order to gain insight into the likely effects of nutrient enrichment, Johnston (1973) had carried out a series of sewage enrichment experiments in large tanks, which could be considered applicable to an open coast situation with modest flushing and an input of well-diffused effluent at a depth of 20 meters. After comparison of the result with the field situation in Scottish coastal waters, he concluded: "considerable stress is exerted on the marine environment if normal nutrient levels are increased more than threefold." In this situation there was an abundant growth of some marine plants. At a ten-fold enrichment only a few species could survive. Such a situation would, according to Johnston, "not be found over a significant area in UK waters." To this he added that in enclosed bays the situation would be much worse. Also the simultaneous discharge of toxic material would cause a largely different situation, although there was very little experimental work on such promoter-inhibitor situations. Johnston concluded his contribution in an optimistic way:

"There is a wide choice of bays and estuaries where sea water is compounded with all sorts of effluents. Fortunately, at the same time, this area also offers a great wealth of marine life and rich stocks of commercial fish and shellfish as significant material for study, adding great social and commercial impact to many aspects of research on the North Sea as a right and proper outlet (under enlightened control) for domestic and industrial wastes of the people round its shores."

According to Postma (1973), the supply of nutrients seemed beneficial from the perspective of enhanced productivity and, consequently, more food for marine organisms, including commercial fish. He warned, however, that also negative effects were possible. These included "disturbance of the oxygen balance and undesirable changes in the composition of the populations of marine organisms, including phytoplankton." For the North Sea in general the oxygen balance was not yet endangered, but accumulation of organic material could occur in marginal areas such as estuaries. Postma presented an example of anaerobic conditions which had occurred near Texel after a heavy spring bloom of *Coscinodiscus concinnus*. Interestingly, in the original report of the event, published by Roskam (1970), the situation was assessed as an example of a natural event. As to the phytoplankton composition, Postma referred to the relatively high abundance of nitrogen in river water. In the central and southern North Sea nitrogen was more rapidly exhausted than phosphorus, but in coastal waters phosphate might be a limiting factor. Even more important might be the relatively low silica concentrations in Dutch coastal waters, which seemed to be limiting the diatom population (Postma 1973).

Van Bennekom et al. (1975) wondered to what extent the nutrient enrichment had caused enhanced phytoplankton growth and an increase in the frequency of algal blooms. In the previous 10-year period several mass occurrences of algae had been reported, which, according to the authors, could also be due to the increased number of observers. Gieskes and Kraay (1975) discussed the effects of nutrient loading on the phytoplankton spring bloom in more detail. They presented a description of the development of the phytoplankton spring bloom in Dutch coastal waters, and stressed, first of all, that in the nearshore area most influenced by the Rhine plume, this bloom started only in April, whereas in more offshore waters this was already the case in February. The reason was the high turbidity of the nearshore area, which counterbalanced the effect of higher nutrient loads. Because of the reduced consumption of nutrients in the nearshore area, Gieskes and Kraay argued, the effects of Rhine nutrient loading on primary production would be perceptible over a larger area. The higher the turbidity, the larger this area would be (Gieskes and Kraay, loc.cit.). Because in early spring grazing by zooplankton was limited, they concluded that the early phase of the phytoplankton bloom was predictable and determined primarily by incident radiation, turbidity and depth of the water column. On the basis of their measurements from the period 1971–1975, Gieskes and Kraay (loc.cit.) also concluded that diatoms determined the algal crop at the start of the growing season. After depletion of silicate in April, *Phaeocystis poucheti*, which needs no silicate, became dominant and might remain so until phosphate became limiting, "in spite of the greatly increased

P loading of the eastern part of the Southern Bight in the last decades" (Gieskes and Kraay, loc.cit.).

At the 1975 ICES North Sea Fish Stocks Symposium, Hagmeier (1978) presented data from measurements carried out at Helgoland since 1962. The average annual course of phytoplankton development showed an inverse relation with salinity, indicating the influence of enrichment by freshwater sources. Although annual average phytoplankton stocks showed high fluctuations, Hagmeier (loc.cit.) concluded that phytoplankton (as well as phosphate content) "have a distinct tendency to rise in Helgoland waters." Reid (1978), at the same symposium, presented North Sea data from the Continuous Plankton Recorder (CPR, see also 2.5.2). Although the CPR is primarily aimed at collecting zooplankton, changes in phytoplankton standing stock can be assessed by the so-called "greenness" of the sampling silk (Reid, loc.cit.). The CPR data showed that the standing crop of phytoplankton in all areas of the North Sea had increased since 1958. Of the total crop, diatoms had decreased "drastically" and *Ceratium* species showed only minor changes, which led to the conclusion that "an unidentified component of the phytoplankton is increasing" (Reid, loc.cit.). The patterns of change were generally comparable over the whole of the North Sea. Reid (loc.cit.) discussed five possible causes of the observed changes, namely pollution, climatic change, biological succession, phytoplankton composition and sampling artefacts. The increase in nutrients could, according to Reid (loc.cit.), affect productivity in the southern North Sea, but seemed unlikely to be the cause, since a recent ICES report had concluded that the North Sea was relatively unpolluted (see also 3.4.1). Changes observed in the North Sea, both for phytoplankton and zooplankton, showed the same pattern in the North Atlantic (see also 2.5.2), which made it plausible that climatic changes were responsible. Reid (loc.cit.) mentioned the reporting of changes in atmospheric pressure belts, caused by changes in solar energy. Also long-term phytoplankton succession might be a possible cause, but there was very limited understanding about this. In this respect microflagellates, which were not identifiable in CPR samples, could be the "unidentified component" responsible for the increase in total phytoplankton stock.

Gieskes and Kraay (1977) investigated one specific area sampled by the CPR in more detail, the so-called "Ow rectangle," in the eastern part of the Southern Bight (figure 3.2). Also these researchers were faced with the question how to explain the increased phytoplankton stock since the mid of the 1960s, taking into consideration that *Phaeocystis* numbers had even decreased during that period. They stated that it was "tempting" to agree with the suggestion of Reid (1978) that some uncounted organisms, possibly micro-flagellates, were responsible for the observed increase. Although fluctuations in the Ow rectangle were comparable to those in other areas of the North Sea, Gieskes and Kraay (1977) observed one important difference: whereas in the southern North Sea as a whole the colour had increased by 30%, it was 70% for the Ow rectangle. The time of the year in which the increase had taken place was the same as in the other areas, namely spring and early summer. The mean silk colour value for the period 1968 to 1972 was 1.7 times that of the period 1948–1952 for the whole south-eastern North Sea. In the Ow rectangle there was a 2.4 fold increase. According to Gieskes and Kraay (loc.cit.), this value corresponded well with the estimated increase of suspended particulate organic matter in the Dutch North Sea coastal area, as documented by de Jonge and Postma (1974) (see further below). To this they added, however, that also in the western part of the Southern Bight and in the Channel region the colour value had doubled, compared to the early fifties. They concluded: "It is possible that the increased fertilisation of this area has stimulated phytoplankton and zooplankton in the eastern part of the Southern Bight. However, the natural long-term variability can still be recognised clearly."

Fig. 3.2. Map of the North Sea. C1, C2, D1, D2, Ow = ICES rectangles

Primary and secondary production

In the foregoing, it was documented how nutrient levels in the North Sea had increased as a result of increased anthropogenic inputs. It was also shown how difficult it was to relate these changes to changes in phytoplankton developments. Even more difficult, however, was the establishment of the link between increased phytoplankton biomass and increased primary production as, for example, worded by Reid (1978):

"A major change appears to have taken place in the composition and abundance of North Sea phytoplankton within the last decade. How these changes may have affected phytoplankton productivity is impossible to assess, since there are no time-series of phytoplankton production measurements in the North Sea."

Postma (1978) stated that direct measurements of primary production before 1965 were lacking, and he therefore made an attempt at an indirect estimate. He calculated that in the Southern Bight 30% of the present primary production was caused by the eutrophying influence of the Rhine River. Postma (loc.cit.) also referred to the results of the CPR and stated that it could be expected that regions not under direct influence of the Rhine River "benefit in one way or another from the increased nearshore productivity." "Much more evident," Postma continued, "is the influence of this increased productivity on the Wadden Sea" (figure 3.2). He referred in this respect to the publication by de Jonge and Postma from 1974, in which the results of a comparison of organic matter content of the water of the western Dutch Wadden Sea between 1950 and 1970 were given. De Jonge and Postma (1974) showed a considerable increase (factor 2–3) of yearly average concentrations of phosphate and particulate phosphorus in both the western Wadden Sea and the adjacent coastal North Sea over this period. They also established a linear relationship between particulate phosphorus and total suspended matter, and concluded that the amount of suspended matter in the Wadden Sea had doubled between 1950 and 1970. According to de Jonge and Postma (loc.cit.), the increased supply of phosphorus from the Rhine River had probably caused increased primary production in the North Sea which, in turn, had caused an increased import of organic matter into the Wadden Sea. Already in 1954 Postma had postulated that the Wadden Sea accumulated organic matter produced in the adjacent North Sea (compare 2.3.3), and that the imported material was rapidly decomposed, as a result of which concentrations of nutrients in the Wadden Sea were usually higher than in the adjacent North Sea. De Jonge and Postma (1974) assumed that in the Wadden Sea itself primary production had not increased because already in 1950 phosphate had not limited phytoplankton growth. This was confirmed by Cadée and Hegeman (1979) on the basis of a comparison of chlorophyll values from the periods 1951–1953 and 1974–1976. Postma (1978) compared these results with the developments in the Southern Bight, and concluded that a considerable part of the assumed primary production in the North Sea would be mineralised in nearshore waters. According to Postma (loc.cit.),

"The estuaries would then be the principal beneficiaries from the increase in North Sea productivity. In other words, the nursery grounds for fish would be the main receivers of extra food. Whether the fish stocks themselves benefit from this state of affairs depends, of course, on whether food is a limiting factor here. The answer must be given by the marine biologists."

Several attempts to provide such answers were given by the participants in the ICES North Sea Fish Stocks Symposium. But the presentations about changes in benthos did not provide many indications of increases. McIntyre (1978) stated:

"all we can say from the meagre benthic information on the North Sea is that there is no conclusive evidence of long-term change."

Boddeke (1978) comprehensively discussed the possible causes of observed changes in the stock of the brown shrimp (*Crangon crangon*) in Dutch coastal waters. The main factors influencing the stock were, according to Boddeke, decrease of nursery grounds, pollution and eutrophication, climatological changes and changes in fishing techniques and intensity. With regard to eutrophication, Boddeke was very pertinent about its positive effects. He first of all referred to the comparatively high densities of brown shrimp around a pipeline, discharging organic waste in the eastern Dutch Wadden Sea. He then showed, on the basis of distribution plots of brown shrimps in different parts of the Dutch coast, that in 1969–1974 shrimp distribution in the southern part had become more extended, i.e. the shrimps were distributed over a wider area than before. Boddeke (loc.cit.) explained this as being caused by increased Rhine nutrient loads and the fact that the Rhine water was discharged more concentrated and with higher speed into the North Sea, which was caused by the construction of an approach channel to the Rhine. As another important factor for the recovery of the shrimp population, Boddeke (loc.cit.) mentioned the introduction of new fishing techniques, through which the by-catch of undersized shrimps had been strongly reduced.

The strong increase of North Sea fish stocks since the 1960s, the central issue of the ICES Symposium, could, according to model studies by Ursin and Andersen (1978), not be explained by increases in phosphorus inputs. The latter could explain only 5% of the observed increases.

In the synopsis of the symposium, chairman Hempel (1978b) concluded:

"Regarding the effects of hypertrophication and toxic pollution, the present level, as described in relevant reports, does not call for immediate concern. If anything it seems likely that the increased discharge of nutrient salts and organic waste into the North Sea had a small indirect positive effect on the fish stocks."

He also stated that there was insufficient knowledge to be able to make estimations about nutrient accumulation in any part of the North Sea, but that it supposedly remained in the shallow southern part, particularly the Wadden Sea.

3.2.6 The Baltic Sea

Contrary to the North Sea, eutrophication in the Baltic Sea had already in the 1960s been subject to increasing scientific investigations. According to Paulsson (1972), continuous measurements since the 1890s had shown that the oxygen contents in bottom water had gradually decreased, and in the late 1960s hydrogen sulphide in the deeper parts had caused extinction of the bottom fauna. Also indications of effects on the cod population had been observed, and several countries had intensified research into the possible causes. The specific eutrophication problems of the Baltic Sea were, according to Dybern (1972) and Fonselius (1972a), caused by the very limited water exchange with the North Sea (on average the water is exchanged every 21 to 24 years) and the pronounced halocline, the division of relatively fresh- and saltwater masses. Organic material, produced in the surface layer, sinks to the bottom, where its degradation consumes oxygen. There is, however, little oxygen transport from the upper layer to the more saline and, therefore, heavier bottom water. Both Dybern and Fonselius provided data on the oxygen content of the bottom water of the Landsort Deep, which had dropped from 2.5 to around 0.5 ml/l since 1900. At the same time, the phosphate concentrations in the deeper parts of Landsort Deep had increased from about 1 micromole in 1950 to more than 3 micromole in 1970. According to Dybern (1972), the increase of phosphate values coincided with increased outflow from land, as a result of the use of washing-powders and artificial fertilisers. Fonselius (1972a) also referred to the use of synthetic detergents as the main source of phosphate. A rough phosphorus balance, elaborated by the ICES Working Group on the Baltic, showed that some 7,000 to 20,000 tonnes of phosphorus annually entered the Baltic on a net basis (Dybern 1972). Compared to the 300,000 to 400,000 tonnes present in the deep layers, this could be considered as small. Dybern (1972) concluded:

"It is therefore at present very difficult to evaluate the influence of the increased phosphorus outflow during the last few years on the oxygen conditions in the Baltic deep basins, although it seems quite certain that many coastal waters are negatively influenced."

Fonselius (1972a) discussed in detail the possible effects of increased nutrient levels on primary production in the Baltic. He illustrated the dangers of anoxic conditions of the bottom water, which could start a "vicious circle of fertilisations." Such conditions might enhance the release of nutrients, which could, in turn, fertilise the relatively nutrient poor surface water and trigger plankton blooms. The organic material, produced by the latter, would settle out to the deeper waters and aggravate the poor oxygen situation which, in turn, would cause the release of more nutrients. Fonselius considered phosphorus to be the limiting factor for primary production in the Baltic. According to Dybern (1972), less than 20% of the sewage entering the Baltic was biologically treated and none had phosphorus removal. He considered this a prerequisite for controlling phosphorus discharges.

The Soviet-Swedish symposia on pollution of the Baltic

During the period 1971–1975 three international scientific symposia on the pollution of the Baltic Sea were held. The main reasons for organising the first symposium (Stockholm 1971) was, according to Paulsson (1972), "the difficult interpretation of the research findings and the fact that the scientific problems could only be solved through international co-operation." In his introductory speech to the Symposium, organised by the Soviet Academy of Sciences, the Royal Swedish Academy of Engineering Sciences and the National Swedish Environment Protection Board, the president of the Royal Swedish Academy of Sciences, Brohult, referred to "alarming reports" about decreasing oxygen content, which had first appeared ten years ago (Brohult 1972). He stated: "On many occasions during recent years, legitimate demands have been made for new measures and stricter laws against the pollution of the Baltic Sea," and also: "The responsible authorities have frequently consulted scientists to get a relevant analysis of the pollution problem as a basis for the making of decisions" (Brohult 1972).

Fig. 3.3. Map of the Baltic Sea.

The symposium, in which also a Finnish observer delegation participated, mainly dealt with methodological aspects of pollution research, such as statistics, modelling and analytical methods, predominantly related to nutrients and organic pollutants. But there were also contributions on effects of PCB and DDT on fish populations, the distribution of radionuclides and oil pollution. A comprehensive overview of the eutrophication situation in the Baltic, highlighting the specific features of this brackish water body, was given by Stig Fonselius from the Fishery Board of Sweden. He presented long time-series of oxygen, phosphorus and primary production from various monitoring stations in the Baltic (Fonselius 1972b). Because of the restricted water exchange between Baltic Sea and North Sea and the permanent stratification, the Baltic Sea can be regarded as a "nutrient trap." Organic material from the upper euphotic layers sinks to the bottom where it is mineralised, thereby consuming oxygen. Oxygen depletion and hydrogen sulphide formation in the deepest parts are a natural phenomenon, and are periodically relieved through water inflow from the Belt region, which washes out the anoxic water from the deep basins and, at the same time, releases the accumulated nutrients. Fonselius (loc.cit.) presented time-series (1950–1971) from the Gotland Deep (depth -240m) (figure 3.3), showing alternating periods of oxic and anoxic conditions. Time-series from 1900 to 1979 from the northern central basin at a depth of 150–170 m, showed a continuous decline from 3 mg O_2/l in 1900 to zero in 1970. Fonselius (loc.cit.) especially pointed to the cycling of nutrients between deep water and surface water, which he termed the "driving force" of eutrophication in the Baltic.

According to Fonselius (loc.cit.), phosphorus was the limiting factor for primary production in the Baltic Sea and especially in the Gulf of Bothnia. He argued that nitrogen could hardly be the limiting factor, if both phosphorus and nitrogen were simultaneously depleted because in the surface water there would always be some ammonia available, which could act as a nitrogen source. Fonselius (loc.cit.) presented two data series of dissolved phosphorus, both showing a steady increase from the end of the 1950s onwards. In the Landsort Deep mean values from 100–400 meter depth had increased from 1 µmol/l in 1955 to 3 µmol/l in 1970. In the surface water of the central basin, concentrations had increased from about 0.1 µmol/l to on average 0.3 µmol/l in 1970. On the basis of Swedish input data, Fonselius inferred phosphorus inputs for the whole Baltic and arrived at a figure of 17,000 tonnes P per year. Fonselius stated that similar models could be constructed for nitrogen, but that "more work is certainly needed before we will be able to understand the nitrogen cycle in the sea." Time-series of primary production were available for the 1960s only. The series from the lightship "Finngrundet" in the Gulf of Bothnia, measured at depths of 0 to 15m, showed about a threefold increase from 1961 to 1968. Measurements at other locations were too few to be able to allow for trend analysis. Fonselius used the "Finngrundet" data to compute an average annual primary production in the Baltic Sea of about 26 million tonnes of carbon. The average annual value of some 57 g C/m^2 was, according to Fonselius, high for an oligotrophic system like the Baltic. This high value could be the result of fertilisation by domestic and industrial wastewater, or derive from natural sources.

The second Soviet-Swedish symposium on the pollution of the Baltic was held two years later in Riga. This time, participants from all Baltic states participated. The Symposium dealt with hydrography, hydrochemistry, biological effects of pollution, modelling and abatement techniques. Nutrient and eutrophication related issues were well represented, but the number of presentations dealing with oil and hazardous substances had increased, compared to the first symposium. Again Fonselius presented an elaborate contribution about nutrients and eutrophication (Fonselius 1976). This time, the emphasis was on the limiting factor for primary production. According to Fonselius, nitrogen supply in the oceans was considered the limiting factor, and, in cases of large phosphorus increase, this might also be the case in inland and brackish waters. For the Baltic, Fonselius stated, it had been claimed that phosphorus would limit primary production in the Gulf of Bothnia. One of the reasons for this claim was the fact that there were more accurate methods for the routine determination of phosphorus. Fonselius (loc.cit.) stated in this respect: "Nitrogen analyses are more complicated and all inorganic nitrogen parameters have to be analysed in order to understand the system. Therefore very few complete nitrogen analyses have been carried out until quite recently. All attention has been directed towards phosphorus." Still, Fonselius considered it hardly possible that nitrogen could be the limiting factor in the Baltic because, generally, sufficient nitrogen was present in the surface water, which was hardly ever depleted to zero. Moreover, contrary to phosphorus, for which there were no additional sources, nitrogen could be supplemented by rainwater and by nitrogen fixation. According to Fonselius, in Sweden phosphate was removed from sewage, so as to stop the increase of inputs to coastal waters, which might result in phosphorus becoming the limiting factor. Moreover, in the process of phosphorus removal also heavy metals were being removed from the sewage effluent. With regard to the removal of nitrogen, Fonselius remarked: "efforts in this direction will probably face overwhelming difficulties and be to no avail, due to the other, external nitrogen sources."

In the recommendations of the Symposium it was stated: "It was commonly felt that even if the deterioration of the water of the Baltic deep basins is a natural process, pollution causing increased oxygen demand may be of importance" (Valeskaln and Hannerz 1976). A number of issues were considered urgent for intensification of scientific and technical efforts, among which "the elaboration of methods for evaluation of the degree of eutrophication of waters of the Baltic and its separate areas, and methods for evaluation of the condition of the Baltic ecosystem in terms of its eutrophication." Also the water exchange with the North Sea and the effects on the status of the Baltic were considered important for future studies.

At the third Soviet-Swedish Symposium on the pollution of the Baltic (Stockholm, 1975) Fonselius again presented an overall overview of the nutrient and eutrophication situation in the Baltic (Fonselius 1977). He started his presentation by referring to recent publications about the limiting factor for primary production and concluded: "it had not been possible to give a definite and final answer to this problem." Fonselius reasoned: "Probably the main difficulties are that different factors may be production limiting in different parts of the Baltic and also during different parts of the productive seasons." Another difficulty was that turnover times for nutrients were not known (Fonselius, loc.cit.). He presented updates of time-series of oxygen concentrations from different monitoring stations in the deep parts of the Baltic, showing a continuation of the pattern of alternating periods with and without oxygen depletion. Time-series of the development of concentrations of different nutrients from the bottom to the surface, generally showed low phosphate values in the surface waters in summer and accumulation of phosphate in the deep parts. In most summers, nitrate in surface water was depleted to zero. Also zero values for nitrate were found in stagnant, hydrogen sulphide containing basins, which was explained by denitrification. Still, many uncertainties were connected to this process, which made Fonselius conclude that intensive research into the nitrogen cycle was necessary (Fonselius, loc.cit.).

The leading role of Sweden in pollution abatement was made clear in a contribution about Swedish legislation with regard to sewage treatment (Hartwig 1977). During the period 1965–1976, the percentage of the population connected to biological sewage treatment plants had increased from 35 to 95%. From 1970 to 1976 additionally tertiary treatment (phosphate removal) had increased from practically zero to 65%.

In the Recommendations of the Symposium it was stated that the participants had been informed about possible pollution influences from the North Sea, and studies of the effect of water exchange between the North Sea and the Baltic Sea were considered highly desirable (Valeskaln and Paulsson 1977). With regard to future scientific co-operation, reference was made to the work of the Interim Commission of the Helsinki Convention (see also Annex 1 and 3.4.2) and harmonisation with the work of this Commission was emphasised.

AMBIO special issue on the Baltic Sea

In 1980 a special issue of Ambio about the Baltic Sea system was published, covering human activities, natural values, pollution and international co-operation. In a general overview of human impact on the Baltic ecosystem, Leppäkoski (1980) listed the main pollution problems of the Baltic: eutrophication, accumulation of toxic substances and oil pollution. He stated that many data about pollution relevant parameters had been collected in local pollution studies, but that in only a few cases the ecological significance of the data had been assessed. The increasing nutrients inputs were, in Leppäkoski's view, "one of the essential Baltic problems." The visible effects were mostly local and comprised decreased water transparency due to increased phytoplankton growth and proliferation of green algae. Also the bottom fauna had changed in such a way that species numbers had decreased and biomass had increased. With regard to fish, Leppäkoski (loc.cit.) mentioned observations that the relative importance of economically important fish species seemed to decrease. Because the Baltic is an oligotrophic system, increased nutrient inputs could, according to Leppäkoski, be beneficial up to a certain limit. Excess eutrophication could, however, cause acceleration of anoxic conditions in stagnant bottom waters and heavy blooms of algae, and pose a severe threat to the system (Leppäkoski, loc.cit.). "Unfortunately," Leppäkoski continued, "whereas a number of local case studies have well documented this subject in various coastal areas, there is not so much direct evidence of the biological effects of eutrophication in the Baltic proper." Available data from waters around Gotland and Öland (figure 3.3) indicated an increase (up to threefold) of benthic biomass above the halocline in the past 50 years. Also zooplankton summer biomass had increased since the 1950s. Leppäkoski considered this an indication of a higher production level in the upper layers, which was the result of "creeping eutrophication." The presence of oxygen free bottom water had, according to Leppäkoski (loc.cit), "been one of the central points in discussions of the Baltic problem." Although the stagnation itself could be regarded as a natural phenomenon of stratified waters, and also examples existed of oxygen free periods from earlier centuries, periods of oxygen depletion and hydrogen sulphide formation "seem to be longer and more frequent." According to Leppäkoski, "There is little doubt that man can play an important role in these processes." But, whatever the cause, oxygen deficiencies had serious impact on the bottom fauna, causing elimination of all macroscopic life, thereby creating "enormous biological deserts in the deepest parts of the Baltic proper." The area affected in the 1960s and 1970s was estimated at some 100,000 km^2, which is about one quarter of the Baltic Sea. For the near future, Leppäkoski expected that joint scientific efforts would result in a more comprehensive assessment of the effects of the pollution of the Baltic. But, Leppäkoski (loc.cit.) concluded, even if the Baltic Sea is only a "minute appendix to the world ocean [....] it is still large and complicated enough, never to be fully understood as a natural and man-influenced system."

The Ambio special issue also contained a contribution about land-based inputs to the Baltic Sea (Pawlak 1980). The article was based upon a comprehensive survey, carried out under the auspices of ICES (see also 3.4.1), supplemented with recent data. The period covered was 1972–1977. Of the different sources, land runoff and rivers were particularly significant. The total annual average river flow is some 430 km^3, which is 2% of the total volume of the Baltic Sea. Because organic matter and nutrients were generally monitored in rivers and in direct discharges of domestic and industrial wastes, the inputs of these parameters could be estimated most reliably. For the whole Baltic, the annual input of organic matter was estimated at some 1.4 million tonnes (expressed as BOD_7; compare Box 1), of nitrogen 309,000 tonnes and of phosphorus some 26,000 tonnes. The riverine contributions varied between area and substance, but were between 66 and 90% of total inputs. For toxic substances no overall input figures could be given. Only for the Gulf of Bothnia estimates for heavy metals and oil were presented. Pawlak stated that in the past decade there had been a trend for the Baltic countries to restrict the use of toxic substances and to construct industrial and municipal biological sewage treatment plants with tertiary treatment. The latter reduced BOD_7 and P loads by 90% and N loads by 40%. On the basis of Swedish data it could be concluded that this had caused an overall decrease of phosphorus inputs along the Swedish coasts of 50%. Municipal and industrial nitrogen inputs had decreased by 35 and 22% respectively, but the overall loads had increased by 10%, which was caused by increased nitrogen runoff to rivers (Pawlak, loc.cit).

The Marine Pollution Bulletin special issue on the Baltic Sea

On the occasion of the publication of the first overall assessment of the state of the Baltic within the framework of the Helsinki Convention (see further 3.4.2), a special issue of the Marine Pollution Bulletin was published (Kullenberg 1981). Relevant for eutrophication were contributions about the physical oceanography of the Baltic (Kullenberg and Jacobsen 1981), the oxygen and hydrogen sulphide conditions (Fonselius 1981), phosphorus (Nehring 1981) and nitrogen (Gundersen 1981).

Fonselius concluded that the decrease in oxygen concentrations in several parts of the Baltic since the beginning of the century, had most probably been caused by two large salt-water inflows, the first one during World War I and the second in 1951. These inflows had reinforced stratification and, consequently, oxygen depletion below the halocline. The first inflow had caused continuously declining oxygen values in the bottom water of the Gotland deep, but this water was renewed in 1932. The recovery from the last inflow seemed, according to Fonselius, to be very difficult. He mentioned increasing anthropogenic inputs of organic material as a possible cause, but stated also that it was presently not possible to differentiate between human influences and natural causes. Nehring (1981) showed significantly increasing phosphate winter concentrations in the surface waters of the southern Gotland Sea since 1969. However, in the Gulf of Bothnia phosphorus concentrations seemed to be declining (see also above). There was a close correlation of the winter phosphate values with salinity, which made Nehring conclude "that the eutrophication appears to be related to hydrographic processes." With regard to phosphorus in the deep layers, Nehring pointed to the Gotland Sea, where phosphate accumulation had been documented. This phenomenon had, however, only been found in the central Baltic and not in the Arkona and Bornholm seas, nor in the Gulf of Finland or the Gulf of Bothnia. Nehring concluded: "At present it is not possible to say with certainty whether the accumulation of phosphate observed in the deep water of the central basins is mainly due to natural causes or is a consequence of increasing pollution."

A description of the distribution and cycling of nitrogen compounds was given by Gundersen (1981), who concluded:

> "As judged from data which can be considered reliable, and which have been collected during the last 10–15 years, it does not appear that the increasing degree of pollution of the Baltic is reflected in the nitrogen picture of the Baltic as a whole. The most important conclusion to be drawn from this study is that all the various biological processes which, in the so-called nitrogen cycle, participate in the transformation processes of the diverse species of nitrogen seem to be healthy and in balance."

Kullenberg (1981), in the editorial to the special issue, presented a number of overall conclusions. With regard to primary production he stated that, even though it was difficult to compare data, there had been an increase in some parts of the open Baltic and a clear increase in coastal waters. Different factors (mainly phosphorus, nitrogen and light) acted as limiting factor in the various sub-areas of the Baltic Sea, also depending on the season. The average annual primary production was about 100 gC/m^2 which was, according to Kullenberg, "rather in agreement with the average." In some coastal waters, especially in archipelago areas, the discharge of nutrients had caused local eutrophication. Generally, the dominating source of nutrients in the open Baltic was the natural circulation, but "it is conceivable that the human input has an effect on the production also there." Kullenberg finally stated: "On the basis of present knowledge one can clearly conclude that the Baltic is not dying."

3.2.7 Estuaries and nutrients

Several international scientific symposia about marine pollution had been held since the end of the 1950s, but none had dealt exclusively with eutrophication. The first to do so was the "International symposium on the effects of nutrient enrichment in estuaries," convened 1979 in Williamsburg, Virginia, USA, in the framework of the Chesapeake Bay Programme. Eutrophication was one of the principal areas of focus of this five-year research programme because of the recognition "that nutrient loading into the Chesapeake Bay system had been and was the source of serious and extensive damage to parts of the estuary – and that the resident population is projected to double within a few decades" (Neilson and Cronin 1981). The editors of the conference proceedings expressed the hope that its value

> "may appear in many forms – in improved management in the Chesapeake Bay area and in other estuaries, in stimulation of further advance in comprehension of estuaries and the questions of adequate and excessive enrichment, in guidance for the expenditure of sufficient funds for prevention of serious problems without wastage because of ignorance, and in improved knowledge among managers, scientists and citizens concerned with achievement and maintenance of the wisest balance in our environment" (Neilson and Cronin 1981).

The papers, published in the proceedings, mainly dealt with estuaries in the United States. Also several cases from Australia had been presented, but none from Europe. About half of the papers from the proceeding were invited review papers, in which attempts of the authors to translate scientific knowledge into applied management were clearly recognisable. At the same time, however, the problems associated with these attempts became obvious, mainly because of the poor state of scientific knowledge concerning effects of enrichment of estuaries, compared to freshwater systems.

The relevance of freshwater studies

Schindler (1981) reviewed the relevance of eutrophication studies in lakes for estuaries. He referred to the so-called "Vollenweider model" which relates phosphate loading and depth to the eutrophication status of a lake. Schindler (1981) stated: "Due to the apparent complexity of mixing processes in estuaries, the simple, one-element models which have proven useful in lakes are unlikely to be applicable, even when limitation by a single element occurs." According to Schindler (loc.cit.), studies had shown that nitrogen was the limiting element in North American east coast estuaries, but he also wondered whether nutrient limitation was significant at all. No background data existed on the basis of which it could be concluded whether P or N limitation had developed as a result of anthropogenic inputs, or whether nitrogen had been limiting in pristine conditions as well. Also McErlean and Reed (1981), who gave a review of indicators of estuarine overenrichment, warned for applying freshwater methodology to estuaries. According to these authors, the measurement methodology of freshwater systems had reached a high degree of sophistication. Marine and estuarine eutrophication, however, had only in the last decades become a concern, and researchers and managers had attempted to apply freshwater concepts and methodologies to these systems, assuming basic similarities. McErlean and Reed seriously questioned "the acceptance of processes or preconceptions which have not been validated for estuarine areas and which may have significant economic penalty if misinterpreted." They concluded that the application of freshwater indexes and indicator techniques to estuarine and coastal areas had been done with "varying degrees of success" because of a lack of a widely accepted definition of an estuary, a basic lack of knowledge of nutrient limitation and cycling and possible fundamental differences between estuaries and other water bodies.

Classifying estuarine eutrophication

Jaworski (1981) presented a proposal for an estuarine specific, qualitative classification of eutrophication stages, and attempted to relate these classes to the anthropogenic nutrient loading of various estuaries. He also included the North Sea in this comparison, using data of James and Head (1972), which were based mainly upon the 1969 ICES inventory (2.5.1). In order to make nutrient loadings between the systems comparable, Jaworski (1981) expressed them as inputs per m^2 per year and inputs per m^3 per year by dividing the amounts of anthropogenic nutrient inputs by the surface area and volume of the receiving water body. For the North Sea he arrived at very low relative input levels of both nitrogen and phosphorus because he used the whole surface area, or volume of the North Sea. On the basis of these normalised data, Jaworski (1981) concluded that estuaries had a much higher loading than the North Sea, and that the higher the phosphorus loading, the more "excessive" the eutrophication conditions were. He furthermore related N/P ratios to the eutrophication status of the investigated estuaries, and found an almost equal share of phosphorus and nitrogen limited systems. On the basis of his findings he proposed a "permissible" phosphorus loading, under which excessive eutrophication could be prevented (Jaworski, loc.cit.).

Also Ryther and Officer (1981) attempted to specify the eutrophication status of estuaries. They did so by describing the characteristics of "good" and "bad" algae. A typical beneficial type of alga was, according to Ryther and Officer (loc.cit.), one that grows fast and can outcompete undesirable species. It decomposes rapidly and is also a good food source for zooplankton or benthos. Because of these characteristics such species do not often cause nuisance blooms. Undesirable algae, on the other hand, do not have these features and may develop nuisance or toxic blooms." Ryther and Officer (loc.cit.) prepared a ranking of seven taxonomic algal groups, ranging from beneficial to undesirable. Centric diatoms were identified as the most beneficial group, and bluegreen algae the least beneficial. Because, generally, species from all these categories appeared together in the water column, they tried to answer the question which would be the main factors determining their relative proportions. They discussed the relevance of high temperature, low salinity, availability of organic nitrogen, toxic contaminants and low silicon levels, and concluded: "Except in extreme cases, it seems unlikely that any single factor normally controls species composition of phytoplankton." It was unfortunate, Ryther and Officer concluded, that these factors frequently tended to coincide in estuaries, due to human enrichments.

Remineralisation and nutrient cycling

"If all we know is the amount of nitrogen entering an estuary from sewage or any other source, we have not learned anything that is very useful. In order to assess the importance of nutrient inputs in sewage, it is necessary to compare them to others in the nutrient budget to see how much of the primary production it might sustain, how the input compares with recycled nutrients that are already turning over in the system, and how it compares with other inputs and losses."

On the basis of this rationale, Nixon (1981) comprehensively discussed the relevance of recycling of nutrients in estuarine and coastal waters. According to Nixon (loc.cit.), there had been an "increasing appreciation" of the complexity of the processes involved in the remineralisation and recycling of nutrients in coastal marine ecosystems in the last 30 years. The classical concept of the recycling of nutrients via the decomposition of dead organic matter by bacteria and fungi, had turned out to be much more complicated (compare figure 3.4).

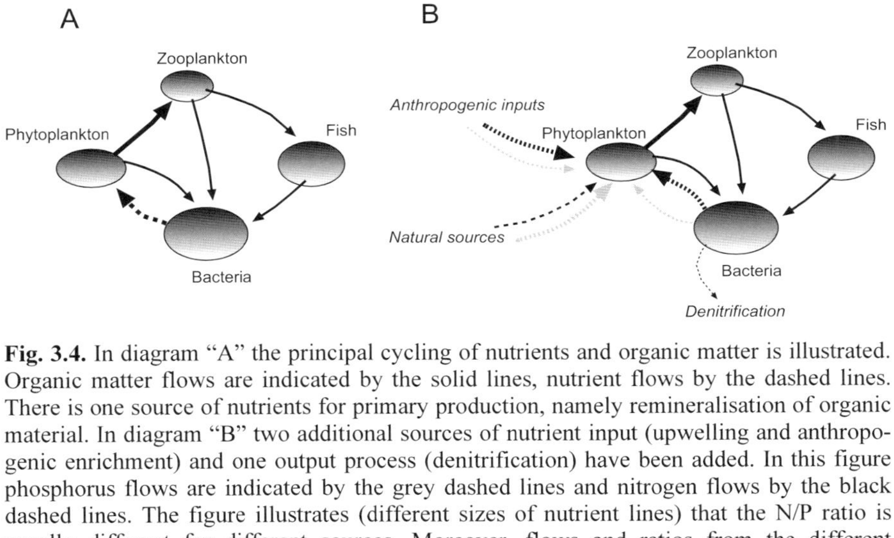

Fig. 3.4. In diagram "A" the principal cycling of nutrients and organic matter is illustrated. Organic matter flows are indicated by the solid lines, nutrient flows by the dashed lines. There is one source of nutrients for primary production, namely remineralisation of organic material. In diagram "B" two additional sources of nutrient input (upwelling and anthropogenic enrichment) and one output process (denitrification) have been added. In this figure phosphorus flows are indicated by the grey dashed lines and nitrogen flows by the black dashed lines. The figure illustrates (different sizes of nutrient lines) that the N/P ratio is usually different for different sources. Moreover, flows and ratios from the different sources and sinks will change with time. It is, therefore, extremely difficult to predict the amounts and proportions of nutrients available for primary production.

Not only benthic bacteria, but also pelagic bacteria were involved, as well as zooplankton, protozoa, animals and plants, and it was not clear which group played the most important role. Moreover, there were large differences in duration of the regeneration and uptake processes, both within and between species groups and for different substances. On the basis of his own experimental work, Nixon (loc.cit.) stressed the importance of benthic remineralisation, which was responsible for 25–50% of the consumption of primary production and, consequently, for a large flux of inorganic nutrients to the overlying water. This flux, however, usually contained the elements N and P in a ratio of far less than 16, which Nixon (loc.cit.) explained by denitrification losses. This could, according to Nixon (loc.cit.), explain the observation that nitrogen is commonly the nutrient most limiting primary production in estuarine and coastal systems. If denitrification was a widespread phenomenon in coastal waters, it might be a major sink in the global nitrogen budget. The fact was, however, that eutrophication problems were increasing in estuaries, which was, according to Nixon, a "dramatic warning that anthropogenic nutrient inputs can overwhelm the recycling and remineralisation processes in coastal waters."

3.3 The relative importance of marine eutrophication

In the foregoing sections of this chapter, basic elements of marine eutrophication research have been described on the basis of a selection of publications from the scientific literature. In this section an attempt is made to evaluate the importance of marine eutrophication research, relative to marine pollution research in general. The basic assumption is that the outcome of such an evaluation is, to a certain extent, indicative of the estimation by the scientific community of the relevance of marine eutrophication as a pollution problem. First, in Sect. 3.3.1, an analysis is made of the contents of the Marine Pollution Bulletin (MPB) over the period 1970–1980. Second, in Sect. 3.3.2 the journal Ambio will be analysed for publications related to marine eutrophication for the period 1972–1980. The choice for these two periodicals was made because they both deal with pollution research. In Sect. 3.3.3 two scientific textbooks about marine pollution, both published in 1976, will be analysed for their coverage of the marine eutrophication theme.

3.3.1 The Marine Pollution Bulletin

Table 3.1 contains the results of the analysis of the MPB. Of the reports, news items, reviews and letters published, only the reports have been used for the analysis. The results clearly show that throughout the 1970s "metals" and "oil" were, with an average share of more than 20%, respectively 30%, the by far predominant categories. At the end of the 1970s the number of reports dealing with organohalogens (mainly PCBs, DDTs and other organochlorines) slowly started to increase. From 1970–1980 only 15 out of 621 contributions dealt with nutrients and eutrophication, which clearly shows the limited importance attached to this topic. Of these 15 reports, 11 dealt specifically with nutrients and eutrophication cases around the world, namely an inventory of phosphate and phytoplankton in inshore British waters (Sykes and Boney 1970), an analysis of nutrient discharges from British North Sea estuaries (Head 1970), the effects of nutrients on Kaneohe Bay in Hawaii (Johannes 1971), nutrient enrichment by sewage discharges causing increased growth of green algae in the Clyde area (Perkins and Abbott 1972), high nutrient levels and oxygen depletion in Kingston Harbour (Jamaica) (Wade et al. 1972), abatement measures for the Great Lakes (Sanger 1972), distribution of nutrients in the North Sea (Folkard and Jones 1975) (see 3.2.5), eutrophication of an estuarine mudflat near Dublin (Fahy et al. 1975), eutrophication of an estuary in Tasmania (Buttermore 1977), blooms of the alga *Cladophora* in Bermuda (Bach and Josselyn 1978) and eutrophication of the Adriatic Sea (Degobbis et al. 1979). The other four articles also addressed other pollutants. Nearly all 15 cases dealt with the eutrophication of estuaries or enclosed bays. The only papers concerned with eutrophication of sea areas were those by Folkard and Jones (1975) for the North Sea and Degobbis et al. (1979) for the Adriatic Sea.

Table 3.1. Papers published in the Marine Pollution Bulletin in 1970–1980, according to subject categories

	1970	'71	'72	'73	'74	'75	'76	'77	'78	'79	'80	Total	%
Organic waste	3	1	1	0	3	0	1	2	2	2	0	15	2,4
Industrial waste	1	1	1	1	1	0	1	1	3	2	2	14	2,3
Dumping (sewage sludge/ dredge spoils)	1	1	3	3	1	0	0	0	0	0	0	9	1,4
Nutrients/ eutrophication	**3**	**1**	**2**	**1**	**2**	**2**	**0**	**1**	**2**	**1**	**0**	**15**	**2,4**
Oil/detergents	20	12	11	13	12	21	19	9	29	21	24	191	30,8
Radionuclides	2	1	0	1	0	1	1	1	3	1	0	11	1,8
Metals	0	5	6	13	11	12	21	19	14	18	13	132	21,3
Organohalogens	3	1	1	4	4	3	5	5	2	10	7	45	7,2
Pollution general	4	6	0	1	3	2	4	2	2	1	5	30	4,8
Thermal	2	1	0	2	0	0	0	3	1	2	1	12	1,9
Microbiology/ consumer safety	1	0	2	0	1	1	0	6	1	2	4	18	2,9
Biotoxins	1	0	0	1	0	0	0	0	0	0	1	3	0,5
Solids	0	2	3	2	2	0	1	2	1	2	5	20	3,2
Species/habitats	0	0	1	3	1	0	2	0	2	2	1	12	1,9
Methodology	11	10	4	4	1	1	1	2	5	5	2	46	7,4
Policy/ legal/ management	4	3	3	1	1	1	0	4	1	1	1	20	3,2
Other	3	2	3	1	1	4	1	2	4	6	1	28	4,5
Total	59	47	41	51	44	48	57	59	72	76	67	621	100,0

3.3.2 AMBIO

Ambio deals with the environment in general and, consequently, the number of contributions related to marine pollution is much lower than in the Marine Pollution Bulletin. The results for Ambio have therefore been summarised for the period 1972–1980. In this nine-year period 60 articles dealing with marine pollution were published. The scoring, according to the same categories as for the MPB, is given in Table 3.2. For comparison the totals of the MPB are also shown in this Table. The AMBIO results for the category "Oil and detergents" show a high conformity with the MPB. This is also the case for the sum of the categories "Metals" and "Organohalogens" (about 30%), although the individual scorings for these groups show that Ambio paid relatively more attention to organohalogens and the MPB to metals. Only two contributions in Ambio were about eutrophication, supporting the conclusion drawn above that marine eutrophication was of limited importance as a subject of marine pollution research. For Ambio this is, however, somewhat surprising because this journal had a strong Scandinavian focus and, as demonstrated in Sect. 3.2.6, eutrophication was considered an essential problem of the Baltic Sea.

Table 3.2. Absolute and relative number of papers about marine pollution in Ambio and the Marine Pollution Bulletin (MPB), classified according to subject categories

	AMBIO 1972–1980	%	MPB 1970–1980	%
Organic waste	1	1.7	15	2.4
Industrial waste	5	8.3	14	2.3
Dumping (sewage sludge/dredge spoils)	0	0.0	9	1.5
Nutrients/eutrophication	**2**	**3.3**	**15**	**2.4**
Oil and detergents	16	26.7	191	30.8
Radionuclides	0	0.0	11	1.8
Metals	8	13.3	132	21.3
Organohalogens	9	15.0	45	7.3
Pollution general	8	13.3	30	4.8
Thermal pollution	0	0.0	12	1.9
Microbiology/consumer safety	0	0.0	18	2.9
Biotoxins	0	0.0	3	0.5
Solids	0	0.0	20	3.2
Species/habitats	2	3.3	12	1.9
Methodology	3	5.0	46	7.4
Policy/management/legal	3	5.0	20	3.2
Other	3	5.0	28	4.5
Total	60	99.9	621	100.1

3.3.3 Eutrophication in marine pollution textbooks

Marine Pollution

Under the editorship of Johnston of the Marine Laboratory in Aberdeen, eight experts contributed to the textbook "Marine Pollution" (1976). In the preface, Johnston wrote that the book "should prove helpful to those in government, local authority, industry and environment protection generally who have to tackle pollution problems." Johnston also expressed the hope "that senior delegates to international organisations dealing with marine pollution will use this work (and others) to establish for themselves a deeper understanding of the subject rather than merely get by with snatches of generalities from their advisers" (Johnston 1976).

Next to sections dealing with mechanisms in marine pollution, biological response to pollutants, dispersion, heavy metals, oil, seabirds and pollution and legal aspects of pollution, an elaborate chapter by Topping was dedicated to the effects of discharges of sewage into the sea (Topping 1976). Topping discussed the role of nutrients in the marine system, the effects of nutrient enrichment and the fate and effects of viruses and bacteria from sewage. As main sources of nutrient enrichment of N and P, he mentioned sewage wastes. Agricultural runoff and waste from food processing industries were considered small, compared to sewage inputs. With regard to the role of N and P in biological systems Topping stated that, unlike in freshwater systems where P appears to be limiting, the limiting factor for primary production in the sea appeared to be the amount of available nitrogen. In this respect, Topping also referred to experimental evidence for the fact that P is regenerated faster as a result of which there will always be a small but persistent supply of P. He also mentioned recent work of Ryther and Dunstan about the relevance of the N/P ratio (see 3.2.2) and discussed the effects of deviations from the 15 to 1 ratio for species diversity. Topping concluded that unbalanced nutrient supply might lead to changes in species composition, generally at the expense of the more sensitive species. In an assessment of the general effects of sewage enrichment, Topping made a clear differentiation between coastal waters and the open ocean. He presented results of calculations of the effects of sewage inputs on oxygen, nitrogen and phosphorus levels in 10 km, 20 km and 50 km stripes along the North Sea coast. The calculations were based upon input data collected by ICES (3.4.1). For the oxygen levels in the North Sea, he concluded that, under the theoretical assumptions that no re-aeration would take place and that no breakdown of organic matter would occur in rivers and estuaries, there would be only three areas (outer Thames, Netherlands and Germany) where oxygen levels could be depressed to low levels, but that these would be confined to a 1 km small strip. Topping therefore stated: "In practice, there is little evidence of significantly low O_2 levels in the North Sea outside the influence of the major rivers and upper reaches of estuaries." Levels of nitrogen and phosphorus were high in the southern North Sea only, but the N/P ratio in these enriched areas did not differ much from background values, which Topping took as an indication that enhanced growth rather than changes in diversity might be expected. According to Topping, the nutrient enrichment in the coastal strip was a clear indication of man's influence, but "It is much less easy to produce an equivalent assessment of biological response." He referred in this respect to the results of the Continuous Plankton Recorder (see 3.2.5) and underlined the need for differentiating between natural and man-made causes. Contrary to the open sea, the signs of pollution were most obvious in many fjords and estuaries, mainly expressed by oxygen depletion and reduced species diversity. Topping finally pointed to the limited knowledge concerning pollution. He concluded:

"We are beginning to learn the extent of the present marine pollution both for sewage and industrial inputs; if we do not want the problem to be exacerbated, perhaps uncontrollably, we need to take action now while we apparently have time on our side. The solution to these problems will require considerable thought by experts in many fields, philosophers as well as practical men, a lot of money and perhaps a change in our attitude towards a life style compatible with survival, or recovery perhaps, of the land and sea and their non-human populations."

Marine Pollution: Diagnosis and Therapy

The original version of this textbook by Sebastian Gerlach was published in 1976 under the title "Meeresverschmutzung; Diagnose und Therapie." Ten topics were covered, namely domestic waste, industrial waste, dumping, oil pollution, persistent dangerous substances, mercury, lead, other heavy metals, organochlorines and legislation. In the chapter "domestic waste," Gerlach described the biological degradation of organic material discharged into the sea. He stated that in open coastal areas with sufficient and well-mixed water masses, the breakdown of domestic waste could take place without considerable environmental impact. "Also in a marine area with strong tidal currents like the Öresund between Denmark and Sweden," Gerlach continued to say, "it is debatable whether it is necessary to build elaborate treatment plants in order to bring about the breakdown of organic matter." He mentioned the different Swedish and Danish positions, the first being in favour of complete treatment, the second regarding this as unnecessary. According to Gerlach, problems would only occur where sewage was discharged into narrow fjords and estuaries. Sewage was the main source of phosphorus, and agriculture the main source of nitrogen inputs. Through biological sewage treatment only one third of the phosphate would be retained. For the North Sea as a whole about 15% of phosphorus inputs would, according to Gerlach, derive from riverine sources, the large majority coming from the Atlantic. It could not be stated with certainty that the North Sea as a whole showed effects of increasing eutrophication and it had, as yet, not been possible to relate the higher fish yields to higher amounts of nutrients (3.2.5). For the Dutch and German coastal zone the picture was quite different, with an equal share of riverine and marine nutrient loading. Gerlach pointed to the increased discharges of the Rhine River from 3000 tonnes P in 1932 to 30,000 tonnes in 1970 and the high P concentrations in the southern North Sea and the German Bight. It was, however, unfortunate that so few long time observations of nutrient concentrations and phytoplankton existed, and that the effects of eutrophication could only be traced on the basis of a few examples. Gerlach referred in this respect to the Dutch Wadden Sea and the Helgoland area (3.2.5). According to Gerlach, phosphorus normally limited growth of phytoplankton. In the Dutch Wadden Sea the phytoplankton production had doubled and also in summer phosphate was abundant. Because planktonic algae need nutrients with a P:N:Si ratio of 1:15:7, and silicon loading had not increased, the latter nutrient was now limiting. Also in the Helgoland area, increased phosphorus concentrations had been observed, but, due to large interannual variations, an increasing trend for phytoplankton growth could not yet be established.

Gerlach comprehensively discussed the situation in the Baltic Sea for which, in his opinion, it had been proven without doubt that the upper layer was eutrophied, but for which it was still an open scientific question whether this could influence the oxygen situation in the deeper layers. He referred to the changes in the water exchange between North Sea and Baltic Sea, caused by changes in wind climate in the North Atlantic Ocean. Should it be the case, Gerlach stated, that natural causes would change the Baltic so much that, compared to this, anthropogenic eutrophication would only play an insignificant role, then the measures to clean sewage, which were presently carried out by the Baltic states, would not be necessary. Treatment plants would only have to be planned in accordance with local hygienic requirements, or with requirements for maintaining the self-purification capacity of surface waters. Gerlach concluded that from the perspective of marine pollution, eutrophication was not a problem that could reach global dimensions. The perspective of providing raw materials would be more important.

Gerlach also described effects of local eutrophication on species diversity. Generally, the original vegetation was replaced by more robust species, and average biomass was higher than before. There were also species profiting from this. With regard to the occurrence of large plankton blooms, for example *Coscinodiscus*, or large numbers of jellyfish on beaches, Gerlach warned that these were not automatically related to marine pollution, as some might want us to believe. He also pointed to natural changes in population sizes. Man might only partly be responsible as a trigger, and in many cases it seemed to be that a coincidence of factors was responsible for extraordinary events. It would be an important task to register such events as a key to causal understanding.

3.4 Marine eutrophication and official bodies

Whereas in the foregoing sections the focus was on how the academic scientific community viewed marine eutrophication, this final section will deal with the official perception of the issue. How did official scientific advisory bodies assess marine eutrophication and how far had the issue progressed within relevant science-policy networks? These questions will be investigated for the ICES-Osparcom network and the ICES-Helcom network, dealing with the North Sea and the Baltic Sea respectively (2.6.2). First, in 3.4.1, the activities of ICES, which is the scientific advisory body for both Osparcom and Helcom, will be evaluated. Next, in 3.4.2, the first official assessment of the status of the Baltic Sea by Helcom will be addressed. In addition to these international assessments, Sect. 3.4.3 addresses the comprehensive analysis of environmental problems of the North Sea, carried out by the German Council for Environmental Affairs. This report proved to be of high relevance for international political developments with regard to the protection of the North Sea, which are comprehensively covered in Chap. 4.

3.4.1 ICES

In 2.5.1 it was described how, at the end of the 1960s, the International Council for Exploration of the Sea (ICES) became involved in marine pollution issues, resulting in 1969 in the publication of a report on the pollution of the North Sea (ICES 1969) and, in 1970, a report on the pollution of the Baltic Sea (ICES 1970). At the beginning of the 1970s several international developments in the field of marine pollution stimulated ICES to continue this type of work. At the 1970 FAO Conference "Marine Pollution and Sea Life" (2.5.2) it was recommended that pilot regional monitoring exercises be carried out and that ICES should organise these for the North Sea (ICES 1974a). In 1971 the third meeting of GESAMP (2.5.2) had proposed to carry out baseline studies, and that the IOC, together with ICES, should cooperate with regard to the North Sea. In 1971 the IOC agreed that baseline studies on marine pollution should be carried out, starting with regional studies (ICES, loc.cit.). In this section, activities of ICES, related to marine pollution in the 1970s, will be described, with specific emphasis on marine eutrophication.

Pollution of the North Sea

In 1971 ICES established a working group with the remit of carrying out a study of the pollution of the North Sea. The study consisted of three main parts, namely an inventory of inputs, a baseline survey of metals and organochlorines in fish and shellfish and a baseline survey of trace metals in water. The results, together with an assessment of the Working Group, were published in 1974 (ICES 1974a). The input data were collected by means of a questionnaire, which was circulated in 1972. The Working Group considered information on inputs "an essential part" of the study of pollution of the North Sea, and stated that, as a result of greater public awareness and the impact of the 1969 report (2.5.1), a more comprehensive study of pollutant inputs was now feasible (ICES, loc.cit.). In the questionnaire, North Sea countries were requested to provide information on sewage discharges, industrial discharges, dumping and atmospheric deposition. On the basis of the sewage input data it was calculated that, on a daily basis, 7.4 million m^3 of sewage entered the North Sea, of which about half into the southern North Sea. According to ICES (loc.cit.), most of the sewage was not treated at all. In Sweden, one third of all sewage was treated biologically, in some other areas this was only done for major discharges, for example in the Rotterdam and London areas, where 61, respectively 70% of the sewage received secondary treatment. Only a very small proportion of sewage received tertiary treatment: 3% in Norway, 8% in Sweden and less than 1% in Germany and England (ICES, loc.cit.).

The annual oxygen demand of the sewage was estimated at more than half a million tonnes BOD. This amount had not caused problems with seawater oxygen levels outside estuaries and fjords, which was, according to the Working Group, "hardly surprising" considering, for example, that a one km^2, one meter deep sea area contains 70 tonnes of oxygen. The yearly inputs of nutrients from the sewage were estimated at some 200,000 tonnes of nitrogen and 36,000 tonnes of phosphorus. It was unclear how much of the nutrients were sedimented or removed by biological activities in estuaries and fjords. The Working Group stated:

"Although no effects have been noted which might indicate a danger of eutrophication arising in the North Sea, there is some evidence of increases in nutrient levels in the waters of the southern North Sea in recent years and these may be linked with increasing amounts of sewage discharged into the sea."

The introductory remarks of the report were more specific about eutrophication: "The Working Group has received information that the levels of phosphate and nitrate in the waters off the Netherlands coast have doubled in recent years and has agreed that an urgent check is necessary to establish whether the increase applies to other areas of the North Sea." Because the effects of nutrients, suspended material and organic matter introduced by the sewage were "not well understood," two study groups were established, one for the Southern Bight area, in which the UK, Belgium, The Netherlands and Germany participated, and one for the Kattegat/Skagerrak area, consisting of representatives from Denmark, Norway and Sweden. Depending upon the results of the work of these groups, further work on P and N budgets would have to be initiated.

Monitoring

In 1974 the Working Group on the Pollution of the North Sea was disbanded and a new group established, the Working Group on Pollution Baseline and Monitoring Studies in the North Atlantic and NEAFC[2] areas. The new group was to conduct a baseline study in the part of the North Atlantic which had not been surveyed yet and to carry out monitoring in the North Sea (ICES 1977b). The Oslo Commission (see 2.6.2) had, at its first meeting in October 1974, decided that ICES would be invited to co-ordinate a baseline study in the Oslo Convention area. The substances the Commission wished to be measured were, in order of priority, organohalogen compounds, metals (lead, mercury, cadmium, zinc, copper and chromium), petroleum hydrocarbons, nutrients (nitrates, ammonia, total N, phosphate and total P) and coliform bacteria. For pragmatic reasons, first priority should be given to fish and shellfish quality for human consumption (ICES 1977a). The survey was carried out in 1975, supplementing the 1972 survey. The Oslo Commission had also suggested that baseline studies should include monitoring of sensitive food chain processes, for example primary production indices (ICES 1975). The ICES Advisory Commission on Marine Pollution (ACMP, see further this section) had studied this proposal and concluded that the methodology for measur-

[2] Northeast Atlantic Fisheries Convention

ing primary production had not been developed sufficiently, and also that it would be hard to relate changes in primary production to contaminant inputs. ACMP therefore advised not to include this parameter in baseline studies (ICES, loc.cit.).

Pollution of the Baltic Sea

In 1971 a joint Working Group of ICES and the Scientific Committee on Oceanic Research (SCOR) was established with the aim of studying the inputs of pollutants to the Baltic, carrying out a baseline study of contaminants and co-ordinating a scientific programme for the study of distribution and fate of pollutants (ICES 1977a). The Group was to work in close co-operation with the North Sea Working Group because of the need for intercalibration and standardisation of methods, and because further improved knowledge was needed of the water exchange between the two seas (ICES 1974a). In 1972 the Working Group decided to circulate a questionnaire amongst the Baltic countries, requesting information on the amounts and sources of discharges into the Baltic Sea (ICES 1977c). The aim was to obtain more detailed information than was contained in the report from 1970, which had been compiled by the ICES Working Group on the Pollution of the Baltic (see 2.5.1). The ICES/SCOR Working Group considered information about inputs of pollutants a prerequisite for the determination of measures to protect the Baltic, as well as for applied and fundamental research. It was furthermore stated that such information would be necessary to be able to develop mass balances and numerical models. Data on nutrient inputs were considered important for the evaluation of the risks of eutrophication.

On the basis of the information received, the Working Group was able to calculate for the year 1972 an estimate of total BOD and nutrient inputs from sewage and industrial discharges (see also 3.2.6). A total BOD of some 1.1 million tonnes per year was calculated, of which about 750,000 tonnes resulted from industrial sources. Total yearly nitrogen inputs were calculated for sewage only and amounted to some 77,000 tonnes. Total phosphorus inputs were estimated at 33,700 tonnes per year of which 27,000 came from sewage. According to the Working Group, about 40% of direct and 20% of indirect sewage inputs were not treated at all and about one third of all sewage inputs was treated biologically. Only 3% of the direct and 15% of the indirect discharges received additional treatment. This was mainly the case in Sweden and Finland. It was concluded that in some cases more information had become available, mainly on sewage and industrial waste, but detailed figures on oxygen demanding substances and nutrients were still limited. Little information had become available for metals and hardly any for pesticides and PCBs. According to the Working Group, this was caused by a lack of national investigation programmes, a lack of national co-ordination and the fact that monitoring programmes had only started recently (ICES 1977c).

Under the auspices of the Working Group also a baseline survey was carried out to determine levels of toxic substances in living resources and in the environment (ICES, loc.cit.). The study was also intended to parallel the North Sea baseline study. The substances covered in the survey, which was carried out in 1974, were mercury, lead, cadmium, zinc, copper, BHC, Dieldrin, DDT, PCBs and oil.

In 1973 the Working Group elaborated proposals for a research programme into pollution problems of the Baltic Sea (ICES 1974b). In the introduction to the report, the Working Group emphasised that it had dealt with problems of the Baltic as a whole and not with local problems. According to the group, in many cases "local problems cannot be understood without a clear view of the behaviour of the Baltic as a whole." In the report first a brief overview was given of the oceanographic characteristics of the Baltic, followed by a listing of the main sources of pollution. As most significant pollutants, the Group identified toxic substances that appear in the food chain, eutrophicating substances and oil spills. In the discussion about how to approach the problem of assessing the present and future situation of the Baltic, the Group stressed the limited understanding of the system, stating: "We are in fact not able to answer even rather simple and fundamental questions. [....] Much effort is needed before we can say that we reasonably well understand a complicated ecosystem such as the Baltic." The latter was the basic purpose of the research programme, outlined in the report. The Group considered it important that preventive actions should not be delayed, despite the incomplete knowledge and underlined that "it is obvious that some preventive measures will have to be taken before anything like final results from research are available." For the understanding of human influence on the Baltic, more knowledge was needed about physical oceanography (particularly transport and exchange processes), chemical oceanography (distribution, transfer and accumulation of substances), marine biology (in particular the interplay between abiotic and biotic factors) and marine geology (particularly soil erosion, beach processes, sedimentation and exchange processes between the sediment-water interface). For the interplay of the physical, chemical, biological and geological processes, a modelling approach according to the energy circuit approach, developed by Odum (2.3.2), was proposed and elaborated. In discussing the various aspects of basic processes needed for the model, much emphasis was put on nutrient cycling, and the work of Fonselius (a member of the Working Group) was used extensively (see also 3.2.6). On the basis of the discussions, the Group elaborated a number of issues for which research objectives should be formulated. It concerned, among others, the mechanisms of exchange processes with the North Sea, the decisive mechanisms that determine oxygen, nutrient and organic matter budgets in the deep waters and the influence of different policies on these budgets, as well as the significance of eutrophication for food production for human use. It should also be investigated what kind of predictive models could be developed for use in management. The scientific research, necessary to provide answers to these questions, was formulated in a comprehensive international programme. An important part of the programme was carried out in 1977 in the so-called Baltic Open Sea Experiment (BOSEX).

The Advisory Committee on Marine Pollution

The Advisory Committee on Marine Pollution (ACMP) was established in 1973 as a subsidiary body of the ICES Council, with the aim of giving "advice on marine pollution and its effects on living resources and their exploitation to member governments and any intergovernmental body for the control of pollution which may request such advice." The members of ACMP were responsible to the ICES Council only, and the composition of the Committee was not determined by national representation. In its second meeting, ACMP decided that it would produce annually a report "summarising the advice which had become available through the Council's activities in a form suitable for submission to member countries and to co-operating international organisations" (ICES 1974b). The first ACMP report was published in 1974. In the second ACMP report (1975) marine eutrophication was addressed. The Committee made a difference between hypertrophication, the negative effects of which were visible in, for example, the Oslo Fjord and the American Great Lakes, and eutrophication, the effects of which were neutral or even beneficial. According to ACMP, "From quantitative considerations, the effects of added nutrients to the open sea are not likely to be harmful at all" (ICES 1975). The Committee stated that there were no simple rules by which to predict the transition from eutrophication to hypertrophication, although the latter was, in many cases, associated with reduced species diversity. ACMP proposed that investigations be carried out concerning the eutrophication in areas where nutrients were added and referred in this respect to the Southern Bight, where nutrient levels had increased twofold since 1962 (ICES 1975). In the same annual report the Committee addressed toxic dinoflagellate blooms, which had, in the last decade, caused mussel toxicity on the east coast of the United Kingdom. It was recommended to develop and execute a programme for observing mussel toxicity (ICES, loc.cit.).

In 1976 ACMP produced a report for the Oslo Commission (Oscom) and the interim Helsinki Commission (Helcom). In the report to Oscom it was stated that in recent years ICES had been much concerned with the problems of hypertrophication in certain areas, but that it was unlikely that such phenomena were significant outside coastal waters (ICES 1977a). An international programme for the study of eutrophication processes was announced. In 1977 a report to the Oslo, the interim Helsinki and the interim Paris Commissions was compiled. In this report eutrophication was not addressed (ICES 1978). In the 1978 and 1979 ACMP reports some relevant results of the BOSEX programme were mentioned (ICES 1979; 1980). Intercomparison exercises had made clear that there were acceptable correlations for phosphate, but very poor ones for nitrate and ammonia. ACMP furthermore noted that very little was known about the interaction between chemicals and biota, for example in relation to primary and secondary production (ICES 1979). The results of BOSEX also showed that the distribution of nutrients and other variables was very patchy and that "this imposes very great difficulties in the interpretation" (ICES 1980).

Apart from the methodological aspects in relation to BOSEX, eutrophication issues were not discussed in the 1978 and 1979 reports. During this period, ACMP was mainly dealing with the preparation and development of baseline surveys for metals and for organochlorines in fish and shellfish, including intercalibration exercises for these substances, as well as input studies. In the 1980 ACMP report, however, red tides were again addressed (ICES 1981). In 1976 a working group on red tides and eutrophication had been convened, which had studied the possible relationship between unusual plankton blooms and eutrophication, but follow-up studies had not been carried out. In the discussions also reference was made to the results of the international conferences on toxic dinoflagellate blooms (3.2.3), and it became clear that much more effort was needed for the study of factors responsible for major phytoplankton blooms and for improving the capability to predict such blooms (ICES 1981). At the meeting it was suggested to continuously monitor those localities where blooms frequently occurred.

3.4.2 The Helsinki commission

In 1978 the (then interim) Helsinki Commission (Helcom) initiated a review of "the existing data on the parameters, substances, and the processes relevant to, and affected by, pollution in the Baltic Sea and to provide an assessment of the present state of pollution and its effects" (Melvasalo et al. 1981). ICES was requested to provide assistance. The report "Assessment of the Effects of Pollution on the Natural Resources of the Baltic Sea" was compiled and edited by an editorial board, consisting of representatives of the Helcom Scientific and Technical Working Group and ICES. The assessment was meant to be a baseline for the Baltic Monitoring Programme which started in 1979. In 1980 the overall conclusions from the Report were approved by ACMP (see also 3.4.1) and in 1981 by Helcom. In the overall conclusions a differentiation was made between changes due to natural causes alone, changes partly due to human activities and changes due to human activities alone.

Natural changes

The irregular intrusions of high salinity water into the deep basins of the Baltic were regarded as being caused by meteorological processes over northern Europe. Also the distinct layering of the Baltic water into less haline surface water, deep water with higher salinity and bottom water with the highest salinity, was seen as a natural feature of the Baltic Sea system. The deep water was renewed by a more or less continuous inflow through the Danish Straits, whereas the bottom water was only renewed by inflows of water with a high enough density to replace the bottom water. Also the recurrent periods of oxygen depletion in the bottom waters were regarded as natural phenomena. Furthermore, since the beginning of the century, temperature and salinity of the deep and the bottom water (compare 3.2.6) had increased.

Changes due to natural and/or anthropogenic causes

This category addressed changes for which it was unclear whether they had been caused by natural changes, human activities or both. One of these types of changes was the decreasing oxygen content of the bottom water and part of the deep water (compare 3.2.6). Since the beginning of the century it had decreased from 3 mg O_2/l to zero. The size of the bottom area with reduced oxygen varied from year to year, but in 1975 a maximum of 100,000 km^2 had been observed. The low oxygen values were partly related to increased phosphate concentrations. Since the beginning of the 1950s phosphate concentrations in the deep and bottom waters of the central basin had increased up to threefold: precipitated phosphate is released from the sediment under anoxic conditions. In the surface water of the Baltic proper an increase in phosphate winter concentrations had been measured.

With regard to primary production it was stated that it "may have increased during the last two decades in some parts of the open Baltic Sea and it has clearly increased in coastal waters receiving large amounts of municipal waste water." Because of differences in methods it was, however, difficult to compare data. In vicinities of population centres, changes in phytoplankton species composition and an increase in supply of organic matter to the benthic community had been observed, in addition to increased primary production. Also changes in fish populations in these areas had been found, although it was difficult to distinguish between pollution, fishery and natural causes.

Changes caused by human activities

The section on changes caused by human activities almost exclusively dealt with toxic synthetic chemicals, such as PCBs and DDT, heavy metals and oil. With regard to nutrients the only statement made was that the discharge of nutrients had caused hypertrophication in coastal areas with a low rate of water exchange.

Unresolved issues

For the above described observed changes, the conclusions were based upon "the most widely accepted opinions." However, as stated in the report: "Even though the Baltic Sea is one of the most extensively studied sea areas, it is evident that there are several cases where different opinions regarding the causes for and the effects of these changes still exist." The unresolved issues mainly related to eutrophication. It concerned the causes and effects of eutrophication and its relation to the oxygen depletion of the deep basins, the nitrogen cycle, which was "very complicated and poorly understood," and the limiting role of nitrogen compounds for primary production. According to the report, these issues

> "cannot be resolved without extensive studies on the factors regulating the Baltic ecosystem. For instance, in order to resolve the problem to what extent the primary production in the Baltic Sea has increased and to what extent this is beneficial or harmful, not only longer time-series but also more frequent observations are needed."

In 1981 Helcom adopted a recommendation (Helcom Recommendation 2/8), in which the Governments of the Baltic Sea states were invited to "carefully consider the document 'Assessment of the Effects of Pollution on the Natural Resources of the Baltic Sea' and to take into account the results of this assessment when taking measures towards the abatement of pollution on the Baltic Sea."

In 1979 the Baltic Monitoring Programme (BMP) started with an experimental stage, comprising a limited number of stations and measurements, but with a comprehensive set of mandatory parameters, among which nutrients, phytoplankton primary production, phytoplankton species composition and biomass and soft-bottom macrozoobenthos (BMEPC 1988).

3.4.3 The German Council on Environmental Affairs

In 1977 the German Minister of the Interior requested the Expert Council for Environmental Affairs (Rat von Sachverständigen für Umweltfragen) to prepare a special assessment of the North Sea (Rat von Sachverständigen 1980). The Council, consisting of independent experts, had been installed in 1971 with the remit of periodically assessing the state of the environment. Already in 1976 a special assessment of the Rhine River had been published, and the council itself had already then acknowledged the need for a special report about the North Sea. The report "Umweltprobleme der Nordsee" (Environmental Problems of the North Sea) covered the whole North Sea, although specific emphasis was put on the Wadden Sea and the German Bight. The report not only addressed pollution issues but also impacts by other human uses, such as coastal development and fishery, as well as the legal situation with regard to environmental protection.

Organic pollution and eutrophication were discussed in the chapter concerning the effects of anthropogenic inputs to the North Sea. The Council pointed to the "problematic situation" in the German Bight, south of Helgoland, an area where the sewage sludge of the city of Hamburg was dumped. According to the Council the area was, as a result of thermohaline stratification and the resulting poor oxygen supply to the bottom water, not very suited as a dumping area. Moreover, the bottom contained already a high amount of organic material from river sediments. Also the Wadden Sea received a high organic waste load, but had, by nature, a higher self-purification capacity. Although the Wadden Sea as a whole was not threatened, there were local effects and it was recommended to reduce inputs of organic substances. For most parts of the North Sea, however, a moderate input of easily degradable substances was considered acceptable, provided the oxygen supply was sufficient.

With regard to nutrient enrichment, the Council stated that in the central and northern North Sea no anthropogenic increase of inorganic phosphorus and nitrogen compounds had been found, whereas such was the case for coastal areas, the southern Bight, the inner German Bight and the Wadden Sea. The most comprehensive research had, according to the Council, been carried out in the Dutch Wadden Sea by Postma and co-workers. Not only the increase of Rhine phosphate inputs was responsible for the observed phosphate concentrations in the Dutch Wadden Sea, but also the increased import of organic material from the North Sea (see 3.2.5). Also in the inner German Bight phosphate levels had increased. With regard to nitrogen, it was stated that ammonium levels in the western Dutch Wadden Sea had doubled from 1961–1971, which was attributed to increased Rhine inputs and increased mineralization. The nitrate values had not changed. As a result of the changed concentrations, the P/N/Si ratio had changed in such a way that silicon had become the limiting factor for diatom growth.

The effects of the increased nutrient inputs on primary production were very hard to assess because of a lack of data, and technical problems in the measurement of primary production. For the Wadden Sea, so far, no anthropogenic induced increase of phytoplankton and microphytobenthos growth could be established. Also for the Helgoland area no increase in phytoplankton biomass had been found. According to the Council, an increase in primary production in the coastal area of the Southern Bight had been postulated by Postma, and the increased input of organic matter to the Wadden Sea was seen as a support for this postulation. The Council also addressed blooms of dinoflagellates and stated that a relation between dinoflagellate blooms and eutrophication could not be proven. Reference was furthermore made to the results of the Continuous Plankton Recorder (3.2.5), which did not support a relationship between anthropogenically increased nutrient concentrations and phytoplankton biomass.

The Council concluded that for some areas an increased phosphate level could be determined, and that in the Dutch Wadden Sea ammonium concentrations had increased. An increase in primary production as a result of human activities might, according to the Council, be assumed for the southern North Sea. The observed phenomena were, however, not considered a threat to the North Sea ecosystem. On the contrary, the increased primary production could have positive effects for fisheries. At the same time, the Council stressed that the distribution of nutrients in the North Sea was very inhomogeneous. In regions with stratification and limited oxygen exchange, such as parts of the inner German Bight, increasing nutrient concentrations could lead to ecological changes, such as massive algal blooms and oxygen deficits. It was, therefore, recommended to reduce the nutrient loads of rivers and to avoid inputs of sewage and sewage sludge into critical areas.

3.5 Summary and conclusions

In this chapter a description and analysis have been given of several cases of marine eutrophication during the period 1950–1980. The first scientific report of marine eutrophication dates from 1954 and concerns a case from the US East coast, documented by Ryther and Dunstan (1954). Also in Europe, in the Oslo Fjord, problems with marine eutrophication occurred already in the 1950s and were reported by Føyn (1960). The analysis of the cases addressed in this chapter has focused on determining the main themes in the study of marine eutrophication, the relative importance of marine eutrophication research and the valuation of marine eutrophication as a (potential) pollution problem. The conclusions with regard to these issues are given in Sects. 3.5.1, 3.5.2 and 3.5.3 respectively. In Sect. 3.5.4 these conclusions will be used to discuss the potentials of marine ecology to contribute to the management of the marine eutrophication problem.

3.5.1 The main themes in the study of marine eutrophication

The main themes in the study of marine eutrophication in the 1960s and 1970s, identified in this chapter, are:

- Inputs, sources and levels of nutrients;
- The limiting factor for primary production;
- Nutrient ratios;
- The impact of eutrophication on the occurrence of plankton blooms;
- The effect of increased nutrients on primary and secondary production;
- The impact of increased nutrient levels on the oxygen situation in bottom water.

This selection is based mainly upon the (potential) relevance of the issues for management and politics. The separate elements are discussed in more detail below.

Inputs, sources and levels of nutrients

In both the North Sea and the Baltic Sea research into inputs, sources and levels of nutrients played an important role. Especially ICES was active in the early investigations into the amounts of pollutants discharged into the North Sea and the Baltic Sea. For both seas, most data were available for sewage inputs. On the basis of national information, total inputs of organic material and inorganic nutrients from sewage were calculated. Other important sources investigated were the Rhine and Elbe rivers. For the Rhine, data on nutrient inputs were available as of the 1950s. Some interesting differences between the USA and Europe can be noted. For the North Sea and Baltic Sea, sewage was generally mentioned as the main source of nutrient inputs, with the emphasis on phosphorus from detergents. In US publications both sewage and agriculture were mentioned as important sources, the latter being the main source of nitrogen, resulting from the increasing use of fertilisers. Given this fact, it is surprising that there was relatively little attention for nitrogen in marine eutrophication research in Europe, especially since Vollenweider had already in 1970 emphasised the importance of nitrogen inputs from diffuse (agricultural) sources (see 3.1.1).

Also in the studies of concentrations of nutrients in the North Sea and the Baltic Sea the emphasis was on phosphorus. One plausible reason for the emphasis on phosphorus may be that, as stated in several publications, the analysis of phosphorus compounds was much easier and more reliable than the analysis of nitrogen compounds. Also the fact that the phosphorus cycle is much simpler than the nitrogen cycle may have played a role in the observed difference. Finally, as clearly indicated at the symposium "Nutrients and Estuaries" (Neilson and Cronin 1981), marine eutrophication research was influenced by freshwater eutrophication concepts, in which phosphorus played the dominant role. Already one practical implication of the emphasis on phosphorus can be noted: in Sweden, 65% of the sewage treatment plants had, by 1976, facilities for the removal of phosphorus, and by the end of the 1970s reductions in phosphate levels were documented for the Bay of Bothnia.

The limiting factor

The limiting factor for primary production was, without doubt, the central theme in marine eutrophication research in the 1970s. The reason for this is the direct relevance for management, namely the ability to decide which nutrient to remove from inputs, so as to reduce or prevent possible negative effects of increased primary production. The limiting factor was, however, also the most controversial issue. On the basis of their research findings on the US east coast, Ryther and Dunstan (1971) concluded that nitrogen was the limiting factor for phytoplankton growth. In the North Sea coastal zone and the German Bight, Dutch and German scientists assumed phosphate to be the limiting factor. In the UK, however, nitrogen was considered the limiting factor. Topping (1976) referred in this respect to the fact that phosphorus is regenerated faster than nitrogen. For the central North Sea, Schott and Ehrhardt (1969) assumed nitrogen to be the limiting nutrient. In the first half of the 1970s it was phosphorus that was considered to limit primary production in the Baltic Sea but, in the course of the decade, doubts increased. In the first assessment of the state of the Baltic (1981), P, as well as N were considered possible limiting factors, and the limited knowledge of the nitrogen cycle was regarded a major knowledge gap.

Nutrient Ratios

Closely related to the discussion about the limiting factor for primary production were the deliberations about the consequences of changes in the ratios between the main nutrients N, P and Si. This discussion centred around the relevance of the N/P ratio of 15 to 16, the so-called Redfield ratio, which is the average composition of these elements in plankton cells, and which is also the average ratio in which dissolved N and P can be found in seawater. Ryther and Dunstan (1971) were very pertinent about the limited value of the Redfield ratio for evaluating effects of nutrient inputs to coastal waters. Interestingly, this conclusion from their frequently cited article was hardly reflected in eutrophication studies in other parts of the world. Both Van Bennekom et al. (1975) and Lucht and Gilbricht (1978) compared the N/P ratio in the North Sea Southern Bight, respectively German Bight, with the Redfield ratio of 15–16. Van Bennekom et al (1975) concluded that, despite increasing P levels, this nutrient could still be limiting because the N/P ratio was higher than 15. Lucht and Gilbricht (1978) used the value of 16 as the transition value from P to N limitation. With regard to silicon, Van Bennekom et al (1975) had found that, due to the increasing inputs of nitrogen and phosphorus, silicon had become limiting for diatoms in the Southern Bight.

Plankton blooms

The two main fields of investigation were changes in occurrence of plankton blooms and changes in the composition of the plankton stock. Already in the 1950s and 1960s, increased amounts of phytoplankton had been observed in fjords and embayments, receiving high nutrient inputs. For larger bodies of water, such as the North Sea and the Baltic Sea, the relationship between increased nutrient levels and increasing amounts of phytoplankton appeared not to be so straightforward. On the basis of an evaluation of data from the Continuous Plankton Recorder, Gieskes and Kraay (1977) had found indications for increasing amounts of phytoplankton in the southeastern part of the Southern Bight, an area which is most directly under influence of the Rhine River. However, also in the offshore North Sea the phytoplankton stock had increased. Generally, large-scale climate changes were assumed to be responsible for the observed changes. Hagmeyer (1978) assumed an increasing tendency for phytoplankton concentrations at the monitoring station at Helgoland, but according to Gerlach (1976) and the analysis of the Rat von Sachverständigen (1980) no significant increase could be substantiated. In the Baltic Sea, increases in phytoplankton had been found in coastal waters but not in the central parts. Changes in the composition of the plankton stock could only be demonstrated by the data from the Continuous Plankton Recorder. These could, however, not be related to changes in nutrient levels. Generally, meteorological forcing was assumed to be the cause of the observed changes in phytoplankton and zooplankton species composition.

Toxic dinoflagellate blooms were an emerging issue in the 1970s. Two international conferences, both held in the United States, were dedicated to this theme. The anthropogenic input of pollutants was considered an important factor in the occurrence of such blooms, however, not because of the nutrient enrichment, but because of the impact of chelating substances on metal toxicity. A world-wide increase in frequency and intensity of blooms could not be substantiated. Also in the North Sea blooms of toxic algae and increased nutrient inputs were linked, although it was acknowledged that the effect of increased observations should be taken into consideration.

Primary and secondary production

Data indicating increasing primary production were only available for one station in the Baltic Sea for the period 1961–1968. Otherwise, insufficient data were available to allow for trend analyses. For the North Sea, no primary production data prior to 1965 were available. On the basis of the increased nutrient inputs to the southern North Sea, Postma (1978) estimated that 30% of the primary production in the Southern Bight was caused by nutrient inputs from the Rhine. De Jonge and Postma (1974) considered the increased import of organic material from the North Sea to the Wadden Sea in 1950–1970 as indirect evidence of increased primary production in the North Sea. Also indications for increases in secondary production were scarce. One of the final conclusions of the Symposium "North Sea Fish Stocks – recent changes and their causes" from 1975 was that it seemed likely that the increased nutrient inputs to the North Sea had caused a small, indirect positive effect on fish stocks (Hempel 1978b). For the Baltic Sea, Leppäkoski (1980) referred to benthos data from Gotland and Öland, according to which biomass above the halocline had increased up to threefold in the past 50 years.

Oxygen

The effect of eutrophication on the oxygen content of seawater was one of the main issues in Baltic Sea eutrophication research. One of the problems encountered was the complicated hydrographic situation in the Baltic, with three stratified layers of water. For the deepest parts, oxygen depletion was regarded a natural phenomenon, and time-series since the beginning of the 1950s showed several periods of oxygen deficits, alternating with periods with levels above zero. For the central basin a continuous decline of oxygen levels in the deep water since the beginning of the century had been observed, but it was unclear whether this was due to natural or anthropogenic causes. The possible effects of discharges of nutrients and organic material to the North Sea were discussed by Topping (1976), who concluded that there was little evidence for low oxygen levels in areas outside rivers and estuaries. Also Gerlach (1975) was of the opinion that in open coastal areas with well-mixed water masses, the breakdown of organic wastes would cause few environmental problems.

3.5.2 The relative importance of marine eutrophication research

On the basis of a comparison of the relative number of publications about marine eutrophication in the journals Marine Pollution Bulletin and Ambio, it is concluded that the importance of marine eutrophication research, relative to marine pollution research in general, was low in the 1970s. Oil pollution, heavy metal pollution, and the effects of organochlorines, were by far the most important research topics. The limited role of marine eutrophication as a marine pollution issue is also substantiated by the fact that it was not covered in Goldberg's global textbook "The health of the oceans" (Goldberg 1976). Neither was this the case at the 14th European Marine Biological Symposium (Helgoland 1979), which was dedicated to the theme "Protection of life in the sea." The majority of the presentations at this symposium dealt with heavy metal pollution and oil pollution (Kinne 1980). The third international scientific Wadden Sea Symposium (Ribe, Denmark, 1979), dedicated to the theme of environmental problems of the Wadden Sea region, did not address eutrophication issues either (Tougaard and Helweg Ovesen 1981). Also indicative of the modest role of marine eutrophication research from 1970 to 1980 is the fact that the first international symposium dealing with eutrophication of marine waters, was held at the end of the 1970s and was dedicated to estuaries only (Neilson and Cronin 1981).

3.5.3 The assessment of marine eutrophication

Although, as shown above, there were considerable differences within the scientific community about the causes of marine eutrophication, there was broad consensus that adverse effects of nutrient enrichment were limited to certain embayments and estuaries, and that in the open sea there was a capacity for the breakdown of organic wastes. Several researchers even underlined the principal positive effects of increased nutrient inputs for secondary production in the form of fish and shellfish.

For the North Sea this picture was confirmed by assessments, carried out by official scientific advisory bodies, in this case ICES and the German Environment Council. In the Baltic Sea, however, as concluded on the basis of an evaluation carried out under the auspices of Helcom, marine eutrophication was considered one of the main pollution problems. This difference is reflected in the monitoring programmes developed in the second half of the 1970s. The Joint Monitoring Programme (JMP) of the Oslo and Paris Commissions (Osparcom) that had started in 1979 and that covered the Northeast Atlantic Ocean, only dealt with mercury, cadmium and PCB in organisms, as well as mercury and cadmium in seawater. The Baltic Monitoring Programme (BMP) that had also started in 1979, was much more comprehensive and included also eutrophication relevant parameters, such as nutrients, primary production and macrozoobenthos. Also the international scientific assessment progressed faster in the Baltic: within six years after the signing of the Helsinki Convention (1974) a comprehensive assessment of the quality status of the Baltic Sea had been carried out and published. With regard to eutrophication phenomena, such as oxygen deficits in the deep layers, the report concluded that it was still largely unknown to what extent these were natural or human induced.

3.5.4 Marine ecology and marine eutrophication policy

As clearly illustrated in this chapter, the increasing engagement of ecologists in environmental issues and the mounting expectations of policy makers towards ecology to support the management of environmental problems, described in Chap. 2, are also valid for the marine eutrophication case. As concluded above, marine eutrophication was, by the end of the 1970s, not yet on the political agenda, at least not in the North Sea countries. But should the issue become politically relevant, what might realistically be expected from marine ecologists regarding their support of public policies? In Sect. 1.4, four contextual factors were listed which may influence the use of science in policy-making. It concerns complexity and uncertainty, consensus within the scientific community, different time frames between scientific and political processes and dealing with values. Although, generally, confidence was expressed by scientists about the ability of science to meet the challenge of environmental problems, there were also more modest views. Leppäkoski (1980), for example, felt that the Baltic Sea, although a "minute appendix to the world ocean" was still too large and complicated to be ever fully understood as a natural and man-influenced system (3.2.6). Leppäkoski's assessment of the limits of ecological research may be related to the results of research programmes that had become available, showing that ecosystem processes were more complex than originally thought (see for example 3.4.1 and 3.4.2). Even the development of basic monitoring programmes proved to be very time consuming. The dilemma of ecology's aspirations and limitations is very well illustrated by Kinne during the 14th European Marine Biology Symposium (1979). In his opening address he presented the following statement about ecology:

"Ecological research has become the essential fundament and means for providing the knowledge and concepts necessary for restricting, adjusting and controlling man's logarithmically increasing impact on nature, for protecting life – in the sea, on land and in freshwater – and for sound management of species, environment and living resources. In short: ecological research and its wise application have become the basic prerequisites for civilised human life on this planet to continue beyond the next few decades or centuries" (Kinne 1980).

In the summary session, a few days later, Kinne closed the symposium with the following words:

"The often heard argument 'We cannot act unless ecologists present hard facts on which our action can be based' is unrealistic. Ecologists may in fact never be in the position to provide the kind of solid data presently asked for by politicians. Ecosystem dynamics may turn out to be too complex and too unpredictable to bring sufficient light in its machinery for reliable forecasting of impact consequences at the ecosystem level" (Kinne, loc.cit).

At the end of the 1970s also the second contextual factor, consensus within the scientific community, seemed to be potentially problematic for the successful use of science in marine eutrophication policies. As concluded above, there was much controversy about the limiting factor for primary production, and it was precisely this item that was considered most relevant for marine eutrophication control policies. The contextual factors time and dealing with values were of less importance. The cases described were, with the exception of some locally relevant events, not serious enough to raise substantial public concern and, thus, political pressure.

In addition to the above mentioned contextual factors, which relate to inherent differences between the scientific and political processes, also the institutional or structural setting in which science is transferred to politics is relevant (1.4). In Chap. 2 the development of science-policy networks for the control of marine pollution was described. As shown in the current chapter, in the course of the 1970s marine eutrophication had become a relevant topic in the science-policy networks for both the North Sea and the Baltic Sea.

In the following chapters, it will be investigated to what extent the contextual and structural factors were relevant for the use of marine ecology in marine eutrophication policies for the North Sea and the Northeast Atlantic Ocean.

4 The politics of marine eutrophication

"As is common in the sciences, short-term horizons can turn cycle into crisis, periodicity into problem." (Machlis 1992)

In 1981 the Danish National Agency for Environmental Protection (NAEP) published the results of a five-year study into the water exchange processes and eutrophication in Danish waters. The NAEP, which had been established in 1972, considered the then existing knowledge insufficient as a basis for monitoring and pollution abatement policies. Therefore, in 1973 a research programme, the so-called Belt Project, was started, which lasted until 1978 (Kampmann 1981). In the summary and conclusions chapter of the final report of the Belt Project it was stated:

"The water exchange of the Danish waters is very intensive. The dilution rate is therefore high, and the possibilities for decomposition of discharged substances are good. In areas heavily loaded with nutrients an increase of the concentrations of phosphorus and nitrogen, as well as the plankton algae production has taken place over the last 30-year period. In the open Danish waters this increase is, however, not reflected by increased turbidity or decreased oxygen concentration. Therefore, it can be concluded that problems related to eutrophication do not occur in open Danish waters" (Ærtebjerg Nielsen et al. 1981).

In the autumn of that same year oxygen deficiency was observed in large areas of the Kattegat, the Belt Sea, the Kiel Bay and the German Bight (figure 4.1). In Denmark the event caused a lot of public concern, and in the autumn of 1981 the Environment Minister informed the Danish Parliament about serious threats to the quality of marine and fresh waters (Jensen 1989). As a first step, NAEP was commissioned to carry out a study into the causes of the event. In 1981 also along the Swedish west coast, especially in the Laholm Bay (figure 4.1), oxygen deficits were observed, and also here a literature study into the causes and consequences was commissioned (Rosenberg et al. 1984). German researchers regarded the oxygen deficits in the German Bight as a clear sign of pollution (von Westernhagen and Dethlefsen 1983).

These events can be regarded as the start of marine eutrophication as an international political issue. As depicted in the previous chapter, marine eutrophication in the Baltic Sea was already on the political agenda of the Baltic countries, especially Sweden. This had, however, not led to large-scale international action and marine eutrophication was still a predominantly scientific issue. In this chapter the development of marine eutrophication as an international political issue will be described. Whereas the foregoing chapters still had a rather general focus, the reason being the importance of the global aspects of marine pollution science, this chapter will concentrate on the North Sea. This is because in the 1980s a unique scientific-political process related to North Sea pollution developed, encompassing all aspects of the science-policy cycle as outlined in Chap. 1. Central elements in this process are the three international political conferences on the protection of the North Sea, held 1984, 1987 and 1990.

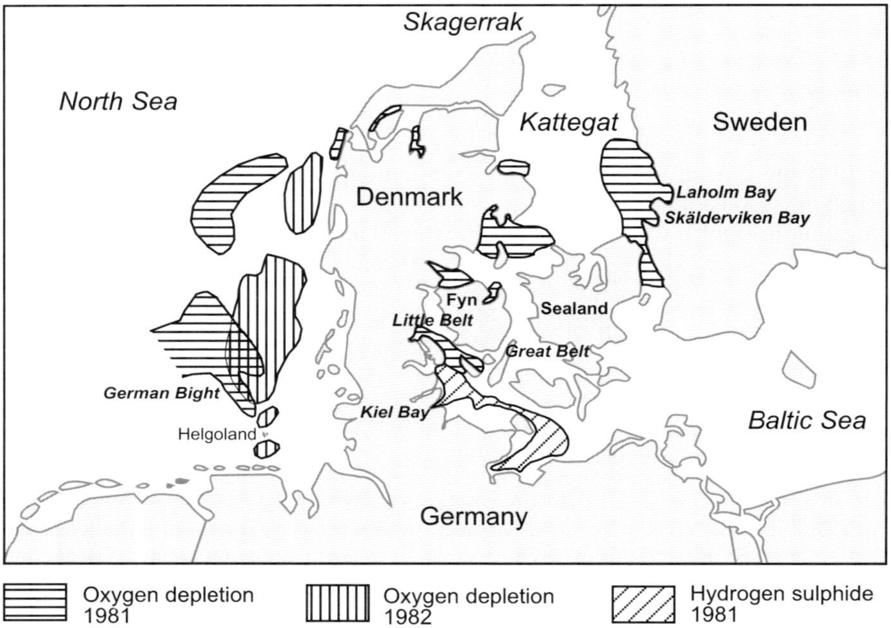

Fig. 4.1. Oxygen depletion in the German Bight, Kattegat and Belt Seas and Kiel Bay. Redrawn from Gerlach (1990)

This chapter addresses all central questions of this study. It concerns the development of the perception of the rational decision-making model, the impact of science on the policy process, the contextual factors that influence the role of science and the structural aspects of the interaction between science and policy. The emphasis of this chapter is on the role of science in political decision-making. More in particular, the following questions will be addressed:

1. Which factors, among which science, have been relevant for the construction of the marine eutrophication problem?
2. To what extent were political decisions on marine eutrophication based upon science?
3. What did politics expect from science in the implementation of political decisions, and has science been able to meet the expectations?

4. How has the science-policy network, in particular the science-policy interface, developed as a result of political developments and which role has it played in the use of science in political decision-making?

This analysis in particular considers the relevance of contextual factors. It concerns the complexity of the problem and the uncertainty of scientific information, the consensus within the scientific community, the difference in time frames between the scientific and political processes, as well as dealing with values.

This chapter has been divided into three main sections covering, respectively, the periods 1981–1985, 1985–1987 and 1987–1990. In 4.1 (1981–1985) the assembly phase of the marine eutrophication problem is addressed. Section 4.2 covers developments during the period 1986–1987, and focuses on political decision-making with regard to marine eutrophication, in particular the decisions taken at the second North Sea conference (INSC-2) in 1987. In Sect. 4.3 the main developments in 1980–1987 are summarised and analysed. Section 4.4 covers the years 1988–1990 and is concerned with activities within the science-policy network, following the political decisions of INSC-2, as well as the preparation of the third North Sea Conference (INSC-3) of 1990.

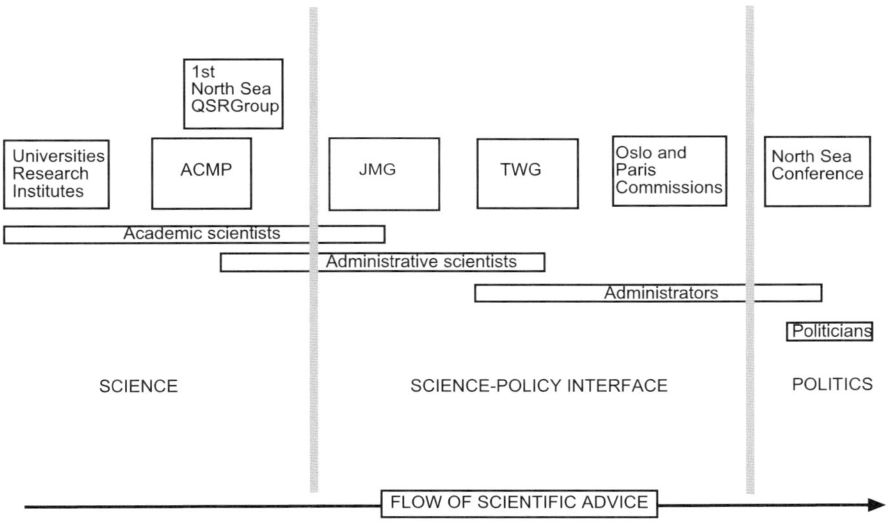

Fig. 4.2. The main elements of the science-policy network in the first half of the 1980s. A schematic indication is given of the type of members of the mentioned elements (Academic scientists, Administrative scientists, Administrators or Politicians), the main field covered (Science, Policy and Management or Politics) and the direction of the flow of scientific advice. Acronyms: see List of Acronyms

For each of the above mentioned periods, the development of and the activities within the science-policy network will be described and analysed, in accordance with the flow of scientific information as assumed in the rational policy-making model (chapter 1). The flow of scientific information through the science-policy network is schematically shown in Fig. 4.2. This figure also indicates the various "actors" in the science-policy process, which will be addressed in this chapter. Generally, each section starts with a description of events, followed by scientific analyses, activities of working groups at the science-policy interface and, finally, of political developments.

4.1 1981–1985: The assembly of a problem

How was it possible that during three decades of increasing awareness of marine eutrophication (chapter 3), the issue had hardly been noticed at the national and international level, whereas, as will be shown in this section, it became an internationally acknowledged problem within less than four years after the 1981 oxygen depletion events? Environmental problems can also be seen as social constructions (Machlis 1992; Hannigan 1995). Machlis (1992) concluded:

> "For all scientific problems, it is through human perception that the challenge to conventional knowledge is raised, through human values that it is accorded importance or ridicule, and through the culture of science that the problem is organized and solutions pursued."

Human values indeed play a central role in whether an issue is perceived as a problem or not. A clear example was already given in Sect. 2.7 for the discharge of sewage of the city of The Hague. Hannigan (1995) has identified three elements in the construction of environmental problems, the assembly, the presentation and the contesting of the problem. For the first element, which includes the discovery, the problem definition and the establishment of the main parameters, science is considered the central forum. The mass media play a dominant role in the presentation of the problem and politics are mainly responsible for invoking action, mobilizing support and defending ownership. In this section, the assembly of the marine eutrophication problem will be investigated, including the main events and actors in this process. The main issue addressed is the importance of science in the assembly of the problem, relative to other factors. First, the scientific developments in 1981–1985, relevant for marine eutrophication, will be described. It concerns the oxygen deficiency events in the Skagerrak-Kattegat area (4.1.1) and in the German Bight (4.1.2). In addition, the situation in Dutch coastal waters is addressed (4.1.3), so as to illustrate national differences in the awareness of marine eutrophication as a problem issue. How the marine eutrophication situation was valued by international scientific advisory bodies, which are part of the science-policy network, is the subject of Sect. 4.1.4, in which the analyses of the ICES Advisory Committee on Marine Pollution (ACMP) and the conclusions of the first North Sea Quality Status Report (QSR) are addressed. Finally, Sect. 4.1.5 investigates how international political fora dealt with the issue.

4.1.1 Oxygen depletion in the Skagerrak-Kattegat and Belt areas

Denmark

As depicted above, the year 1981 was characterized by several oxygen depletion events in the Skagerrak-Kattegat and Belt areas. In Denmark, NAEP started an investigation into the extent and the causes of the observed phenomena. The results were published early 1984 in a report called "Oxygen deficiency and fish kills in 1981; extent and causes" (Miljøstyrelsen 1984a). From the collected data the conclusion was drawn that the area with low oxygen values had most probably been a large coherent area with a size of the islands Fyn and Sealand together (figure 4.1). In some cases also hydrogen sulphide production had been observed. The study into the causes of the event focussed on a comparison of several factors with the situation in previous years. Oxygen depletion events had also occurred in the past, mainly in enclosed bays and fjords, and probably a worsening of the situation had occurred. For open waters, it was stated that a clear aggravation had occurred, and that the situation had never been as alarming as in 1981. The NAEP report suggested that the event had been caused by a combination of several factors. As a result of high discharges of nitrogen from land in the previous winter, there had been a strong spring phytoplankton bloom. The winter concentrations of nitrogen had shown an increasing trend in the open waters during the period 1976–1981. The organic matter from the spring bloom had, after remineralisation, caused a high production in the bottom water. The primary production in the bottom layer had been increasing since 1977 and was, in 1981 in the Kattegat, twice as high as in 1952–1960. There were weak winds in the summer of 1981 and, consequently, there was a very stable halocline, separating bottom and surface waters and preventing oxygen supply to the bottom layer. Such situations had also occurred in the past, but had never led to such large-scale oxygen depletion. The main difference between the situation in 1981 and previous situations was the high nitrogen run-off from land, which had been considerably higher in the winter of 1980/81 than in previous years. Generally, the winter concentrations of nitrogen salts had increased in the surface waters of the Kattegat and the Great Belt during the period 1975–1981, reflecting increases in land run-off in the same period. For phosphorus compounds, no such correlations were found. The report furthermore revealed that the nutrient contribution from the Baltic was relatively low, whereas the nutrient-rich North Sea water had a considerable influence on the nutrient situation in the Kattegat and Belt seas. In the course of the 1970s the annual phytoplankton primary production in the Great Belt had doubled and, according to the report, the increase in the second half of the 1970s was directly related to the increased nitrogen concentration in the winter half year.

The environment minister had also requested NAEP to investigate to what extent the event had been caused by man-made activities, and which preventive measures could be taken. The report was unequivocal with regard to the first issue:

"Our coastal waters and marine waters are influenced by the inputs of nutrient salts from our own land area, from neighbouring land areas, via the atmosphere and from bordering sea areas, the North Sea and the Baltic Sea. Also in the latter mentioned seas (in the North

Sea only the coastal waters) an increased nutrient loading can be observed, originating from the land areas that border them."

Specific reference was made to the bad water quality of many of the Danish enclosed fjords and bays, and it was emphasised that communities and city councils should be aware of this and, on the basis of the regional water quality standards in force, take measures to reduce the nutrient inputs from the Danish land area. Moreover, it should be investigated which sources could not be properly controlled by the water quality planning and for which measures, based upon central initiatives, might be necessary. The agricultural sector was called upon to more actively reduce losses of nutrients and to improve the storage capacity of manure.

But not only Danish action was considered necessary. It was stated:

"From the Danish side the national efforts for reducing the leaking out of nutrients must be supplemented by an active effort in the international framework – in the international sea conventions and in the European community – to stimulate reductions of nutrient inputs by all countries that contribute to the burdening of our sea."

The report finally summed up a series of measures, necessary for improving the knowledge about nutrient sources, transport and budgets, as well as for improving monitoring. Also proposals were given for improving knowledge of processes in the sea, among which nutrient conversion (especially denitrification), uptake and release of nutrients, as well as toxic algae.

Sweden

In the autumn of 1981 oxygen depletion events also occurred along the Swedish Kattegat coast, most pronounced in the Laholm Bay and the Skälderviken Bay (figure 4.1). Upon request of the Swedish National Environment Protection Board, a group of 16 Swedish scientists carried out a comprehensive literature study with the aim of evaluating the events. The study covered both the Swedish North Sea and Baltic Sea coasts and was published in 1984 (Rosenberg et al. 1984). The final conclusion of the report was:

"there are several symptoms to suggest that a process of eutrophication is established in the Baltic, Sound and Kattegat, as well as in particular coastal areas of the Skagerrak. Although some connection with other large scale changes cannot be dismissed, eutrophication may be considered a major contributory factor for the developments described above" (Rosenberg et al., loc.cit.).

The developments referred to, were increased levels of nutrients, decreasing oxygen levels, increased benthic biomass, an increased occurrence of filamentous macroalgae and a decrease of *Fucus vesiculosus*. There was no direct evidence for increased primary production because of too short time-series and different methods, but the observed increase in nutrients "makes such an increase plausible in the Sound and certain areas of the Baltic." It had not been possible to establish a direct correlation between eutrophication and increased landings of commercial fish, which had occurred in the past decades, but neither was it possible to dismiss the possibility that eutrophication had contributed to the increased landings. The report was unequivocal about nitrogen being the nutrient usually limiting primary production in the Skagerrak, Kattegat, Sound and the Baltic proper, and phosphorus being the limiting nutrient in the Bothnian Bay.

In 1985 a paper, summarizing the results of the study, was published in the Marine Pollution Bulletin under the title "Eutrophication – the future marine coastal nuisance?" (Rosenberg 1985). In this paper Rosenberg stated that, contrary to heavy metals, chlorinated compounds and oil, eutrophication of marine waters had received little attention and that there was little literature about this subject. He believed that there were good reasons that eutrophication would become, in the near future, "a common hazard in many coastal areas in many parts of the world." Rosenberg referred in this respect to the increased atmospheric deposition and riverine inputs of nutrients, especially nitrogen. The main sources were combustion of fossil fuels and agriculture. According to Rosenberg (loc.cit.), there was increasing evidence of nitrogen being the critical limiting nutrient for phytoplankton throughout most of the year, and he referred, among others, to the work of Ryther and Dunstan (3.2.2). Inorganic phosphorus would be particularly important for the growth of bluegreen algae in hyposaline areas. Referring to the recent events of oxygen depletion, which had aroused public concern, Rosenberg underlined: "It would, however, be preferable to have earlier and less catastrophic warnings of such large scale disturbances."

As a result of the 1981 events a research programme was started in Sweden, which was concentrated principally on the nitrogen cycle, including an assessment of the extent of denitrification. The primary goal of the programme was to enable "predictive advice, based on sound background knowledge, about where to introduce counter measures in potentially eutrophic areas" (Rosenberg, loc.cit.).

Norway

In an editorial in the Marine Pollution Bulletin, Gray and Paasche reflected on the discussion in Scandinavia about the extent, causes and potential cures of marine eutrophication (Gray and Paasche 1984). They referred to the experiences with eutrophication in the Oslo Fjord, where anoxic conditions had been occurring for already a long time (3.2.4). Gray and Paasche also criticized the campaign against phosphate-containing detergents as a measure against the eutrophication of the Oslo Fjord. Such a measure had been very successful in Lake Mjøsa, Norwegians largest freshwater lake, but, according to Gray and Paasche (loc.cit.), could not simply be applied to a marine situation. They referred to the analysis of the oxygen depletion events in Laholm Bay, where nitrogen from agriculture was considered the main cause of the eutrophication effects. They presented several examples of both nitrogen and phosphorus limitations at different salinities and at different periods of the year and concluded:

> "Establishing which overall nutrient is limiting for the inner Oslo Fjord becomes a question of what season one is referring to, which species one is concerned with, and what scale one is interested in. Thus a clearly applied problem of how one should design sewage treatment plants for discharge to the sea still requires basic research to arrive at a solution."

Unfortunately, Gray and Paasche noted, contrary to Sweden, such basic research was not envisaged in Norway.

4.1.2 Oxygen depletion in the German Bight

Large-scale oxygen depletion events in the German Bight in the years 1981 and 1982 were reported by Rachor and Albrecht (1983) and Von Westernhagen and Dethlefsen (1983). Rachor and Albrecht (1983) referred to the 1974 ICES study on North Sea pollution (3.4.1), in which it was stated that, with the exception of some fjords and estuaries, there had been no problems with oxygen in the North Sea. The low oxygen conditions recorded in 1981 and 1982 in areas with a size of several 1000 km^2 made it, according to these authors, clear that it concerned events of a different order than referred to in the ICES study. The low values in 1981 had been measured below the halocline in the outer German Bight, north-west of Helgoland (figure 4.1). In 1982 several German marine research institutes carried out regular oxygen measurements, and again low values were found, this time mainly north of Helgoland (figure 4.1). According to Rachor and Albrecht (1983) the special weather conditions had, both in 1981 and 1982, played the determining role in the development of the low oxygen situations, and the relevance of anthropogenic nutrient inputs could not be answered in a quantitative manner. However, the role of nutrient inputs by man should, in the light of nutrient recycling processes and the summation of oxygen consuming processes, not be considered as irrelevant. The fish and macrozoobenthos mortality, which accompanied the 1982 event, was the main topic in a paper in Ambio by von Westernhagen and Dethlefsen (1983). According to these authors, the oxygen situation had been "rather alarming" and they concluded: " present practice of waste disposal via rivers and seas reflects a short-sighted environmental policy. There is accumulating evidence that the assimilative capacity of the southern North Sea has been surpassed, and for many wastes the ultimate disposal must therefore be on land only."

The events in the German Bight and the Kiel Bay prompted the German Minister of the Interior, also responsible for environmental protection, to initiate a research project, which started early in 1984 (Gerlach 1984). The project encompassed, for both the German Bight and the Kiel Bay, studies into trends in nutrient inputs, trends in nutrient concentrations in coastal waters, trends in primary production in coastal waters, sedimentation of organic matter and effects in the sediment, effects on higher links in the food chain, historic evidence of oxygen depletion and weather impacts on stratification. A comprehensive interim report on the oxygen depletion events was prepared for the first International Conference on the Protection of the North Sea (INSC-1), which was to be held in Bremen in 1984 (Gerlach, loc.cit.) (see further 4.1.5). The report contained descriptions and preliminary results of the sub-projects, conclusions and recommendations from these, as well as a chapter on the problems of monitoring. With regard to nutrient inputs, it was concluded "likely" that phosphorus concentrations in the German Bight had increased as a result of anthropogenic inputs, mainly via the rivers Elbe, Weser, Ems and Rhine.

Although no official monitoring programme had been in operation, it had been possible to detect increases in phosphorus concentrations in the German Bight on the basis of observations at Helgoland where, since 1962, daily water measurements had been carried out by the Biologische Anstalt Helgoland marine research institute (BAH). Some methodological problems were, however, associated with these measurements (Gerlach, loc.cit.). First, pre-1970 nitrogen data were rare, due to technical problems with the analysis of nitrogen compounds. Second, the seasonal variations were larger than the variations on a multi-year time scale. Finally, the monitoring site at Helgoland was, depending on weather conditions, irregularly influenced by the nutrient-rich freshwater plume from the river Elbe, causing a high variability in concentrations. It was, furthermore, concluded that the causal connection between these increased concentrations and the oxygen depletion events was not clearly established, and that further research on this issue was necessary. For precautionary reasons it was recommended to equip sewage treatment plants in the catchment areas of rivers with facilities for the removal of phosphates. Another conclusion was that in summer phytoplankton blooms, nitrogen was more often than phosphorus the limiting factor. It was, therefore, recommended to reduce the input of nitrogen compounds into the sea "as far as possible," by reducing emissions to the atmosphere, reducing surplus application of mineral fertilizers and liquid manure in agriculture and by taking appropriate measures in sewage treatment plants. However, it was stated: "whatever measures will be taken, due to the complex situation in the North Sea and the Baltic, positive effects may only be expected after many years" (Gerlach, loc.cit.).

4.1.3 Dutch coastal waters

In Sect. 3.2.5, the work carried out in the 1970s by the Netherlands Institute for Sea Research (NIOZ) regarding increasing nutrient concentrations in Dutch coastal waters was described. This work had not been continued in the second half of the 1970s, but in the first half of the 1980s several publication appeared, from which it may be concluded that the interest in marine eutrophication was increasing again. In 1982 and 1983, reports on the development of the water quality in the Wadden Sea and the North Sea were published by the Governmental Institute for Water Purification (RIZA). Both were based upon data from monitoring programmes, which had been operational as of 1971 in the Wadden Sea, and from 1975 onwards in the North Sea. Eutrophication was addressed in both reports.

The fourth scientific Wadden Sea symposium, held 1984, was dedicated to the theme of the role of organic matter. These reports will be briefly discussed below for the Wadden Sea and the North Sea respectively.

The Wadden Sea

The report on the quality of the Wadden Sea (De Wit et al. 1982), covering the period 1971–1981, contained an extensive part about eutrophication. In this period, the winter ortho-phosphate concentrations had clearly increased, and this increase could be wholly attributed to the increased inputs from Lake IJssel and the coastal water of the North Sea. The increase had been strongest in the second half of the 1970s. There was no change in chlorophyll concentrations, and it was concluded that the main determining factor for primary production was the light availability. The Biochemical Oxygen Demand (BOD) had unexplicably decreased during the period under consideration.

At the fourth scientific Wadden Sea symposium it was stated (Laane and Wolff 1984):

> ”In a few cases it has already been shown, and it is likely that it is generally true, that an additional input of organic matter through either direct discharge via rivers, polder drainage and pipelines or eutrophication by excessive supply of dissolved phosphate and nitrogen compounds, will cause serious oxygen deficits as well as the development of dense algal mats on the tidal flats.”

For this reason, it was recommended that the governments of the Wadden Sea countries would take measures to considerably reduce inputs of organic matter and nutrients to their coastal waters. Interestingly, the Dutch Minister responsible for Nature Management, also mentioned eutrophication in his opening address. He questioned whether eutrophication was a priority issue, compared to pollution by metals and halogen compounds. According to the Minister, eutrophication had, so far, seldom led to ”large-scale disfunctioning of the ecosystem.” In order to be able to set the right priorities, the precise role of nutrients in the cycle of organic matter would have to be known.

The North Sea

North Sea monitoring data for 1975 to 1982 had been evaluated by RIZA (1983). It was concluded that in those parts of the coastal waters that were directly influenced by the Rhine, a reduction in ammonium concentrations had occurred. In the mouth of the Western Scheldt and north of the Ems estuary an increase in orthophosphate was observed. An increase in chlorophyll concentrations, however, could not be established and, consequently, no conclusions about the eutrophication situation could be given. In the framework of the development of harmonized policies for the North Sea, the Dutch government decided in 1983 upon a comprehensive marine ecological research programme, in which also eutrophication issues would be addressed (Beukema et al. 1986).

4.1.4 International scientific advice

Although the international scientific community was aware of the potential impacts of excess nutrients in the marine environment, it valued marine eutrophication as a pollution problem of minor importance, limited to restricted coastal areas (chapter 3). The oxygen depletion events in the German Bight and the Skagerrak-Kattegat can certainly not be categorized as local incidents, and the question presents itself whether and how the perception of the international scientific community would change as a result of what had happened. Even more interesting is the question how official scientific advisory bodies, which are part of the North Sea science-policy network (figure 4.2), assessed these events, in particular in the light of current political developments. First, the Paris Commission (Parcom) had shown interest in the marine eutrophication issue, and requested ICES to provide advice on the problem of unusual plankton blooms. Secondly, Germany had taken the initiative for an international political conference on pollution problems of the North Sea. The scientific backing for this conference was to be provided by a so-called Quality Status Report (QSR) of the North Sea, to be prepared by experts from the North Sea countries. In the following section, ICES activities with regard to marine eutrophication, as well as the North Sea QSR will be discussed.

ICES and marine eutrophication

Parcom had, so far, not dealt with nutrient issues (see also 3.4.1). Parcom regularly requested ICES for advice, in particular on monitoring issues. Amongst the issues to be addressed by ICES in 1984 was the request "to examine, as a priority issue, the problem of unusual phytoplankton blooms, evaluate the possible causative factors including the role of nutrients and hydrographic conditions, and review the environmental effects" (ICES 1985).

Plankton Blooms. In Sect. 3.4.1 it was described how the ICES Advisory Committee on Marine Pollution (ACMP) valued the issue of marine eutrophication. In the 1970s the main emphasis of ACMP was on red tides, about which in 1976 a study had been carried out by a working group. As stated in the 1980 ACMP report, as yet no follow-up had been given to the work of this group (ICES 1981). The 1981 meeting referenced to observations of mortality in fish farms in Scotland and Ireland and problems with dissolved oxygen in the United States, and it was agreed that ACMP would continue to pay attention to the issue of plankton blooms, especially in relation to hypertrophication and pollution (ICES 1982). During the 1982 ICES statutory meeting, a joint meeting of the Hydrography, the Biological Oceanography and the Marine Environmental Quality Committees discussed plankton blooms. The outcome of this meeting was on the agenda of the 1983 ACMP meeting, together with a paper on exceptional plankton blooms and their implications for fisheries by Parker (ICES 1983b). The latter stated that the interest in "abnormal" blooms had increased as a result of the "apparent increased frequency and scale of effects" in the North Atlantic (Parker 1983). With regard to the causes of the increase, three options were given, namely an increase in observers, increased nutrient sources and long-term changes in oceanographic conditions. As to the observer increase, Parker (loc.cit.) stated that it was true that "much of the new interest derives from the effects of blooms on new activities (mariculture)," but that "there is some evidence that the problems are genuinely more prevalent currently than in previous years." In enclosed systems, increased nutrient loads could exacerbate blooms, but, generally, could not be considered the cause of the blooms. That was, firstly, because nitrogen and phosphorus were not the limiting factors in dinoflagellate blooms, which also developed in nutrient poor waters. Parker mentioned in this respect “conditioning” factors, such as micronutrients and vitamins (see also 3.2.3), but also stated that the role of nutrients and micronutrients in bloom development was not well understood. Secondly, the extent of blooms could not be explained by the presence of point sources of nutrients. Still, it was concluded that ”control over such sources could help to ensure that anomalous natural events do not become disasters.” The biological and chemical conditions, necessary for bloom development, were considered to be of secondary importance, compared to the physical (oceanographic, climatological) conditions. It was also concluded that the changes in bloom incidence might be related to large-scale physical changes, caused by climatic changes.

A special ICES meeting on the ”Causes, Dynamics and Effects of Exceptional Marine Blooms and Associated Events” was held in 1984. The outcome of this meeting was largely consistent with the conclusions of the Parker Report. On the basis of the recommendations of the special meeting, it was decided to install a Working Group on Exceptional Algal Blooms. The terms of reference of the Group were to establish a system of information exchange regarding bloom incidence, to consider how the predictability of blooms could be improved, to consider management proposals for overcoming the effects of exceptional blooms and, finally, to prepare advice to ICES countries about site selection for mariculture (ICES 1985).

Advice to the Paris Commission. In dealing with Parcom’s request, ACMP first of all referred to the 1984 Special Meeting (see above). In addition, ACMP evaluated the results of a discussion by WGMPNA[3] about primary production and nutrients, based upon, among others, review papers on nutrient distribution and trends, nutrient enrichment and primary production in the North Sea and reports on the recent oxygen depletion events in Danish coastal waters. The Working Group concluded: ”much of the evidence presented on the possible role of nutrient enrichment in increasing primary production and inducing exceptional bloom events was either inadequate or contradictory” (ICES 1985). The Group especially criticized that frequently references to nutrient concentrations instead of fluxes were made, although the latter were more relevant for changes in primary production. It was stated, however, that there was sufficient evidence of relationships between enrichment and increase in primary production, especially in inshore waters, ”to merit deeper studies.” The group, therefore, decided to continue the discussions and to prepare a comprehensive overview. In the meantime, it was considered premature to include nutrient studies in the contaminant baseline study.

The outcome of the Special Meeting ”Causes, Dynamics and Effects of Exceptional Marine Blooms and Associated Events” was discussed in more detail in the 1985 ACMP meeting. ACMP concluded: ”there is little evidence in North Atlantic waters for any rising trend in bloom incidence, although the data are very sparse” (ICES 1986). Furthermore, on the basis of available data, it could be concluded that there was little evidence for the existence of large-scale hypertrophication effects in the North Atlantic. ACMP concluded that, generally, the understanding of

[3] Working Group on Pollution Baseline and Monitoring Studies in the Northeast Atlantic (see also 3.4.1)

exceptional blooms and its relation with eutrophication was only possible in relation with a better understanding of primary production in coastal and shelf seas. There was, consequently, "an urgent need to establish long time-series of data on primary production and nutrient fluxes in addition to exceptional blooms incidence." There was also "an urgent need to develop and extend the methods currently applied to the studies of primary production." The ICES Biological Oceanography Committee was encouraged to initiate an appropriate programme.

Oxygen depletion. In the 1982 ACMP report an account was given of the "unusually low" oxygen events in the Baltic Sea and the Kattegat Area (4.1.1) (ICES 1983a). It was stated that the oxygen conditions in the whole Baltic Sea seemed to have deteriorated in 1981, that H_2S had been observed in the Kiel Bay, and that fish mass mortality had been reported for Danish and Swedish coastal waters in the Skagerrak-Kattegat area. Also large dinoflagellate blooms had been observed. Remarkably, no reference was made to the oxygen depletion events in the German Bight. With regard to eutrophication in the Baltic Sea, the ICES/SCOR study group (3.4.1) had agreed that there were still many unanswered questions about the influence of land-based nutrient discharges on the nutrient situation in the open Baltic and that it was, therefore, not possible to give statements about the results of decreasing inputs of nitrogen and phosphorus (ICES 1985). Such questions were regarded "of great economic importance, given the costs associated with reducing point source discharges of these nutrients."

The 1981 oxygen depletion events in the Skagerrak-Kattegat area had prompted the Nordic Council of Ministers to request ICES to establish a forum for scientists, working in the Skagerrak-Kattegat area, to discuss their results. In 1982, the ICES Statutory meeting established the "Working Group on Pollution Related Studies in the Skagerrak and Kattegat" to meet this request. The Working Group finished its work in 1986 and published a report in 1987 in the ICES Cooperative Research Report series (ICES 1987b). The Working Group had discussed a range of pollution issues, but the emphasis was on nutrients and eutrophication. It was concluded that there had been an increase in inputs of nutrients to the Kattegat since the 1930s, and that significant increases in the concentrations of total P and increasing trends of inorganic nitrogen had been observed in this area. Primary production had recently increased in the southern Kattegat and the Belt Seas, but it was emphasised that an accurate assessment of primary production over longer time periods was difficult. Also benthic animal communities had shown changes in biomass and composition during the 20th century, and a decreased vertical distribution of macroalgal communities had been observed in some Swedish coastal areas in the Skagerrak. Furthermore, mass mortality of Norway lobster had occurred, and high numbers of dinoflagellates had been observed in recent years. Eutrophication of the Kattegat and some coastal areas of the Skagerrak was considered to be the main cause of the described events. The effect of other pollutants (heavy metals, organochlorines, radionuclides) was considered of minor importance. Climatic changes could also have an effect, but these alone were not considered to be significant (ICES 1987b).

The first North Sea Quality Status Report

At the end of 1983, the North Sea littoral states decided that a group of experts should co-operate in drawing up the quality status of the North Sea, as a preparation of the first North Sea Conference (see further 4.1.5). The first North Sea QSR (1986 QSR) consisted of contributions by scientists from governmental institutions of the North Sea countries, and was structured according to the themes physical oceanography, inputs, concentrations and ecological effects. The latter comprised the issues fish diseases, accumulation of substances in organisms, effects on fisheries, mammals, sea birds, productivity and other biological effects (Carlson 1986). The group of experts had also prepared a synthesis of the national contributions, together with joint conclusions. It was concluded that no increase had occurred in concentrations of nutrients in the northern and central North Sea, but that in the inner German Bight, especially in the Wadden Sea and the Southern Bight, a clear increase in phosphate concentrations had taken place. For the Southern Bight, it was stated that this was a tentative conclusion, on the basis of data collected between 1961 and 1974. This statement derived from the contribution by the UK, in which it was also emphasised that there were no synoptic North Sea nutrient data after 1974 (see also 3.2.5).

In the part on ecological effects, the section "productivity" dealt with effects of increased nutrient levels on primary production. In the synthesis it was stated that the fact that changes in phytoplankton productivity occurred across the Northeast Atlantic, proved that climatic changes were the cause. It was also stated:

"So far, there appears to be no evidence that anthropogenic nutrients have caused any significant change in productivity in the North Sea, or even in the Southern Bight. There is circumstantial evidence that the organic pollution load may be significant in enclosed and semi-enclosed marine coastal waters" (Carlson, loc.cit.).

With regard to the observed increase in nutrient concentrations, however, "serious concern" was expressed. It was, furthermore, remarked that the influence of climatic changes and changes in nutrient levels were discussed by scientists as "a matter of controversy." Reference was made to the German research project on eutrophication in the North Sea and the Baltic Sea (4.1.2) and the ICES Special Meeting on exceptional plankton blooms (4.2.2). The controversy can be traced back to the differences in the German and British positions.

With regard to the effects of increased nutrient loads, the German contribution referred to the study by Gerlach that would be made available for the Conference (4.1.2). The UK had submitted an extensive paper dealing with the effects of eutrophication on normal and abnormal (toxic) plankton blooms, species diversity, and the anthropogenic contribution to the North Sea nutrient balance. The UK conclusions were almost identical to those from the synthesis, described above. Also other North Sea states had produced scientific reviews on the effects of nutrient enrichment. In the Dutch contribution it was stated, mainly on the basis of findings from the 1970s (3.2.5), that the possible increase of annual algal production might cause enhanced production of phytoplankton and, assuming that this overproduction would only partly be consumed by zooplankton, an enhanced flux of organic material to the sediment and possible oxygen depletion in stratified waters. However, increased zoobenthic production could be of benefit for the benthos and might support larger fish stocks. The Swedish contribution listed the oxygen depletion events described in 4.1.1, and the studies that had been initiated. No general conclusions or recommendations were given. In the Norwegian contribution reference was made to strong phytoplankton blooms in the inner Oslo Fjord, but coastal areas had not been significantly affected by such local eutrophication. It was also mentioned that increased dinoflagellate blooms had affected oxygen conditions in deep waters and fish farms along the south and west coasts of Norway. With regard to the causes, the Norwegian report referred to the 1983 ICES ACMP report, in which it was stated that it was not clear whether natural or anthropogenic causes were responsible (4.1.4). Denmark had, surprisingly, not produced a text on effects of ecological effects of anthropogenic nutrient enrichment.

The above described German-UK controversy not only related to eutrophication, but to the effects of pollution in general (see also De Jong 1986). It was most explicit for fish diseases, an issue about which both countries had submitted comprehensive contributions to the QSR. The British point of view was that fish diseases had also occurred over the past hundred years, and that there was no evidence of a link between pollution and (changes in the prevalence of) fish diseases. The German contribution contained two different scientific points of view. According to Möller, undernourishment was the main factor responsible for fish diseases in the Doggerbank area. Dethlefsen, however, concluded: "many findings speak for a correlation between the type and intensity of the waste water pollution and the frequency of several fish diseases" (Carlson 1986). In the general conclusions on ecological effects, Germany underlined the problem of the large natural variability, and the fact that only in a few cases a causal relationship between pollutants and biological effects had been established. It was concluded:

"Because the natural conditions in the sea pose these principal difficulties, because harmful alterations can therefore under certain circumstances not be recognized in due time, because damages occurred can be irreversible, prudent precautions should be taken and negative anthropogenic influences, especially in near coastal areas like estuaries and the Wadden Sea, should be reduced" (Carlson, loc.cit.).

4.1.5 International politics and marine eutrophication

The previous section shows that marine eutrophication had become a political issue in Denmark, Germany and Sweden and that its magnitude, impacts and causes were being assessed in international scientific advisory bodies. In this final section it will be investigated to what extent national developments and scientific advice have influenced international political fora.

The first international North Sea conference

In June 1982 the German Minister for the Interior, also responsible for environmental affairs, announced that the German Government would take the initiative to organize an international North Sea Conference, with the aim of analysing deficiencies in the execution and enforcement of relevant existing international legal instruments, such as the Paris Convention and the Oslo Convention (see also 2.6.2) (Peet 1984). According to Peet (loc.cit.) the initiative was the official reaction to the report "Umweltprobleme der Nordsee" (Environmental Problems of the North Sea), published in 1980 by the German Council for Environmental Affairs (3.4.3). In the report it was, among others, recommended to base North Sea policies upon the principle of precautionary action. The first International Conference on the Protection of the North Sea (International North Sea Conference, INSC-1), held in Bremen on the 31st of October and the 1st of November 1984, was attended by representatives of all eight North Sea states and the European Commission. At the Conference, three focal areas for joint action were discussed, namely the reduction of pollution from land-based sources, the reduction of pollution at sea and the further development of the Joint Monitoring Programme (JMP) of the Oslo and Paris Commissions (Osparcom).

With regard to the reduction of pollution from land-based sources, the Ministers, in §1 of the Preamble to the Ministerial Declaration, "affirmed their strong support for further binding regulations for black and grey list substances that should be adopted within the framework of the EEC, Parcom and River Commissions concerned, if possible as early as 1985" (BMI 1985). The implementation of the black and grey list substances approach (chapter 2) had, so far, been a very slow and cumbersome process, mainly caused by controversies between the UK and most other North Sea countries, about the application of emission- or immission-based pollution policies, as well as the application of the precautionary approach (compare Peet 1984). These principal differences in pollution policies were reflected in the Ministerial Declaration. In §C8 it was stated: "Emissions normally should be limited at source; emission standards should take into account the best technical means available and quality objectives should be fixed on the bases of the latest scientific data." According to §C7 it was expected that present studies within Parcom into the comparability of the uniform emission and environmental quality objective approaches, would bring results as soon as possible and that, at the latest at the next conference, "political decisions should be considered on the simultaneous and/or complementary application of the two approaches on the basis of the results of the assessment of the scientific, economic and environmental data."

As outlined in 2.6.2, nutrients were not part of the black or grey lists of the Oslo and Paris Conventions, neither were they being monitored in the framework of the JMP. The grey lists of the EEC Dangerous Substances Directive and the Rhine Convention contained phosphorus, nitrite and ammonium, but these instruments were mainly directed at freshwater systems. In §C3 of the Bremen Declaration, it was stated that the substances of the black and grey lists of Osparcom and the EEC should be examined more closely, with a view to including new compounds. This decision was specified in Annex 3 to the Declaration and, according to §4 of this Annex, "The effects of nutrients on the North Sea should be studied intensively. On the basis of the results inclusion in the grey list of the Paris Convention is to be examined."

With regard to the further development of the JMP a number of requirements and objectives was agreed upon and laid down in Annexes 15 and 16 to the Declaration. According to Annex 15, the aim of monitoring should be to provide a basis for:

- The assessment of the state of the marine environment;
- Decisions on measures for the protection of the marine ecosystem against contamination and for the reduction of pollution;
- The evaluation of the effectiveness of measures already taken.

The relevance of monitoring for policy was also put forward in §J6 of the Declaration, according to which

"Further measures for the protection of the North Sea should above all be taken on the basis of data and information to be collected and of their evaluation and assessment. The Joint Monitoring Group (JMG) should consider which data from the Joint Monitoring Programme, other monitoring programmes and current surveys and statistics, might be put together, evaluated and assessed for this purpose."

In Annex 16 a number of specific substances was mentioned, for which it should be examined whether they should be included in the JMP. Amongst these were also nutrients. Furthermore, methods for the collective determination of several monitoring parameters had to be reviewed, among which biochemical oxygen demand (BOD) and chlorophyll.

The Conference did not result in specific decisions and, generally, only intentions and wishes were expressed. Many agreements related to the stimulation of research and monitoring. Eutrophication was not an issue at the Conference, which may be explained by three factors. First, the very recent nature of the oxygen depletion events; second, the fact that only in Sweden, Denmark and Germany eutrophication was a political issue and, third, the scientific controversy about the causes and the extent of eutrophication and eutrophication-related phenomena.

The 1985 Paris Convention Consultation Meeting

In Denmark, the 1981 oxygen depletion event (4.1.1) had given rise to two national studies, the results of which were published in 1984 in reports about oxygen depletion and fish kills (Miljøstyrelsen 1984a) and land based inputs of nitrogen and phosphorus to the inner Danish waters (Miljøstyrelsen 1984b). The release of these reports and the fact that also in 1982 and 1983 oxygen deficiencies and low oxygen levels were measured in the German Bight, the Kattegat and the Belt (Jensen 1990), were the reasons why on 31 May 1985 the Danish Parliament adopted the so-called NPO (Nitrogen, Phosphorus, Organic matter) Action Plan (handlingsplanen) for the reduction of nutrient losses from agriculture, aquaculture and sewage treatment plants (Somer 1988; Christensen 1996). But not only national action was initiated. In 1985 Denmark took the initiative for a special meeting in accordance with Article 9 of the Paris Convention. According to this Article, parties have the right to ask for consultation with other parties in case "pollution from land-based sources originating from the territory of a contracting party by substances not listed in Part I of Annex A of the present convention is likely to prejudice the interests of one or more of the other parties." Nutrients were not contained in part 1 of Annex A (compare 2.6.2), and Denmark was of the opinion that the oxygen depletion events, at least in its open waters, depended to a large extent on the nutrient loads originating from central European rivers (Somer 1988). The consultation meeting was held in November 1985 in Copenhagen and resulted in the following conclusions (NUT 1986):

"I. A number of environmental changes have been recorded in the coastal areas. The changes are larger than expected and they cannot only be explained as natural variation;
II. there are indications that the changes are particularly significant in those areas where the nutrient input from land is dominant;
III. no simple explanation of the observed changes can be given that is ascribable to only a single factor;
IV. the nutrient concentration depends, in addition to local conditions, also on long-range transportation;
V. in the German Bight a general trend of increased nutrient load has been established over the last 20 years. In the same period there has been an increase in the phytoplankton biomass and a significant shift in species composition with respect to increases in the flagellate biomass."

It was, furthermore, decided to establish a new expert working group, which was to report directly to the Technical Working Group (TWG, see 2.6.2), with the tasks, among others:

- To consider which quality objectives would have to be achieved, in order to avoid adverse effects of increased nutrient loads;
- To exchange information on current and planned measures to reduce nutrient loads;
- To transmit to each other the results of national nutrient monitoring and research programmes, and to consider the need for coordinated research projects.

The working area of the group was not the whole North Sea, but was defined as an area 60 miles off the coasts of Belgium, The Netherlands, Germany, Denmark and Norway, including the Skagerrak and the Kattegat and excluding the territorial zone of the United Kingdom, underlining the fact that the UK did not consider eutrophication to be a problem in its waters.

The outcome of the consultation meeting can be valued as a first step towards defining and structuring marine eutrophication as a transnational pollution problem. In the terminology of Hannigan (1995), the meeting can be regarded as the completion of the first phase, the assembly phase, of the construction of the marine eutrophication problem (compare chapter 1).

4.2 1986–1987: Political decision-making

Prior to 1985 marine eutrophication had, with the exception of Denmark, Sweden and Germany, been almost exclusively a scientific issue. Moreover, problems experienced were limited to these three countries. Through the 1985 consultation meeting (4.1.5) marine eutrophication had been recognized as an international marine pollution problem. However, as underlined by Hannigan (1995), the assembly of a problem is not yet a guarantee for political action. The claims identified in the assembly phase need further legitimation in several arenas: the media, the public, science and government. This section investigates whether and how international political action with regard to marine eutrophication developed. More in particular, the following questions will be addressed:

1. What was the role of science in furthering the case of marine eutrophication as an international political issue?
2. Was the formulation of international political decisions with regard to marine eutrophication based upon science?

The first question can be specified into four subquestions related to new scientific knowledge, official scientific advice, the perception of science and the role of the science-policy interface.

New scientific knowledge is covered in 4.2.1, in which an account is given of research results from coastal waters in the Southern Bight.

The relevance of official scientific advice for the construction of the marine eutrophication problem is investigated in 4.2.2. Scientific advice on marine pollution problems was prepared for the second international North Sea Conference (INSC-2), scheduled for 1987. This was done in international conferences, by ICES and through the new North Sea QSR.

Section 4.2.3 deals with the changing perception of the role of science in policy-making, and the relevance of these changes for the construction of the marine eutrophication problem.

In 4.2.4 it is investigated how the newly installed eutrophication working group (see 4.1.5) influenced the political perception of the marine eutrophication problem. Whereas the activities described in Sects. 4.2.1, 4.2.2 and 4.2.3 belong to the science-part of the science-policy network (compare figure 4.2), the eutrophication working group is part of the science-policy interface, and its activities concern the translation of scientific information into policy-relevant material.

The second main question, the relevance of science for the formulation of political decisions, is investigated in Sects. 4.2.4 and 4.2.5, which cover the preparation, respectively the outcome, of INSC-2.

In the analysis of the above questions, specific attention will be given to the relevance of contextual factors. It concerns the complexity of the problem and the uncertainty of scientific information, the consensus within the scientific community, the difference in time frames between the scientific and political processes and dealing with values.

4.2.1 Eutrophication in the Southern Bight

Although, as concluded in 4.1.3, eutrophication was considered a potential problem issue in The Netherlands, clear effects of increased nutrient levels had not been documented for Dutch marine waters. This changed in 1986, when two scientific papers were published, showing an increase in the intensity of *Phaeocystis* blooms and an increase in macrozoobenthos biomass, both in the western Dutch Wadden Sea. These findings not only influenced the Dutch political position with regard to marine eutrophication, but also played an important role in the international scientific discussion on the effects of nutrient enrichment.

Phaeocystis in the Marsdiep

Since 1973, Cadée and Hegeman of the Netherlands Institute for Sea Research (NIOZ) had been investigating several parameters in the Marsdiep (figure 3.2). In earlier publications, for example Cadée and Hegeman (1979), data on phytoplankton seasonal development had been published, but there were no indications of increasing levels. At the 1983 scientific Wadden Sea symposium (see 4.1.3), Cadée had published data on an increase in microphytobenthos primary production for the period 1968–1981 and phytoplankton chlorophyll *a*, the spring values of which had increased in the beginning of the 1980s. He was, however, not very optimistic about the assessment of long-term trends of organic matter, chlorophyll and primary production, considering the large seasonal and year-to-year variability in these parameters (Cadée 1984). In the 1986 article, Cadée and Hegeman stated that until 1984 the seasonal appearance of the colonial phase of *Phaeocystis* had been in accordance with reports in the literature: a high spring peak usually appeared some weeks after the spring diatom peak (see also 4.1.3). However, in 1985 *Phaeocystis* colonies were also observed in winter. During the period of observation (1973–1985), there had been an increase and a broadening of the spring peaks. This was well illustrated by graphs showing the number of days per year with more than 100, respectively more than 1000 *Phaeocystis* cells/ml. The number of days with more than 1000 cells/ml had increased from around 20 during the period 1974–1976 to more than 90 in the years 1983–1985 (figure 4.3). Cadée and Hegeman (1986) comprehensively discussed the observed developments. They pointed to the complicated and only partly resolved life history of *Phaeocystis*, and the fact that the timing and intensity of blooms of this species were very different at different sites. A relation between bloom intensity and temperature could not be established and the authors stated that "It seems justified, but difficult to prove, to relate the *Phaeocystis* increase to eutrophication." Because biomass of macrozoobenthos in the western Wadden Sea had also increased (see below), the authors considered it "natural" to assume a causal relationship between the simultaneous increases of nutrient levels, primary production and secondary production.

The investigation of *Phaeocystis* blooms was also part of the Dutch research programme EON (Ecologisch Onderzoek Noordzee: Ecological Research North Sea) that had started in 1984 in the framework of a harmonization of Dutch North

Sea policies. In 1986 first results were published by, among others, Veldhuis et al. (1986a; 1986b).

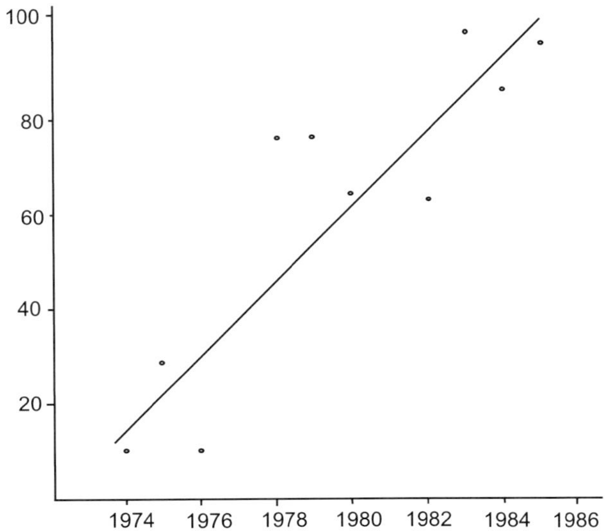

Fig. 4.3. *Phaeocystis* blooms in the Marsdiep. Number of days with more than 1000 cells/ml. From Cadée and Hegeman (1986)

Phaeocystis at Norderney

Increased blooming of *Phaeocystis* was also reported for the East Friesian Wadden Sea. According to Bätje and Michaelis (1986), "unusual amounts of sea foam" had been observed in May and June 1978 on the beaches of several east Friesian islands and "the public suspected that detergents and industrial wastes were the causes." Bätje and Michaelis (loc.cit.) continued saying:

"After its first occurrence in 1978 this phenomenon seems to have become chronic since it has appeared year after year with a different intensity. Therefore the question arises whether the production of *Ph. pouchettii* is enhanced by man-made eutrophication of coastal waters."

Spring blooms of *Phaeocystis pouchetii* had also been recorded in literature dating back to the end of the 19th century and could, therefore, be regarded as a regularly reappearing phenomenon in the southern North Sea. Foam formation had, however, not been recorded for previous decades and it was concluded "Therefore, this phenomenon is evidently a new one and may be regarded as another indication of increased growth of *Phaeocystis pouchettii* populations" (Bätje and Michaelis (loc.cit.).

Phaeocystis blooms in an international perspective

Increased international interest into the effects of increased nutrient loading on *Phaeocystis* may be inferred from a paper, published in 1987 in the journal Ambio by Belgian, Dutch, French, German, British and Norwegian researchers (Lancelot et al. 1987). In the introduction to the paper, entitled "*Phaeocystis* blooms and nutrient enrichment in the continental coastal zones of the North Sea," it was stated that the proliferation of *Phaeocystis* was not a recent phenomenon, and that it was not restricted to the North Sea coastal zone. However, the authors continued to say, "the increasing pressure of human activities on these coastal areas has almost certainly led to the recently observed increase in intensity and duration of *Phaeocystis* blooms in the Southern Bight of the North Sea." With the aim of discussing the behaviour of *Phaeocystis* and its "disquieting increase" a group of European and US scientists met on the Dutch island of Texel in March 1986, and the main results of this meeting were laid down in the Ambio article. Lancelot et al. (loc.cit.) discussed three issues, namely nutrient enrichment, the ecological characteristics of coastal waters and the "peculiar physiology" of *Phaeocystis*. With regard to nutrient inputs, the increased Rhine inputs were mentioned, and reference was made to the increased nitrate and phosphate concentrations, measured in Dutch coastal waters and at Helgoland. In the discussion of the dynamics of enriched coastal systems, the effects of enrichment on the seasonal cycle of phytoplankton were addressed. It was stated that in the enriched North Sea coastal area explosive spring blooms might develop because in the spring period phytoplankton was not controlled by zooplankton. Also the results of a mathematical simulation were given, showing the seasonal development of chlorophyll *a* under different nitrate regimes. These showed, generally, higher spring chlorophyll *a* values for areas with higher nitrate concentrations. The effects of the increased primary production were, however, more difficult to predict, and in this respect especially the dynamics of *Phaeocystis* blooms were discussed. That was because this flagellate often dominated over other flagellate species, and during such periods "the peculiar physiology of *Phaeocystis* colonies strongly influences the working of the whole marine ecosystem." Lancelot et al. (loc.cit.) could not conclude the extent to which the complete *Phaeocystis* bloom could be used as food for zooplankton and benthic organisms, especially since the major part of the *Phaeocystis* primary production was in the form of mucilaginous material, of which it was unknown to what extent and how fast it was decomposed by bacteria. The authors did not expect that this type of primary production would significantly increase fish yield because large parts would not be available for the pelagic food web, due to deposition in the sediment and washing ashore on beaches. Moreover, the mucus could form foam which would wash ashore with onshore wind conditions and might cause "great nuisance" for recreational activities. "*Phaeocystis* blooms," Lancelot et al. (loc.cit.) continued to say, "not only alter the marine environment, but may also have an impact on the atmosphere." With the latter they pointed to the production by *Phaeocystis* of dimethylsulphide (DMS) that might contribute to the acidity of rainwater. Because of these "harmful consequences" it was considered important to know more about *Phaeocystis* blooms, in order to be able to prevent and control them. But the latter would not be a simple undertaking. As a result of the introduction of sewage treatment plants, the aerobic status of most rivers would be restored. This would, however, also reduce denitrification, which could account for as much as 70% of nitrogen removal, as was shown for the Scheldt River by Billen et al. (1985). Because the nitrogen removal by secondary sewage treatment was only 30%, the net result of the introduction of treatment plants without tertiary treatment would, "paradoxically," be an increase in nitrogen inputs into coastal waters (Lancelot et al. 1987).

Macrozoobenthos in the Dutch Wadden Sea

As mentioned above, also secondary production had increased in the western Dutch Wadden Sea. The results of long-term observations into macrozoobenthos species composition and biomass on tidal flats in this area, were published in 1986 in the marine biology journal Ophelia (Beukema and Cadée 1986). According to Beukema and Cadée (loc.cit.), biomass values for most species had approximately doubled over the period 1970–1985. The biomass of short-lived species had increased even more. In discussing these observations, the authors stated: "The parallel increase of nutrient concentrations, algal biomass and production, and macrozoobenthos biomass and production in Dutch coastal waters during the last decades generate strong and positive correlations between these parameters, suggesting straightforward causal relationships." They stressed, however, that some caution was needed in interpreting these correlations because, on the basis of cause-effect relationships, the existence of nutrient and food limitation would have to be assumed. The latter was not necessarily the case in the Wadden Sea, which, also in the 1960s, could not be regarded as an oligotrophic system. On the basis of a literature analysis they did, however, cautiously conclude that limitations did exist in the transfer between the different trophic levels. They furthermore considered the conformity in magnitude of increase in both primary and secondary production (in both cases a doubling) an argument in favour of the cause-effect relationship between the two. Beukema and Cadée (1986) also discussed some alternative explanations for the increased primary and secondary production, among which a decrease of pollutants, such as heavy metals and pesticides, and changes in the concentration of suspended particulate material, but none of these could be linked to increased production. They finally discussed the possible negative consequences of increased production, such as oxygen depletion and increase of macroalgae. But although observed occasionally, such phenomena were not likely to cause large-scale problems in the Wadden Sea because "The system appears robust in treating huge amounts of oxygen-demanding material." They therefore concluded: "if eutrophication affected the Dutch Wadden Sea, the effects so far appear to have been predominantly positive."

4.2.2 Scientific advice on marine eutrophication

In the run-up to INSC-2 several activities took place, which aimed at providing scientific advice to the responsible authorities. It is noted here that not all of these activities were of a purely scientific nature. Not only academic scientists, but also scientists from official bodies and representatives from non-governmental organizations participated. In this section five themes and events are covered. It concerns reports of two scientific symposia, an overview of the state of the art of modelling and monitoring, as well as a description of two official types of scientific advice, provided by ICES and presented in the 1987 North Sea QSR. The section is concluded with an analysis of the scientific advice.

The second North Sea Seminar

The second North Sea Seminar was a scientific conference, organized in 1986 by the Dutch non-governmental organization Werkgroep Noordzee. In the introduction to the proceedings of the seminar, it was stated that the question of whether the status of the North Sea would give reason for concern, would be in the centre of discussions of INSC-2. The aim of the seminar was to provide such an assessment, well in advance of the Conference (Ten Hallers-Tjabbes 1986). Non-governmental environmental organizations had been very active during the period prior to INSC-1, analysing the main problems and formulating recommendations and demands to the Conference. Interestingly, the issue of marine eutrophication had not, or hardly, been a part of their analyses, the focus being mainly on hazardous substances (AKN 1987). The Dutch Werkgroep Noordzee (North Sea Working Group) had also not addressed marine eutrophication in the first half of the 1980s, as can be inferred from their 1984 comprehensive description of pollution problems in the North Sea, in which marine eutrophication was not mentioned (Van Weering and Kramer 1984). But at the second North Sea Seminar, marine eutrophication was one of the issues discussed as a possible reason for concern with regard to the status of the North Sea environment. In the introductory Volume I to the second North Sea Seminar, a brief overview of physical, chemical and biological aspects of the North Sea ecosystem, including nutrients and eutrophication, was presented (Eisma 1986). According to Eisma (loc.cit.) it would not be easy to assess the effects of the increased supply of nutrients to the North Sea. He referred to the low oxygen events in the German Bight and Danish waters and several changes that had occurred during the past decades, such as higher biomass of flagellates and dinoflagellates, lower diatom biomass, an increase of shrimp biomass and an increase of total fish biomass. These could be the result of phytoplankton biomass increase, resulting from increased nutrient supply, but also of a combination of natural shifts, of fisheries, or unknown factors at other trophic levels. More generally, Eisma concluded that "The ecology of the North Sea remains largely obscure because of the complexity of biological relations and the fact that most studies have been limited to commercially important species." He also concluded that "Apart from some self-evident or well studied effects, an assessment of the effects of pollution and other human activities in the North Sea is very difficult to make" (Eisma, loc.cit.).

At the Seminar itself, a presentation on eutrophication in the North Sea was given by Beukema (1986). He discussed nutrient enrichment, primary production, secondary production, changes in species composition, oxygen deficiency and other adverse effects. With regard to nutrients he concluded: "Eutrophication in the sense of elevated nutrient contents is thus a reality in some coastal parts of the North Sea" and pointed to the Dutch coast, the inner German Bight and some British estuaries. The effects of these elevated concentrations were less clear. Beukema (loc.cit.) concluded:

"there is some evidence of increased phytoplankton biomass and production that is specific to the eutrophicated coastal areas and is limited to non-diatoms, exactly as one would expect from the increased load of P and N and unchanged concentrations of Si. However, the causal connections with the enhanced concentrations of P and N cannot be demonstrated exactly. Meteorological conditions appear to play a significant role, resulting in a complex situation."

Environmental Protection of the North Sea

On the occasion of INSC-2 and in the framework of the European Year of the Environment, a scientific conference "Environmental Protection of the North Sea" (WRc Conference) was organized by the British Water Research Centre (WRc) from 24–27 March 1987. In the Editor's preface to the proceedings of the Conference it was stated that this theme had been chosen "in view of continuing political pressure for further measures to protect and improve the quality of the North Sea" (Newman and Agg 1988). Newman and Agg referred to INSC-1, which had called for a major reduction in the discharge of pollutants, and to the publication in 1985 of a proposal by the European Commission to reduce waste disposal at sea. They also mentioned recent campaigns by Greenpeace, in which the focus had been on the UK as a major polluter. The objectives of the WRc were "to produce, as far as present knowledge allows, a definitive appraisal of the impact of potential pollutants on the North Sea," and to deliver this appraisal in time for the preparation of the North Sea QSR (Newman and Agg, loc.cit.).

Five presentations about eutrophication were given at the WRc Conference, illustrating that marine eutrophication was now regarded a "real" pollution issue, like oil and heavy metals. Gerlach (1988) presented a general overview, which focused on the nutrient balance of the North Sea and long-term changes in nutrient inputs and concentrations. In the framework of the German eutrophication research project (4.1.2), nutrient data from Helgoland Reede had been analysed, and Gerlach presented first results of this undertaking. Linear regressions of winter values of phosphate and dissolved inorganic nitrogen showed a 1.6 fold increase for both parameters over the period 1962–1984. During the same period, annual phytoplankton biomass had increased from 9 to 37 μg C/l, mainly caused by an increase in flagellates. It was unclear, Gerlach stated, to what extent the changes had been caused by marine water masses or by river water, but there were no indications that central water masses of the North Sea had higher nutrient concentrations. Moreover, the observed increase was consistent with the results of modelling carried out in the framework of the Dutch North Sea Water Quality Plan (see further below), showing a 50% increase due to anthropogenic loads. Gerlach was even more cautious about the possible effects of the nutrient increase. There were, according to Gerlach, conflicting arguments among scientist as to whether nutrients or light were limiting phytoplankton growth, but "A fourfold increase of phytoplankton biomass at the Helgoland-Reede station between 1962 and 1984 and a shift from diatoms to flagellates requires, however, an explanation."

Several scientists of the Netherlands Institute of Sea Research (Duursma et al. 1988) presented an overview of eutrophication effects observed in the North Sea, most of which were presented earlier in this chapter. They summarized these effects as follows:

- Primary production and algal concentrations in the southern North Sea had doubled in the past decades. They mentioned in this respect the observations in the Marsdiep and at Helgoland, but warned that long-term climatic changes might also play a role. The increase in primary production was less than for nutrient concentrations and the reason might be that insufficient light hampered a further increase of productivity.
- Stocks of benthic fauna and secondary production had increased. Duursma et al. (loc.cit.) first mentioned the situation with regard to zooplankton, for which data from the Continuous Plankton Recorder showed a decrease, although it was less dramatic in the southern North Sea than in the North Atlantic (see 3.2.5). For macrozoobenthos in the western Dutch Wadden Sea, however, biomass and production had approximately doubled during the period 1979–1985 (see 4.2.1), and densities of shrimp in near coastal waters had also increased, as shown by recent work of Boddeke (compare 3.2.5).
- The diversity of benthic species had changed to the detriment of long-lived species. This was illustrated by the work of Beukema in the western Dutch Wadden Sea.
- Total fish catch had increased in the North Sea, especially after the early 1960s. It could not be concluded whether changed fishing techniques or increased fertilization also played a role.
- The regularly observed oxygen deficiencies of North Sea near-bottom waters in the German Bight, and occasionally in the Wadden Sea, seemed to become "critical."

Considering the situation of the North Sea as a whole, it was concluded that there were "serious reasons for concern." Duursma et al. (loc.cit.) made a plea for developing integral management plans for the North Sea, taking account of the various functions, such as fisheries, mineral extraction, transport and waste reception. The Dutch Government was currently developing such a plan. In this respect they stressed that such plans should take due account of the fact that the North Sea "is not just one pool" but consisted of several zones, differing in biotic and abiotic features.

Modelling and monitoring

The possible role of ecological modelling in environmental management was discussed even in the 1960s (see section 2.3.1). Practical examples of the application of models to North Sea pollution appeared in the mid 1980s, albeit firstly physical models. In the first North Sea QSR, only modelling of oceanographic processes was addressed (Carlson 1986), but in the second North Sea QSR also simulations of transport of substances were presented and discussed. Because of the high political relevance of international nutrient transport, the models are described in more detail below.

In the framework of the development of a Water Quality Management Plan for the Dutch North Sea, model calculations were used to determine distribution of pollutants over the North Sea and the anthropogenic fraction of the concentrations. Results of these calculations were presented at the second North Sea Seminar (van Pagee and Postma 1987) and the WRc Symposium (van Pagee et al. 1988). The model showed the spreading of pollutants and nutrients from riverine sources over the North Sea, clearly illustrating that the Rhine influence was confined to the eastern and northeastern North Sea, whereas the Thames influence extended from the southwestern UK coastal waters to coastal waters in the northwestern part of The Netherlands and the German Bight. The Dutch model was a two-dimensional one, and based upon the assumption of conservative behaviour of substances, i.e. that compounds would not undergo changes during transport, as a result of chemical and biological processes. Backhaus and Soetje (1988) presented a three-dimensional model to simulate the transport of pollutants in the North Sea. They stated that models describing physical processes had been developed and applied in the past decades, but that, considering the complexity of the marine environment, biological, chemical and geochemical processes should also be included within models used in environmental management (Backhaus and Soetje, loc.cit). Their own model was a physical one, since, as stated by the authors, "the present modelling act is still in a rather early state with regard to environmental problems," but useful to describe some principal problems. These concerned the three-dimensionality and the stochastic behaviour of physical processes in the North Sea. On the basis of their simulations, Backhaus and Soetje (loc.cit) concluded that models using only two dimensions or average values were likely to produce unrealistic result. They illustrated this by showing time-series of simulated concentrations of substances, which deviated by a factor of two to three from the mean within days to weeks. The deviations were most pronounced in the continental coastal part of the North Sea. According to Backhaus and Soetje (loc.cit), present monitoring activities did not adequately resolve such fluctuations because they had a sampling rate of one to four times per year. They considered the observed fluctuations relevant for management, especially when developing safety margins for dangerous substances. Backhaus and Soetje (loc.cit) anticipated that not only the physical environment, but also non-physical biogeochemical processes might have a pronounced three-dimensional character. Despite the above critique, the results of the Dutch dispersion model were widely used, among others by Gerlach, both in the first meeting of the Parcom nutrient working group (NUT, see further 4.2.4) and in his presentation at the WRc symposium (Gerlach 1988), and by Van der Voet (1987) at the North Sea Seminar, as well as in the second North Sea QSR. Peet (1988) used the outcome of the model simulation to argue that British nutrient inputs did have an effect on eutrophication in the eastern North Sea.

Interestingly, results of biological modelling were also presented at the WRc symposium (Van Pagee et al. 1988), showing the impact of increased nutrient inputs into Dutch coastal waters. This model had been developed in the framework of the Dutch water quality management plan, and was applied to the situation around 1930 and the period around 1980. For the first period, no monitoring data were available, but for the 1980s the model had been calibrated with monitoring data from 1975–1985. The forcing factor was the nutrient input by rivers and the Channel. The main change that had occurred between the two periods was a two to threefold increase in primary production, mainly caused by an increase in non-diatoms. Although the results of this model were labelled as preliminary because of methodological limitations (RWS 1985), they were used in the second North Sea QSR (see below) to substantiate effects of increased nutrient loading, i.e. a doubling of primary production in Dutch coastal waters between 1930 and 1980.

ICES

The results of the first meeting of the Working Group on Exceptional Algal Blooms (WGEAB)[4] were discussed at the 1986 ACMP meeting (ICES 1987a). This group had been installed in 1984 as a result of the Special Meeting on Exceptional Algal Blooms (see 4.1.4). The WGEAB had developed pragmatic procedures for collecting information how to manage bloom effects on mariculture. The Group also proposed additional research into the biology and life histories of bloom organisms. ACMP suggested the formation of a study group to continue the work of the Special Meeting on Exceptional Algal Blooms, which should "suggest research directed towards increasing the knowledge of the role of physical, chemical and biological factors in creating conditions which initiate and sustain the development of specific blooms." ACMP 1986 furthermore emphasised the relevance of primary production studies and the need for an intercalibration workshop on the measurement of primary production (ICES 1987a).

The measurement of nutrients. In the 1986 ACMP report, an account was given of discussions in the ICES Marine Chemistry Working Group (MCWG) on the need for a review of measurement of nutrients in laboratories. Earlier, ACMP had requested MCWG to produce guidelines for the monitoring of temporal trends of nutrients in seawater. MCWG considered it necessary to review both the identification of areas where trend studies could be carried out, and the identification of gaps in the collection of nutrient data (ICES 1987a). In the 1987 ACMP meeting, the results of a questionnaire, distributed by MCWG to laboratories in the Ospar-

[4] In later years this Group was called "Working Group on Harmful Effects of Algal Blooms on Mariculture and Marine Fisheries"

com area, were discussed. One of the aims of the questionnaire was to identify areas suitable for temporal trend monitoring of nutrients. The outcome of the questionnaire made it clear that nutrient data were mainly collected in estuarine and coastal zones, normally at a frequency of 1–12 times per year. Only few of the data collected were available in the ICES nutrient data bank. The results did not provide conclusive information about nutrient trend monitoring (ICES 1988). MCWG had also discussed the quality of nutrient measurements, and, on the basis of this discussion, ACMP concluded that there was a clear need for an assessment of the comparability of nutrient measurements conducted within the ICES area, and for a better quality assurance of nutrient analyses. Rapid action by MCWG for both issues was considered necessary (ICES, loc.cit.). Osparcom had requested ICES to prepare an overview paper on trends in nutrient concentrations in seawater, but for reasons given above, ACMP was concerned that only limited data could be used for such an overview (ICES, loc.cit.).

The North Sea Quality Status Report. The scientific preparation of INSC-2 (4.2.5), was tabled at the 1986 and 1987 ACMP meetings. The 1986 ACMP report referred to the Oceanography Sub-Group that had been established to prepare the oceanography part of the new QSR, and that had evaluated North Sea circulation models (ICES 1987a). In the 1987 meeting the QSR itself was addressed, of which ACMP had received a draft for commenting. Since the QSR had been prepared by national representatives it would, as ACMP noted, "probably reflect a compromise between differing views of the states concerned" (ICES 1988). ACMP commented, among others, the part on algal blooms and proposed that a sentence be added to the report, indicating the difficulties in attributing changes in phytoplankton species composition and increases in primary production to nutrient increases from coastal sources. ACMP referred in this respect to the fact that large-scale changes in phytoplankton species composition had occurred in the Northeast Atlantic, in areas not significantly influenced by coastal nutrient inputs.

The 1987 North Sea Quality Status Report

An important element of the scientific preparation of INSC-2 was the updating of the 1986 North Sea QSR. The work on the new North Sea QSR started in 1986, and was carried out by a scientific and technical working group (STWG). The 1987 QSR was structured in the same way as the 1986 QSR, according to the chapters physical oceanography, inputs of contaminants, concentrations of contaminants and ecological effects (compare 4.1.4). There were two additional chapters on trends and overall assessment (STWG 1987). Contrary to the first one, the new QSR did not contain separate national contributions, but was already a synthesis, prepared by the UK Conference Secretariat on the basis of individual national contributions and contributions of international organizations (ICES, Commission of the European Communities [CEC], Osparcom). The information in the QSR must therefore be valued as "negotiated science," being the result of the outcome of discussions in STWG.

With regard to marine eutrophication, the 1987 QSR was much more elaborate than the 1986 QSR. According to the report, there was "concern" about changes in nitrate and phosphate concentrations that appeared to have occurred especially in the 1970s. At the same time, it was remarked that there appeared to have been no further increase since about 1978. Reference was made to the first meeting of the Parcom nutrient working group (NUT) (see further 4.2.4), at which it had become clear that the inflows from the North Atlantic and the Channel were by far the most important sources of nutrient inputs to the North Sea, but that on a local scale the influence of rivers was important. The increased riverine inputs had caused a doubling of the concentrations of phosphate and nitrogen in the coastal waters of The Netherlands, Germany and Denmark. It was in this respect remarked that the influence of the continental rivers was much more important than that of the smaller rivers along the North Sea coast of the UK. The report also addressed the limiting factor for primary production. It was stated that it was nitrogen that, in the "classic pattern of nutrient cycles," was the limiting factor governing primary production. In Sweden a 30% reduction in phosphate inputs had had only limited impact on plankton blooms, but in some areas the increased nitrate inputs might now have caused phosphorus to be the limiting factor. According to the QSR, this hypothesis had not been proven, and had been extended with the "circumstantial" link with the dominance of *Phaeocystis* and other flagellates in some areas.

Where the QSR was rather unequivocal about the fact that nutrient levels had increased, it was much less straightforward as regards the effects of the increased levels. In the chapter "Ecological Effects," a comparison with freshwater eutrophication was made: "it is believed by some that the recent changes in phytoplankton species composition and the incidence of unusual plankton blooms evidence that something similar to eutrophication in the freshwater is occurring in the sea." It was noted, however, that several alternative causes of the observed phenomena were possible, such as increased awareness (observer effect), the introduction of non-indigenous species or changes in meteorological conditions. With regard to the latter, the results of the Continuous Plankton Recorder (CPR) were given, showing an increase of chlorophyll levels throughout the eastern North Atlantic, which could not be attributed to increased nutrient inputs (see also 3.2.5).

What was not questioned was the fact that there had been changes in plankton species composition. Such had been clearly documented for the Helgoland Reede (see above). Reports from the Netherlands, which were partly based on model computations, indicated a doubling of primary production in a 30 km wide strip along the Dutch coast between 1930 and 1980 (see above). Belgium reported a "marked increase" in the occurrence of *Phaeocystis* blooms in recent years, and in Danish coastal waters there had been "frequent blooms of unusual species." In the final chapter "Assessment of the Status of the North Sea" the issue of effects on plankton populations was summarized as follows:

"Thus in some areas a link between nutrient inputs and plankton blooms, and nutrient inputs and plankton population structure appears possible. There is evidence also that the consequent effects are detrimental, e.g., in the German Bight and in certain Danish coastal waters. More generally, it is apparent that phytoplankton community structures and production, and even sessile algal population structures, have changed over long term periods and large spatial scales, but the causes of these changes are far from clear. If they are linked to changes in coastal nutrient levels, the links are complex and involve other factors, such as meteorology and hydrography."

Also the possible link between increased nutrient inputs and benthos was addressed. From Danish waters periodic mortalities of benthic organisms were reported, which were connected with increased nutrient inputs, followed by oxygen depletion, caused by the decay of phytoplankton blooms. Sweden reported about replacement of macrophytes by filamentous algae, having an effect on the species composition of whole hard-bottom ecosystems. A doubling of biomass of soft-bottom fauna in the Skagerrak had occurred during the last 50 years, and comparable increases had occurred in the Dutch Wadden Sea. The United Kingdom reported an increase in macroalgae over the last 20–30 years on some British North Sea shores, especially in harbour areas. The QSR concluded: "The significance of these changes is often difficult to assess but the severest effects are usually local, and most studies suggest that anthropogenically induced changes are reversible."

With regard to nutrients and dissolved oxygen, it was concluded that over most of the North Sea oxygen levels were usually close to or above saturation. There were, however, extensive areas where occasional oxygen depletion had occurred. In this respect, the German Bight and the Danish coastal North Sea waters were mentioned. Such events were associated with stratification of the water column, and usually the result of the decay of algal blooms.

Analysis of scientific advice

The picture, emerging from the above scientific analyses, is that there was broad consensus about the fact that nutrient concentrations in the North Sea had substantially increased. Several cases had been presented about increased phytoplankton growth and increased secondary production, but it was generally acknowledged that the link with nutrients was unclear, and that also climatic and hydrographic changes had to be taken into consideration. The data of the Continous Plankton Recorder (CPR) had been very relevant in this respect. There was also consensus about the fact that eutrophication phenomena were restricted to confined coastal areas and that there were no problems in the central North Sea. What is striking, is that none of the presentations and analyses by academic scientists gave concrete advice to politicians about the need for an international programme to reduce nutrient inputs to the North Sea. It is true that at the WRc conference NIOZ researchers had underlined the need for a management programme for the North Sea, but not what this would mean in terms of reduction measures. The only reference to a nutrient reduction programme had been given by Somer of the Danish NAEP during the WRc conference, at which he had presented the Danish nutrient action programme (Somer 1988). On this occasion Somer had also announced nutrient reduction initiatives in the preparation of INSC-2.

4.2.3 Environmental policy principles

The scientific discussions preceding INSC-2 not only concerned the analysis of marine ecological information, but also the question whether environmental policies should be based on the precautionary principle or the concept of assimilative capacity. This question is directly related to the application of the Uniform Emission Standard (UES) or the Environmental Quality Objective (EQO) approach, which had already been a major theme at INSC-1 (4.1.5). The UK was the advocate of the EQO approach, whereas most continental North Sea states favoured the application of UESs.

The precautionary principle versus assimilative capacity controversy is discussed in more detail in this section because it is highly relevant for understanding both the political developments and the different valuations of the status of the North Sea ecosystem. The background of this difference already emerged in the 1970s. In Chap. 2 a description was given of the development of waste discharges to the marine environment, the control of which was considered a technically and scientifically manageable practice. In the course of the 1960s and 1970s, however, concern about marine pollution increased and national and international regulations for pollution control became established (chapter 2). The original idea that wastes could be discharged into the environment was challenged, and, according to Walker (1988), "battle lines seemed to be drawn between two groups of people equally concerned to protect the natural environment." The first group "had confidence that man working in harmony with natural processes could deal with pollution." The second group "seemed to be less impressed by observable improvements in the natural environment and much more influenced by the gloomy analysis of 'Limits to Growth'"(Walker, loc.cit). In the UK, the first approach, also termed the more pragmatic and economic approach, became formalized in the early 1970s (Walker, loc.cit). As a scientific response to the growing critique on the practice of marine discharges the concept of "assimilative capacity" emerged in the United States at the end of the 1970s. The "Assimilative Capacity" of a body of seawater was defined as "the amount of a given material that can be contained within it without producing an unacceptable impact on living organisms or nonliving resources" (Goldberg 1981). According to Goldberg (loc.cit) "Recently a mood has developed in countries of the Northern Hemisphere that the oceans are sacrosanct and that any entry of polluting substances is undesirable." He argued that many marine scientists and engineers considered the oceans to have "a finite capacity to receive some societal wastes." On the basis of knowledge from 30 years of marine pollution studies, models could be prepared for the determination of the assimilative capacity of coastal waters. As to the future disposal needs of society, Goldberg stated: "The simplest answer is increased knowledge about the chemistry, physics, biology and geology of the sea" (Goldberg, loc.cit). In the following years, the assimilative capacity concept was adopted by ICES and GESAMP as a basis for pollution control (Pravdic 1985; ICES 1987). During the same period, however, the debate about the concept became more polemic, as can be inferred from several commentaries and comparative studies appearing in the scientific literature (Kamlet 1981; Stebbing 1981; Pravdic 1985; Dethlefsen 1986;

Krom 1986). Important arguments against the concept were the limited knowledge about the complicated marine ecosystem processes, especially with regard to chronic and cumulative effects of pollutants, the limited ability to predict fate and effects of pollutants and the limitations to monitoring. An argument of a very different nature was put forward by Pravdic (1985), who suspected that reluctance by administrators to adopt the concept was caused by the fact that they would become more dependent on science, scientific research results and advice.

ACMP had documented its principles in the 1986 report (ICES 1987a), in which it was stated:

> "There are currently two extreme approaches to controlling the entry of substances to the marine environment. These are, on the one hand, the ultra-conservative approach that demands avoidance of inputs under all circumstances and, on the other hand, the approach that allows almost any input provided it is within certain, often loosely described, constraints as to the rate and quantity."

Both extreme approaches were rejected by ACMP. The aim of international Conventions for the prevention of pollution was, according to ACMP, the protection of the marine environment from pollution. Implicit in the definition of pollution was the fact that controlled input of wastes to the marine environment could be done without causing harm to living resources, or changes that would be unacceptable to society. It was stressed that scientists could advise on whether a particular effect was deleterious, but that society would have to decide whether it was acceptable or not. ACMP's conclusion was that both from a scientific and a societal point of view "there exists a range of contamination levels that do not cause, or are not likely to cause, unacceptable deleterious effects" (ICES 1987a). In the 1987 ACMP Report (ICES 1988) it was noted that, although the article in the 1986 ACMP report had been received positively, there had also been negative views which "deeply concerned" ACMP.

At the WRc Symposium a whole session was dedicated to this controversy. Peet (1988) had made an analysis of national and international environmental policies and concluded that all these policies were based upon the principle of preventing pollution. He concluded that, actually, there were substantial differences in how environmental policies were implemented in practice, the UK applying a more risk tolerating policy and many other countries being more cautious. Von Weizäcker et al. (1988) pointed to the principle problems in the management of large ecosystems. They stated: "Characterizing changes in large ecosystems is a typical example of a question without a scientific answer. [....] In a large ecosystem the chain of cause and effect is continuously broken, at least if we apply a scientifically acceptable standard of proof." They concluded, therefore, that "the larger the body of water, the less likely are EQOs to provide a guide for permissible discharges from an individual source." But they also criticized UESs for not being able to adequately deal with non-point sources, or with a combination of several point sources, all applying UES, but together overloading the environment. With regard to the Precautionary Principle they saw a principle problem, namely to what degree society would be prepared to pay for avoiding risks. The UK policy, presented by Walker (1988), concentrated most of all on the costs of pollution control, favouring the application of the Best Practical Environmental Option (BPEO). With regard to the future of water quality management, Walker was of the opinion that there seemed at present more political sympathy in the UK for the precautionary approach than there used to be.

4.2.4 The science-policy interface: the Nutrient Working Group

The previous sections were concerned with several aspects of what may be termed the science-part of the science-policy network (compare figure 4.2). With the establishment by the 1985 Paris Convention consultation meeting of a special working on nutrients and eutrophication (4.1.5), a science-policy interface for marine eutrophication had been created. Although this nutrient working group, in the following referred to as NUT, was intended to prepare the groundwork for Parcom with regard to marine eutrophication matters[5], it will be shown in this section that the preparation of INSC-2 turned out to be much more important for the activities of NUT than the Parcom developments.

The first two NUT meetings were held 1986 in Berlin and 1987 in Stockholm, both under the chairmanship of Erik Somer of the Danish NAEP. The locations of these first two meetings and the nationality of the chairman illustrated the importance Germany, Denmark and Sweden attached to the new Group. In line with the terms of reference, three main categories of issues were discussed by the meetings (NUT 1986; NUT 1987a):

1. The actual situation with regard to nutrients in the North Sea, including quality objectives;
2. Current and planned measures against nutrient pollution in the North Sea states;
3. The status of national monitoring.

Status of eutrophication in the North Sea

At NUT-1 a comprehensive overview of the eutrophication status of the North Sea was presented by Gerlach (see also 4.2.2). Gerlach concluded that there had been a substantial increase in nutrient loads from anthropogenic sources that had resulted in nutrient concentrations in the German Bight and Dutch coastal waters, which were 50% in excess of "natural" background values. In addition, trend analyses of nutrients in the German Bight by Weichart, who had compared old phosphate data from 1936 with data from a survey in 1978, showed that phosphate concentrations had increased markedly. Also in The Netherlands work was underway to assess natural background values. The meeting furthermore discussed the effects of eutrophication on the basis of a Danish contribution, in which especially the occurrence of algal blooms and the effects of algal blooms were addressed. Denmark also put forward the issue of nutrient limitation and stated that, in general, phosphorus was limiting in spring and nitrogen in summer and autumn. In the general

[5] Parcom was working on amending the Convention, so as to include nutrients in the Annexes (compare 2.6.2). In 1984 such a request had been made by INSC-1 (4.1.5)

discussion it was stressed that it was necessary to look at the combination of nutrients that might affect algal blooms, and that it was necessary to further define and develop appropriate analyses of "nutrient limitation." NUT agreed that "eutrophication phenomena" should be understood to include "primarily an increased frequency and greater geographical coverage of algal blooms, partly of a nature unusual until recently and in some cases toxic." This, in turn, might lead to oxygen depletion, changes in benthic fauna and flora and inconvenience to recreational and touristic activities. With regard to the status of eutrophication NUT, agreed on a number of general conclusions to be forwarded to TWG and Parcom. These were that:

- There had been a considerable increase in nitrogen and phosphate inputs to coastal waters of the eastern and southern North Sea, the Skagerrak and the Kattegat and, consequently, increased nutrient concentrations in these areas;
- The most important sources were municipal and industrial waste water, agricultural losses and atmospheric inputs;
- There was "circumstantial" evidence that the increased inputs were related to "problematic eutrophication phenomena," which had occurred in the mentioned areas.

In NUT-2 several national and international research projects were presented, including a joint European study on *Phaeocystis* dynamics, co-funded by the EEC. Important questions addressed in the studies were, among others, the limiting factor for phytoplankton growth and the role of denitrification. The meeting concluded that a compilation and assessment of the results of the various projects would be of great value for the scientific community and administrators. It was decided that a compendium of ongoing and recently completed research project would be made, aiming at disseminating information resulting from the projects, and identifying areas where more research was needed. During the meeting some topics were identified for which more research was considered necessary, namely the development of models, the development of internationally acceptable assessment methods and a review of nutrient reduction technologies.

Quality Objectives

The development of quality objectives "to be achieved in order to avoid adverse effects in the ecosystems of the area due to increased loads of nutrients," was a specific task given to the group by the Consultation Meeting. However, at NUT-1 only Denmark was able to provide specific information because quality objectives for marine waters were already part of Danish water quality policies. It was, therefore, agreed that all delegations would submit to the next meeting of NUT, maps showing where, for different water bodies, quality objectives would be applied, together with proposals for these objectives and quality standards to achieve the objectives. Only the German delegation reserved its position, given the many different subsystems in German marine waters. At NUT-2 many different national approaches to developing and applying quality objectives were presented and extensively discussed. Denmark proposed to apply qualitative rather than quantitative parameters. Belgium stated that the development of models was the "ultimate objective" of the European study on *Phaeocystis* blooms (see also above). The Netherlands reported about approaches to developing seawater quality standards, using values from the 1930s as a reference. Norway reported that it had just started to develop a eutrophication classification system, based upon nutrient and chlorophyll *a* concentrations. Germany indicated that it had a reservation regarding the development of ecological quality objectives, considering the fact that it was still unclear how large phytoplankton blooms were triggered. It was, therefore, first of all considered necessary to minimize nutrient inputs from all land-based sources. The Dutch and Danish delegations proposed to agree upon a single internationally agreed quality objective for the North Sea. Several delegations pointed out that it would first be necessary to specify the relative importance of nitrogen, phosphorus, silicon and carbon, and that it would be necessary to establish a baseline. The delegation of the United Kingdom emphasised that it would have difficulties to agree upon such a quality objective, especially as the aim of such an objective was unclear. The Belgian delegation also pointed to the large amounts of nutrients already stored in the system and that, hence, a recovery time should be taken into consideration. It was, once again, agreed that all delegations would submit maps to the next meeting, showing where quality objectives would be applied, including proposals for such objectives. The German and Swedish delegations reserved their positions with regard to the proposals for quality objectives, but would submit maps with nutrient concentrations. It was, furthermore, agreed that delegations would submit their views on the proposal for an international quality objective.

Measures against pollution by nutrients

Reports on national measures to reduce nutrient inputs were submitted to NUT-1 by The Netherlands, Denmark, the Federal Republic of Germany and Sweden. These focused mainly on sewage treatment, the banning of phosphorus-containing detergents and agriculture. The presentations made clear that, generally, the emphasis of the measures was on the reduction of phosphorus compounds, and that the reduction of nitrogen was still in a planning phase. The Dutch delegation indicated that the combined effect of all ongoing and planned measures, including the reduction of transboundary pollution of the Rhine river, would result in a 50% reduction of phosphorus inputs to the North Sea, but would have only a marginal effect on nitrogen inputs. Germany gave figures for sewage treatment, which made clear that 90% of the inhabitants were connected to sewage treatment systems with biological treatment, removing 40% of the nitrogen and 30% of the phosphorus. By 1985 9% of the treatment plants had tertiary phosphorus treatment. Germany also presented information on plans to reduce nutrient emissions from agriculture and measures that had been taken to reduce NO_x emissions from large combustion plants. The emphasis on phosphorus policies was confirmed by Sweden, which informed the meeting about developments in the Helsinki Convention framework. In 1986 the Scientific and Technical Committee of the Helsinki Commission (Helcom) had prepared a draft recommendation in which it was proposed that sewage treatment plants of more than 10,000 population equivalents (p.e.) should have phosphorus removal capacity, resulting in effluent P values of 1.5 mg P/l. For larger plants the possibility of improving nitrogen removal capacity should be considered. Denmark had developed, and already partly initiated, the by far most comprehensive and concrete policies. These encompassed, among others, removal of nitrogen and phosphorus from sewage treatment plants by some 80%, an action programme for the control of animal manure, to be complied with by 1990, a reduction of N-fertilizers by 25% and reductions of emissions of NOx from large power plants. An EEC delegation at the meeting gave details about the Commission's activities in the framework of environmental consequences of agricultural activities. In order to evaluate possibilities for coordinated reduction programmes, Denmark had prepared an overview of all national measures. On the basis of this compilation, it was concluded that not all parties were carrying out or planning all possible measures. It was, therefore, decided that for NUT-2, Denmark would prepare a draft coordinated programme, in which the various common actions would be prioritised.

At NUT-2 a Danish proposal for priority actions was presented, but not discussed further because, in the meantime, a draft text for INSC-2 had become available, in which detailed measures for the reduction of nutrients were contained, with the aim of achieving a 50% reduction of inputs of nitrogen and phosphorus between 1985 and 1995 (see further 4.2.5). The meeting was also informed about the Rhine Action Programme, adopted 4 weeks earlier on October 1st, 1987, at the 8th Rhine Ministers Conference. One element of this Programme was a 50% reduction of phosphorus and ammonium inputs by 1995. At the national level several reduction activities were presented, either new or complementary to those already presented at the NUT-1 meeting. Sweden presented its Action Plan for the Marine Environment, aiming at halving the nitrogen inputs to Swedish coastal waters by 1992 and considerably reducing phosphorus inputs. Germany addressed problems that could be expected with the reduction of nitrogen compounds: The elimination at sewage treatment plants was an advanced technique that still needed improvement, and it was therefore considered premature to set standards for nitrogen in effluents. It was recognized that agriculture was one of the major sources of nitrogen inputs, but that the legal basis for measures aiming at reducing these inputs was limited.

Monitoring

As already made clear in 2.6.2, nutrients were not part of Osparcom's Joint Monitoring Programme (JMP). Denmark had prepared a compilation of national nutrient monitoring programmes for the NUT-1 meeting. The compilation clearly showed that there were large differences between the national programmes and that most had started only a few years ago, therefore not yet allowing for trend analyses. On the basis of the evaluation, Denmark proposed that nutrient monitoring should be carried out at monthly intervals. NUT's general opinion with regard to intensifying, harmonizing or extending current national nutrient monitoring can, however, best be described as reluctant. The conclusion was, therefore, of a very general nature: "monitoring of nutrients and relevant hydrographic parameters should be carried out on a routine basis at appropriate intervals." It was, furthermore, agreed that phytoplankton production, biomass and species composition should be monitored regularly along transects.

In the NUT-2 meeting the results of the national monitoring activities of Denmark, Germany, The Netherlands and Belgium were presented. Denmark repeated its statement, made at NUT-1, that it was necessary to sample at least 12 times a year to be able to assess nutrient dynamics. Several contributions discussed the limiting nutrient for primary production, and it became clear that it was not simply a matter of either nitrogen or phosphorus. Dutch data from 1985–86 indicated that in winter, nutrients did not limit algal growth. In summer, nutrients were not the limiting factor in the coastal zone, but at 70 km offshore nitrogen and silicate were potentially limiting. The German data showed strongly changing N/P ratios within one season and according to the report "It is this complicated pattern of either phosphorus or nitrogen shortage at the same time in different areas which makes the interpretation of limiting nutrients so difficult" (NUT 1987b).

Future work

With regard to the future work, the NUT-1 meeting agreed to recommend to TWG and Parcom to extend the geographical area covered by NUT, to encompass the whole North Sea. It argued that nutrient inputs from adjacent sea areas might influence the present area covered by NUT. The group also decided that it would prepare scientific advice on whether to include nutrients in the Annexes of the Convention. An initiative of The Netherlands, NUT-2 identified a number of future activities as a medium term objective of the group, mainly with a view to the implementation of the expected outcome of INSC-2 and the envisaged third North Sea Conference (INSC-3), which would be hosted by The Netherlands. It concerned:

1. The elaboration of a detailed definition of eutrophication problem areas, based upon current water quality and national (and possibly international) quality objectives and standards;
2. Quantification of nutrient inputs to the North Sea;
3. Distribution of nutrient inputs through the North Sea. With the help of models it should be possible to quantify the distribution of nutrients and the input to the problem areas, once these had been established. This implied that NUT would work on the adoption of models, which should include processes such as nutrient transport, nutrient interaction with the sediment and ecological processes, such as algal growth;
4. Expected results of input reduction measures;
5. Assessment of ecological changes, resulting from current reduction programmes;
6. Quantification of necessary further input reductions. Also for this activity it would be necessary to use transport and other models.

Conclusions

When analysing the activities of NUT-1 and NUT-2, it is especially the conclusion formulated by NUT-1 that there was "circumstantial evidence" that the increased inputs were related to "problematic eutrophication phenomena, which had occurred in the mentioned areas," which deserves closer attention. It is about the essence of the marine eutrophication problem: nutrients causing problematic phenomena. The adjective "circumstantial" seems well chosen here. The 1976 edition of the Oxford Illustrated Dictionary defines "circumstantial evidence" as "indirect evidence from circumstances affording a certain presumption." The presumption by the NUT members that nutrients were the cause of the problems, did have important consequences, as will be shown in the following Sect. 4.2.5.

4.2.5 The second North Sea Conference

The Policy Working Group

One of the decisions of INSC-1 (Bremen 1984) had been to hold a second Conference in the United Kingdom, with the aim to review the implementation and effectiveness of the Bremen Conference decisions, and to adopt further concrete measures for the maintenance of the quality of the North Sea. The preparation of INSC-2, scheduled for November 1987, was carried out by a Scientific and Technical Working Group (STWG), responsible for the preparation of an updated QSR (see 4.2.2) and a Policy Working Group (PWG), which had the task to evaluate the implementation of INSC-1 and to prepare the themes for INSC-2. The work of STWG served as scientific input to PWG. As had already been the case in the framework of the Paris Convention (4.1.5), Denmark also took the lead in PWG with regard to marine eutrophication. This resulted in the submission to the second meeting of PWG (9–10 April 1987) of a so-called lead paper, outlining a strategy for the reduction of nutrients in the sea. In this paper, Denmark referred to the work of NUT, which had established "circumstantial evidence" that the increased input of nutrients was related to the problematic eutrophication phenomena that had occurred in the eastern and southern North Sea, the Skagerrak and the Kattegat. According to the lead paper, oxygen depletion had, since 1981, occurred almost annually in Danish North Sea coastal waters and in the Kattegat. It was, furthermore, stated that around 1960 the meteorological conditions were similar to those in the 1980s, and that oxygen depletion events had not occurred in the 1960s. The inputs of nutrients around 1960 were about half of those in the 1980s, and Denmark therefore proposed that the minimum aim should be to reduce both phosphorus and nitrogen inputs by 50%. This proposal was clearly reflected in the background policy paper, prepared for PWG by the secretariat of INSC-2, which was part of the agenda of the Conference (INSC 1987a). With regard to the issue "Inputs of Nutrients," the background paper referred to the conclusion of NUT that an increase of nutrient inputs had occurred and that there was circumstantial evidence for a relation with oxygen depletion events. It was stated that these conclusions were reflected in the QSR, and also that they were supported by the conclusions of the WRc Conference (4.2.2). In the background paper the Danish rationale for nutrient reduction was connected with the following statement: “It was therefore argued that nutrient inputs to the areas displaying serious eutrophication should be reduced substantially (perhaps by 50%).” It is obvious that this proposal differed from the Danish one in that it was clearly limited to certain areas. This position was further supported by another paragraph in the background paper, which stated: “Eutrophication is a particular problem limited to certain parts of the North Sea. It was argued that restrictions on nutrient input should accordingly be concentrated where they were likely to give the best results.” The background paper also addressed the issue of the limiting factor for primary production:

> “Generally nitrate was the limiting nutrient in coastal and deeper waters but phosphorus appears to be limiting in estuaries and in certain enclosed coastal areas during at least part of the year. Appropriate action needed to be directed at both, and to focus on the main sources affecting the particular areas which displayed eutrophication or were at risk from it” (INSC 1987a).

Nutrient reduction

The proposal, prepared by PWG, to reduce both phosphorus and nitrogen inputs by 50%, was indeed adopted at INSC-2. However, the decision also reflected the PWG considerations that this reduction would only apply to eutrophic areas. The 50% reduction agreement, adopted at INSC-2, reads as follows (INSC 1987b):

> “10. take effective national steps in order to reduce nutrient inputs into areas where these inputs are likely, directly or indirectly, to cause pollution;
>
> 11. aim to achieve a substantial reduction (of the order of 50%) in inputs of phosphorus and nitrogen to these areas between 1985 and 1995.”

In order to reach this goal, it was agreed to urgently prepare action plans (§12), to pursue detailed elaboration of possible measures to reduce nutrient inputs within the framework of the Paris Commission Working Group on Nutrients (§12), and to consider actions as listed in Annex E to the Ministerial Declaration to be implemented in national action plans (§14). Annex E contained four categories of measures, namely best available technology for wastewater treatment, the regulation of phosphates in detergents, measures to reduce inputs from agriculture and, finally, discharge licensing for industry. §15 of the Declaration also agreed on appropriate measures to reduce nitrogen oxide emissions to the atmosphere was agreed upon.

It is remarkable that both phosphorus and nitrogen inputs were covered by the decision. There was awareness of the fact that reduction of nitrogen inputs would be much harder to achieve than reduction of phosphorus. For the latter, many measures had already been taken in order to solve the problem of eutrophication in freshwater systems, for example the elimination of phosphorus in sewage treatment and the introduction of phosphorus free detergents. In Dutch coastal waters and in the German Bight there was, furthermore, a focus on phosphorus as the main limiting nutrient for primary production. But, mainly as a result of Danish and Swedish studies, which underlined the role of nitrogen and stressed the need for reducing nitrogen inputs, nitrogen was also included in the 50% reduction decision of INSC-2. Moreover, in Denmark and Sweden decisions had already been taken regarding reductions in nitrogen inputs. It can be concluded that scientific evidence had played the major role in the fixing of the agreement on nitrogen, which was, from an administrative and political point of view, a very unattractive one because of the anticipated difficulties in implementation. As a comparison the situation in Chesapeake Bay is mentioned (compare 3.2.7). Here, a controversy had arisen between state and federal agencies, favouring the reduction of phosphorus as a means of reducing negative eutrophication effects in the Bay and, on the other hand, scientists who stressed the need for reducing nitrogen inputs as well (D'Elia and Sanders 1987). In this case, however, scientific advice had not been followed up, most probably because of the political unattractiveness of nitrogen reduction.

Hazardous substances

But INSC-2 was not only relevant from the point of view of reducing nutrient inputs to the North Sea. It was also agreed to reduce by 50% the inputs via rivers and estuaries of substances that are "persistent, toxic and liable to bioaccumulate." Also this reduction would have to be achieved within the time frame 1985–1995. Interestingly, this decision was taken within the framework of the Precautionary Principle, as stated in §1 of the London Declaration:

> "accept the principle of safeguarding the marine ecosystem of the North Sea by reducing polluting emissions of substances that are persistent, toxic and liable to bioaccumulate at source by the use of best available technology and other appropriate measures. This applies especially when there is reason to assume that certain damage of harmful effects on the living resources of the sea are likely to be caused by such substances, even when there is no scientific evidence to prove a causal link between emissions and effects ('the principle of precautionary action')."

In §VII of the Preamble of the London Declaration the adoption of the Precautionary Principle was worded as follows:

> "Accepting that, in order to protect the North Sea from possibly damaging effects of the most dangerous substances, a precautionary approach is necessary which may require action to control inputs of such substances even before a causal link has been established by absolutely clear scientific evidence."

It is obvious that the advocates of the UES approach had gained an important victory, notwithstanding the fact that in §XV of the Preamble the use of both the UES and the EQO approach was underlined. In six subparagraphs it was, among others, decided to reaffirm both approaches as set out in the Bremen Declaration and to ensure that quality objectives, based upon the latest scientific findings, should form part of strategies to control inputs of hazardous substances.

There are some important differences between the 50% hazardous substances reduction decision and the 50% nutrient reduction decision. The first was taken under the political motivation of a precautionary approach and supported by all participants in the conference. Such was not the case for nutrients. There was one more important difference with the 50% reduction agreement for hazardous substances, namely the condition that the reduction effort for nutrients would only be necessary in case these would cause pollution. This decision reflected the UK position, which considered that negative eutrophication symptoms did not occur in UK coastal waters. This opt-out option, which actually meant that the decision did not apply to all parties, was probably the main reason why a decision on nutrient reductions had been possible at all.

Scientific knowledge

A specific section of the Declaration dealt with the enhancement of scientific knowledge and understanding. Here it was agreed "to endorse the need for further development of harmonized methods for monitoring, modelling and assessment of environmental conditions at national and international levels." In order to achieve these aims, it was decided to establish a joint Task Force of ICES and Osparcom. In Annex G to the London Declaration the rationale for establishing the Task Force, its objectives and its work programme were specified. It was stated that during the preparation of the QSR, it had become clear that there were still shortcomings in data for certain contaminants, in particular with regard to trends in inputs and the link between contaminant levels and environmental changes. It was, therefore, considered necessary to develop a coordinated scientific programme to provide more consistent and dependable data, in order to be able to establish links between contamination and effects with greater confidence. According to Annex G, "Such knowledge is needed not only as a basis for further decisions but also to show the effectiveness or otherwise of measures already taken or planned." The objective of the Task Force was "To carry out work leading, in a reasonable time scale, to a dependable and comprehensive statement of circulation patterns, inputs and dispersion of contaminants, ecological conditions and effects of human activities in the North Sea." The main elements of the working programme of the Task Force were:

1. Agreement on substances and parameters to be measured, including monitoring methodology;
2. A quality assurance programme for sampling and analysis;
3. More and better quality data;
4. Special programmes in specific areas (e.g. the Wadden Sea, the Kattegat and British estuaries);
5. Development of models for assessment and management purposes;
6. Research to fill gaps in knowledge of causal mechanisms needed for the interpretation of data (e.g. impacts on marine ecosystems, indicators of biological

change, fish diseases, nutrient enrichment, contaminant dispersion and sediment movement).

The decision to establish a Task Force must be valued as remarkable because there were already actors within the science-policy network with this responsibility. The most notable is ICES and its working groups. But also within Osparcom scientific advice was being prepared by working groups. The most plausible reason for the decision to install a Task Force is the dissatisfaction by several parties with the functioning and/or opinions of the existing groups. Interestingly, the grounds for this dissatisfaction must have been quite different. For some continental states the position of the ICES ACMP, especially its adherence to the assimilative capacity principle and dissatisfaction with the functioning of Osparcom working groups, may have been reasons to support a new group, in which countries would be represented on a national basis. On the other hand, seen from the position of the UK, the Task Force can be regarded as support for strengthening the scientific basis for pollution policy and management.

4.3 Recapitulation: marine eutrophication constructed

In the introduction to Sect. 4.1 the question was asked how it had been possible that marine eutrophication had become an internationally acknowledged pollution problem within less than four years after the 1981 oxygen depletion events. A question that can be added now is how was it possible that internationally agreed measures to reduce nutrient inputs had been agreed upon only two years after the issue had entered the international political agenda. According to Hannigan (1995), “successfully contesting an environmental claim in the political arena requires a unique blend of knowledge, timing and luck.” Hannigan (loc.cit.) also stressed the relevance of disasters to open up “political windows,” and furthermore stated that “society’s willingness to recognize and solve environmental problems, rests primarily upon the claims-making activities of a handful of ‘issue entrepreneurs’ in science, the mass media and politics.” The five factors mentioned by Hannigan, knowledge, timing, luck, disasters and entrepreneurs, can all be found in the process that started with the 1981 oxygen depletion events and ended with the 50% nutrient reduction decision of INSC-2.

The construction of the international marine eutrophication problem was triggered by several oxygen depletion events that occurred in the Skagerrak, the Kattegat, the Danish Belt seas, the western Baltic Sea and the German Bight, during the period 1981–1983. The impact on the public was great, especially in the Denmark and Sweden, because the effects of the oxygen depletion were tangible in the form of dead fish and other marine species. Danish researchers also claimed that international nutrient transport was responsible for the oxygen depletion events, which was the reason why Denmark took the initiative for a special consultative meeting of the Paris Convention (Copenhagen, 1985). This meeting resulted in the establishment of a special working group on marine eutrophication issues, the NUT group (NUT). With the establishment of NUT, the international marine pollution science-policy network had been extended with a body that can be regarded as an intermediate between science and politics. Denmark, Sweden, Germany and The Netherlands were the most active participants in the first two meetings of NUT, which took place in 1986 and 1987. The involvement of The Netherlands may be seen in the light of increasing scientific proof of eutrophication effects in Dutch coastal waters, in particular in the western Dutch Wadden Sea, but also in the framework of the development of a Dutch management plan for the North Sea. Denmark provided the first chairman of NUT, as well as most of the scientific input.

The original tasks of NUT, which mainly related to the exchange relevant information about eutrophication related matters, were placed in a different perspective as a result of the preparation of INSC-2. Through INSC-2, the work of NUT received political relevance, and Danish civil servants played a central role in transferring the results of NUT from the administrative to the political realm. The Danish entrepreneurship was successful in two respects: First, it was acknowledged that marine eutrophication was an international problem; Second, the 50% reduction decision applied not only to phosphorus but also to nitrogen. Both claims originated from Danish researchers and the Danish administration. This underlines the observation by Hannigan (1995) that, with regard to international environmental problems, it is not the international epistemic community that is most important, but that "the centre of gravity for scientific claims-making on specific issues tends to reside in a specific nation." Indeed, as shown in 4.2.2, the international scientific community, especially ICES, was much more cautious about the scope and causes of phytoplankton blooms. This also underlines the difference between academic science and regulatory science, as put forward by Jasanoff (1990) (chapter 1). In academic science, consensus building is slow, whereas regulatory science must respond to immediate political demands. According to Lambright (1995), different processes operate in regulatory or policy-relevant science: “Debates among scientists take place often through the media rather than scientific journals. Consensus-forming processes are speeded up by special mechanisms, and actions are taken by policymakers on the basis of what may be very tentative agreements based on limited data.” A central feature of policy-relevant knowledge is, according to Lambright (loc.cit.), that everything is done with more urgency. He therefore speaks of “accelerated science.” What happened in the run-up to INSC-2 is a clear example of the application of instantly available knowledge. There were limited data available and targeted research programmes had not yet been finalised at the time the political decisions were taken. Both Lambright (loc.cit.) and Jasanoff (1990) have pointed at the dangers of the application of instant knowledge. Lambright (1995) has worded this as follows: “Science by press conference may speed up its use but can also burn the provider and policy users if the information proves faulty.” For the case of CFCs and the ozone layer depletion, Lambright (1995) concluded that the use of “accelerated science” had been successful. An important factor, according to Lambright (loc.cit.), had been the creation of a participation mechanism for the scientific community, in which the new knowledge had been discussed before it was applied in decision-making. In the case described, the science used was deemed credible by the scientific community, in contrast to the preparation of the INSC-2 decisions. The scientific information applied, derived mainly from Danish research, had not, or only to a limited extent, been subject to discussions within the international scientific community. International discussions about marine eutrophication, carried out by ICES in the framework of the preparation of the North Sea QSR and in scientific symposia, had not even addressed the relevance and extent of international nutrient reduction measures, let alone agreed upon the credibility of the information used. Moreover, as argued by ICES (see 4.2.2), the scientific basis for monitoring nutrients and primary production was fully insufficient. This task was explicitly mentioned by INSC-2 as part of the enhancement of scientific knowledge.

Not only was the scientific underpinning of the 50% nutrient reduction decision poor. There was also political controversy because the United Kingdom considered marine eutrophication not to be a problem in its waters, and underlined the importance of other causative factors for eutrophication-related symptoms, most notably climatic changes. For that reason, the national obligation to reduce nutrient inputs by 50% only applied for those discharges, which were likely to cause pollution. On the eve of the implementation of the decisions of INSC-2, some formidable tasks awaited the science-policy network: new scientific knowledge was needed to strengthen the scientific credibility of the 50% reduction decision, to identify in which parts of the North Sea nutrient inputs would cause pollution, including the question how to define pollution and to monitor nutrient concentrations and eutrophication effects. How the science-policy network dealt with these questions, and which role new knowledge played in finding relevant and practicable answers, are central themes in the remainder of this study.

4.4 1988–1990: Towards the third North Sea Conference

INSC-2 had, contrary to INSC-1, resulted in a number of very concrete political decisions. The responsible Dutch minister Smit-Kroes even valued the conference as an "historical event" (De Jong 1987), and she proposed a third International North Sea Conference (INSC-3), to be held in The Netherlands early 1990. The activities of the science-policy network in 1988–1990 were, therefore, determined both by the implementation of the decisions of INSC-2 and by the preparation of INSC-3. The two main questions addressed in this section are directly related to these two different types of activities. The first question is whether science has been relevant for the fine-tuning of the political decisions of INSC-2. The second question is whether new scientific knowledge has influenced new political decision-making. In addition, it will the investigated how the science-policy network, in particular the science-policy interface, dealt with these two tasks.

But not only political and scientific developments determined the activities within the science-policy network. In 1988 two catastrophic events occurred, which caused much public concern and, consequently, more pressure on the science-policy network. It concerned a bloom of the toxic alga *Chrysochromulina polylepis* along the Swedish and Norwegian North Sea coasts and an epidemic of the harbour seal in the North Sea. These events will be described in 4.4.1.

The further development of the knowledge basis and its relevance for the policy process will be investigated in 4.4.2, providing an overview of several scientific publications relevant to marine eutrophication that became available after INSC-2. It concerns review articles and the results of national studies that had been initiated, following the oxygen depletion events of the beginning of the decade.

Section 4.4.3 is concerned with the activities of the science-policy interface. Following the establishment of NUT in 1986, the science-policy interface had been further strengthened by the installation of the North Sea Task Force (NSTF). In 4.4.3 developments in these groups will be covered, as well as relevant activities of ICES working groups. The focus of the analysis in 4.4.3 will be on how the science-policy interface was organized, how it dealt with the tasks given to it by INSC-2 and how it prepared INSC-3. Of particular relevance will be the question whether and how the science-policy interface applied new knowledge in its activities, and whether new scientific findings would lead to amendments of these INSC-2 decisions. After all, the INSC-2 decisions were based upon instantly available knowledge, and several questions about marine eutrophication remained after INSC-2, such as whether a 50% nutrient reduction would be sufficient to protect the North Sea from adverse eutrophication effects, or which parts of the North Sea were most vulnerable to excess nutrient loading. At the national level, Denmark had already started a nutrient reduction programme with much more ambitious goals, and in The Netherlands such was being discussed seriously. The United Kingdom, on the other hand, regarded marine eutrophication as not a problem in its waters.

In Sect. 4.4.4 an overview of political developments is given. It concerns the impact of INSC-2 as well as the preparation of INSC-3 (The Hague 1990).

4.4.1 The 1988 catastrophes

The "killer alga" Chrysochromulina polylepis

On May 9th 1988, mortality of rainbow trout in fish farms along the Swedish west coast was linked to a bloom of the toxic alga *Chrysochromulina polylepis*. The algal carpet spread in a westward direction along the Norwegian southern coast and reached its peak around June 2nd. By this time the algae had spread as far as the coastal area between Stavanger and Bergen (Berge et al. 1988). Accompanying the spreading of *Chrysochromulina* were large-scale fish kills, especially of caged trout from Norwegian and Swedish fish farms. About 100 tonnes of trout from Swedish and 500 tonnes from Norwegian farms were killed in the first 14 days of the bloom, representing a value of about 5.4 million USD. Not only had trout been killed, but also high numbers of invertebrates, macroalgae and wild fish (Rosenberg et al. 1988; Underdal et al. 1988). A synopsis of the events by Rosenberg et al. (1988) in Ambio carried the title "Silent spring in the sea," illustrating the impression the incident had made. According to Rosenberg et al. (loc.cit), the incident "provoked a major consolidated reaction among Scandinavian scientists and research ships were directed to investigate the hydrography, algal distribution and immediate ecological effects." Rosenberg et al. (loc.cit.) pointed to a land runoff in the preceding winter that had been higher than average, resulting in high nutrient concentrations prior to the spring bloom. Furthermore, the surface water temperature had been 2 °C higher than average. During the bloom, however, the nutrient concentrations were not conspicuously high. Rosenberg et al. (loc.cit.) concluded that the ecological causes of both the bloom and the production of toxins were not known. They pointed, however, to the many local and large-scale changes that had been observed in the Kattegat and Skagerrak, which might be attributable to eutrophication. They stated to be "convinced that man's continuous pollution of the seas during the last decade has put certain marine ecosystems in a state of disorder." Even with a drastic reduction of discharges, Rosenberg et al. (loc.cit.) continued to say, significant ecological disturbances in the sea would continue to happen in subsequent decades.

The event not only attracted interest of scientists in Scandinavia. Two scientific workshops dedicated to the event were held, the first organized by ICES, the second by the Commission of the European Communities. The aims of the ICES workshop (28 February–2 March 1989) were to amalgamate relevant observations on toxicology, physiology and toxicity of *Chrysochromulina polylepis*, to describe the environmental background associated with the bloom, and to evaluate the effects of the bloom on mariculture and on the marine ecosystem (Skjoldal and Dundas 1991). The workshop aimed to determine the role anthropogenic nutrient enrichment might have played in the development of the bloom. According to the scenario in the workshop report, anthropogenically loaded water from the southern North Sea was transported into the Skagerrak-Kattegat area through the so-called Jutland current. Also local land run-off in this area and transport of nutrients from the Baltic contributed to the nutrient loading. The latter two factors were higher than average, due to a high precipitation in winter and a high outflow of Baltic water. Another exceptional factor, caused by specific meteorological conditions, was an effective and relatively long-lasting stratification of water masses in the area, allowing the *Chrysochromulina* bloom to develop. As a result of the anthropogenic loading, the surface water in the Skagerrak-Kattegat area was not nutrient depleted after the spring bloom, and the underlying water had a high N/P ratio. During the mixing of the two layers, there was an upwelling of water with a high N/P ratio, causing phosphorus deficiency, which, as had been shown experimentally, may cause *Chrysochromulina* to become toxic. It was, therefore, cautiously concluded that anthropogenic nutrient load to the affected water masses had played a role in the development of the toxic bloom (Skjoldal and Dundas, loc.cit.).

In October 1989 a special scientific workshop was held in Brussels, in the framework of the Environmental Research and Development Programme of the Commission of the European Communities (CEC). The following reasons for organizing the workshop were given (Lancelot et al. 1990):

"The occurrence in May/June 1988 of a large and unexpected bloom of *Chrysochromulina polylepis* along the coasts of Denmark, Sweden and Norway and the catastrophic consequences it has had on fisheries and mariculture have drawn attention to the potential danger of North Sea eutrophication and to the unpreparedness of the European countries to take measures for avoiding re-occurrence of such accidents in the future."

The aim of the workshop was "to identify the lack of knowledge to predict ecological processes involved and to assess preventive actions to be taken." According to Lancelot et al. (loc.cit.), rational management of the coastal environment would only be possible with integrated approaches, coupling ecological modelling and economic assessment. The main questions addressed in the session on ecological processes, the central theme of the workshop, were whether available knowledge would allow the development of algal bloom models, and to what extent models would be able to predict the occurrence of exceptional blooms.

Reid (1990) presented an overall picture of phytoplankton dynamics, and he stated that physical processes that helped to enhance stability, were of prime importance for blooms to form. High land runoff would improve stable conditions and provide nutrient pulses. An analysis of blooms, which had occurred in the last half century, revealed, according to Reid (loc.cit.), that no increasing trend was present, but that, rather, there were periods with higher numbers of blooms. He also stated that our knowledge of phytoplankton was still in its "infancy." Nielsen and Richardson (1990) presented an overall account of the 1988 *Chrysochromulina* bloom and concluded that, in general, the onset of phytoplankton blooms was controlled by the physical environment. Because the marine ecosystem fluctuated in response to changes in meteorological conditions, and was also influenced by human impact on geochemistry, occurrences of blooms in terms of extent and timing would be difficult to predict (Nielsen and Richardson, loc.cit.). Lancelot (1990) described an international EEC sponsored research programme, aiming at developing a model, which would allow the prediction of *Phaeocystis* development in response to terrestrial nutrient input. She concluded that such a task was attainable, especially because *Phaeocystis* blooms were a recurrent event. However, for *Chrysochromulina* this would be much more difficult because it was a non-recurrent, exceptional event.

The epidemic of the harbour seal

A second catastrophic event in 1988 was the high mortality of harbour seals (*Phoca vitulina*), which occurred in several areas of the North Sea during the period April to December (Reijnders 1992). Although not related to marine eutrophication, the event is mentioned here, since it had an effect on marine pollution policies in general, as will be shown later. In the Wadden Sea about 6,000 animals died, reducing the population by 60%. In the Skagerrak-Kattegat area the population was halved from 6000 to 3000 individuals. Also along the Norwegian North Sea coast and in the Wash area mortality occurred, whereas this was not the case in the northeastern parts of the UK.

4.4.2 Marine eutrophication reviewed

During the period following INSC-2, several comprehensive review papers, specifically dealing with marine eutrophication, were published, indicating the increasing scientific interest in the issue. Not only interest in marine eutrophication as such had increased, but in marine pollution in general. In the second half of the 1980s several research projects dealing with marine pollution, including factors relevant for marine pollution, such as meteorology, hydrology and climate, had been initiated. Among these were the German ZISCH Project (Zirkulation und Schadstoffumsatz in der Nordsee: Circulation and pollutant turnover in the North Sea) in which, in 1984–1989, circulation and contaminant fluxes in the North Sea were studied, and the British National Environment Research Council (NERC) North Sea Project (1988–1991), in which several studies of marine processes were carried out as a basis for the development of prognostic environmental quality models.

Increasing scientific interest in marine eutrophication can also be inferred from publications in the Marine Pollution Bulletin (MPB) and Ambio. In the 1970s the percentage of papers published in these journals, dealing with marine eutrophication, was 2.4% for the MPB and 3.3% for Ambio (see 3.3.1 and 3.3.2). During the period 1981–1990, 44 papers about marine eutrophication were published in the MPB, which is 5.3% of all articles (reports and baseline studies) that appeared in this period. The increase of marine eutrophication contributions in Ambio was much more pronounced. Whereas from 1972–1980 only two marine eutrophication papers appeared, this figure was 24 in the 1980s, equalling almost 25% of all contributions dealing with marine issues. Of the 24 contributions, 13 were part of the special issue on marine eutrophication of May 1990 (see further this subsection). An interesting difference between the MPB and Ambio was that the cases covered by the MPB came from all over the world, whereas the Ambio focus was almost exclusively on marine eutrophication in the Baltic and the Skagerrak-Kattegat area. It is obvious that the Ambio emphasis had been caused by the oxygen depletion events, which had started in the beginning of the 1980s, and by the 1988 *Chrysochromulina* event. The low attention in the MPB for marine eutrophication in the North Sea area probably reflects the position of UK scientists and officials that marine eutrophication was not a large-scale marine pollution problem (compare also Clark 1987). Still, an increase in attention for marine eutrophication was obvious for the MPB, reflecting a world-wide interest in the issue. This is confirmed by the editorial to the Ambio special issue in which it was stated:

"the recent awareness of marine eutrophication as a serious coastal issue is not confined to northern Europe and Scandinavia. From around the rim of the Mediterranean, and from increasing numbers of bays and estuaries along the coastlines of North and South America, Africa, India, southeast Asia, Australia, China and Japan have come increasing reports of noxious (and sometimes toxic) algal blooms, anoxic bottom waters, and fish kills" (Nixon 1990).

Another clear indication of increasing global interest in marine eutrophication was the publication by GESAMP (2.5.2) of a report specifically dealing with this topic (GESAMP 1990).

In the remainder of this section, the results of three literature reviews about marine eutrophication and algal blooms, as well as the outcome of two national eutrophication studies, all published in 1988–1990, will be described and analysed.

Pollution of the North Sea: an assessment

Under this title, a comprehensive volume was published in 1988 under the editorship of Salomons, Bayne, Duursma and Förstner (Salomons et al. 1988). Although not specifically dealing with marine eutrophication, but with marine pollution in general, this volume is covered here because it was, as stated on the back cover, "the first modern review on the fate, distribution and effects of pollutants in the North Sea." In the preface to the book the editors stated: "This preface is being written at a time of exceptional public interest in the North Sea, following media headlines on toxic algal blooms, the mass mortality of common seals and concern over pollution levels." It should be noted here that the preface was indeed written after these events had taken place, but that the contents of the book had, as is usual for scientific papers, been submitted at least one year earlier. The editors underlined the fact that the book was of a multinational character, "expressing remarkable consensus amongst the scientific community as to the vulnerability of the North Sea, and its finite capacity to assimilate waste." They also touched upon the problem of ecosystem complexity, and stated that the theory of ecosystem structure was not yet advanced enough to allow detailed tracing of cause and effects. To this they added that it had been argued that "in systems at this level of complexity predicting catastrophic events may be inherently impossible." For the editors this implied a reduction of the possibilities of irretrievable damage to occur by reducing Man's impact and, in the meantime, "to accelerate the pace of scientific research, in order to identify the most sensitive areas and processes within the North Sea, coupled with careful monitoring to detect change, both as deterioration and recovery."

The general problem of the complexity of the North Sea ecosystem, signalled by the editors, also appeared in several of the individual contributions. However, after reading these individual contributions, the "remarkable consensus amongst the scientific community," expressed by the editors, seems a somewhat too positive judgement. The separation of the scientific community into a pro-assimilative capacity camp and a pro-precautionary principle camp is clearly traceable, the Anglo-Saxon contributions generally belonging to the first and the continental scientists to the second camp (compare 4.2.3). Examples in favour of a precautionary approach can be found in papers about the ecosystem (De Wolf and Zijlstra 1988), natural events (Zijlstra and De Wolf 1988) and the German Bight (Dethlefsen 1988). These authors underlined the complexity of the ecosystem and the associated problem of separating natural and man-induced events, and, for this reason, supported precautionary or no-regret policies. Contributions by Stebbing and Harris (1988) about the role of biological monitoring and by Livingstone et al. (1988)

about biological effects measurements, were generally more positive about the possibilities of inferring and predicting assimilative capacity.

"Pollution of the North Sea" also contained an extensive review contribution about nutrients and eutrophication by a German, a Belgian and a Dutch author (Brockmann et al. 1988). Because of the controversy between the UK and continental scientists about the causes and effects of eutrophication-induced events, it is unfortunate that no British scientist contributed to this paper. The emphasis of the paper was on the distribution of nutrients in the North Sea. The comprehensive description of nutrient concentrations in the various areas of the North Sea illustrated the large regional differences, mainly caused by the different hydrographic regimes. In the section on eutrophication, Brockmann et al. (loc.cit) pointed to the problem of predicting eutrophication effects on the basis of external nutrient enrichment because of the "complex coastal ecosystems, including variable plankton populations and patchy benthic communities, all subject to the influences of hydrodynamic and other physical conditions." The only thing that could be stated, Brockmann et al. (loc. cit) continued to say, was that systems, which were naturally subject to conditions promoting intense algal developments or oxygen depletion, were particularly sensitive to additional external inputs of nutrients. In the final section on future research needs, Brockmann et al. (loc. cit) expressed the expectation that research efforts to study nutrients and eutrophication effects would increase in future. They referred to the new Parcom nutrient working group (NUT), to interdisciplinary activities of ICES, to the continuation and intensification of national research and monitoring programmes and the international *Phaeocystis* project (see also 4.4.1). They underlined the need for considering all relevant parameters for the understanding of eutrophication processes, and pointed in this respect to the role of microphytoplankton and microheterotrophs and flux measurements. Brockmann et al (loc.cit) also expected that the role of modelling would (and should) increase, in order to increase the understanding of ecosystem processes and to be able to predict consequences of man-induced changes.

Eutrophication in the North Sea

The first comprehensive literature review about eutrophication in the North Sea appeared in 1988. The 100-page document was published in 1988 by Nelissen and Stefels, two students of the Netherlands Institute of Sea Research (Nelissen and Stefels 1988). The review was structured according to hydrography, distribution and fate of nutrients, eutrophication phenomena and season-dependent susceptibility of the foodweb for eutrophication. Nelissen and Stefels (loc.cit) underlined the need for a system approach to eutrophication, and questioned in this respect the usefulness of nutrient concentrations as an indicator of the status of the system. They identified the coastal area along the continent, the so-called continental coastal water mass, stretching from the Channel to the Skagerrak, as most influenced by eutrophication. The main reasons given by Nelissen and Stefels (loc.cit) were that land-derived inputs remained confined within the area as far as the "exit" in the Skagerrak. Moreover, nutrients would be recycled constantly within this system. They estimated that about half of the nutrients present within the area were of anthropogenic origin. With regard to assessing the effects of anthropogenic eutrophication, Nelissen and Stefels (loc.cit) stated that they were "aware of the complicating 'naturalness' of the processes involved, in contrast with toxicity effects of most other man-made pollutants." They concluded that the effects of eutrophication ranged from positive to negative, and that a "harmless increase in biomass may end up in mass mortalities." In vulnerable areas, such as the German Bight, little flexibility was left and it was therefore necessary to reduce nutrient inputs (Nelissen and Stefels, loc.cit).

Eutrophication of the North Sea and the Kiel Bay

Between 1984 and 1988, German marine research institutes had carried out a comprehensive research project into the causes of the 1981 oxygen depletion events in the German Bight and the Kiel Bay (see also 4.1.2). In the second half of the 1980s already several results of the project had become available, and in 1990 an English version of the final report of the project was published (Gerlach 1990). Gerlach had integrated the results of the 22 subprojects in sections about long-term developments of weather conditions, oxygen deficiencies and nutrient concentrations and inputs, nutrient processes, such as deposition and denitrification, and the effects of changes in these parameters on phytoplankton development. In summarizing the results, Gerlach firstly demarcated the eutrophication problem area in the North Sea as the belt of the continental coastal water, extending 50 to 100 km from the coastline. It was to this area that annually some 160 km^3 of freshwater were discharged by the continental rivers. Changes in nutrient concentrations in the German Bight could be documented as a result of the measurements, carried out since 1962 at Helgoland. Phosphate winter concentrations at this station had increased by a factor of 1.7 during the period 1962–1975, after which no further increase was recorded. In 1973–1984 winter concentrations of dissolved inorganic nitrogen had increased by a factor of 1.4. According to Gerlach (loc.cit), it was "reasonable" to assume that these changes were the result of increased nutrient inputs by rivers and atmosphere. He, however, also mentioned the fact that changes in phosphate concentrations had occurred in the western English Channel, which could not be correlated to changes in anthropogenic inputs.

The Helgoland monitoring data also showed an increase of overall phytoplankton biomass by a factor of two to three, between the beginning of the 1960s and the end of the 1970s. This increase had been caused mainly by an increase in biomass of flagellates, which had occurred between 1971 and 1978. Changes in phytoplankton and zooplankton had, as documented by the Continuous Plankton Recorder data (see 3.2.5), also occurred in parts of the North Sea not influenced by anthropogenic nutrient discharges. So far, Gerlach continued to say, no correlations had been found between exceptional algal blooms and exceptional nutrient discharges. Therefore, he considered it necessary to study both the relations between phytoplankton and nutrients and phytoplankton and hydrography/meteorology. According to Gerlach (loc.cit.), these relationships were further complicated by zooplankton grazing.

In the final chapter, entitled "Consequences," Gerlach asserted that a halving of the riverine phosphorus inputs could restore the situation of 1970 in the German Bight. He considered phosphorus to be the limiting nutrient here, whereas in offshore waters and in the Baltic, nitrogen was limiting primary production in spring. As a result of a reduction of nutrient inputs, undesirable phytoplankton blooms would occur less frequently and over smaller areas than during the 1980s (Gerlach, loc.cit.).

Phytoplankton of the North Sea and its dynamics: a review

A central question in the study of eutrophication and its effects was whether increased anthropogenic nutrient inputs to the North Sea had caused an increase of phytoplankton blooms and/or changes in phytoplankton species composition, or whether other factors, such as climate and weather, were the main causative factors. As discussed in the foregoing parts of this chapter and in the previous Chap. 3, the negative effects of such changes could be oxygen depletion as a result of the decay of large amounts of dead phytoplankton, and the occurrence of nuisance and toxic blooms. Because there were quite different opinions within the scientific community about the causes and seriousness of such events, the main results of a review article in the Netherlands Journal of Sea Research, jointly written by scientists from the UK, The Netherlands, Belgium and Germany (Reid et al. 1990), will be briefly discussed here. All contributors were, or had been, directly involved in eutrophication related research: Reid of the Plymouth Marine Laboratory was working with the CPR (see 3.2.5), Lancelot of the Free University of Brussels with *Phaeocystis* (see 4.2.1 and 4.4.1), Gieskes of the University of Groningen with phytoplankton development in Dutch coastal waters (see 3.2.5) and both Hagmeyer and Weichart, Biologische Anstalt Helgoland, respectively Deutsches Hydrographisches Institut, were involved in the German eutrophication research project (see above).

One of the main conclusions of the review was that there was no evidence of an increasing trend in the frequency of plankton blooms, with the possible exception of *Phaeocystis* (Reid et al., loc.cit.). With regard to this possible exception, reference was made to, among others, Cadée and Hegeman (1986) and Lancelot et al. (1987) (see further 4.2.1). Other important conclusions were that the North Sea consisted of several sub-regions with characteristic floras, that eutrophication was clearly visible in the continental coastal waters but not in the offshore North Sea, and that eutrophication had not increased since the end of the 1970s, whereas microflagellates had continued to do so.

Reid et al. (1990) also stressed the complexity of phytoplankton dynamics and the inadequacy of survey coverage of the North Sea, both in time and space. An exception was the CPR, but the authors expressed the hope that recently started research projects would improve the situation. With regard to future research work, one recommendation deserves special attention. Reid et al. (loc.cit.) asked for a guarantee that the few long time-series of phytoplankton observation, namely the CPR, the Marsdiep series and the Helgoland series, be continued, as an essential contribution to understanding long-term variability and climatic changes, and the evaluation of anthropogenic impacts.

The Swedish eutrophication research programme

In May 1990 a special issue of AMBIO was published, dealing with the results of the Swedish marine eutrophication research project. This project had started in 1983, following the 1981 oxygen depletion events in the Skagerrak-Kattegat (see 4.1.1). The programme focused on two different eutrophication impacted areas, namely the Stockholm archipelago, mainly influenced by discharges from municipal sewage treatment plants with tertiary treatment, and the Laholm Bay (figure 4.1), mainly influenced by nutrient effluents from agricultural land (Rosenberg et al. 1990). Contrary to the above presented results from North Sea research projects and analyses, the conclusions were quite straightforward. In terms of nutrient demands of phytoplankton, both in the Baltic and the Kattegat discharges of nitrogen dominated over discharges of phosphorus. Despite this fact, nitrogen was generally found to be the limiting factor for primary production. Nitrogen-fixation did not seem to be a substantial source of nitrogen: there were summer blooms of cyanobacteria in the Baltic proper, but these were an exception and limited by phosphorus. The paradox of nitrogen as the limiting factor, despite relatively high nitrogen inputs, was explained by the reduction of nitrogen concentrations by denitrification and the release of phosphorus from sediments. Both processes were enhanced by low or zero oxygen situations (Rosenberg et al., loc.cit.). The question was raised how to restore the marine environment, and what kind of a "balanced ecosystem" was desirable. Rosenberg et al. (loc.cit.) mentioned in this respect the wish for a high fish catch and fairly good bottom conditions. According to the authors, conditions in the 1950s and the 1960s seemed to have been satisfactory for the Baltic and the Kattegat respectively. For both areas, it was argued that nitrogen loading in the respective periods had at least doubled and it was, therefore, proposed to achieve a reduction of nitrogen inputs by at least 50%.

Analysis

The scientific publications described in this section show some remarkable similarities. The most obvious are the emphasis on the complexity, dynamics and variability of the marine ecosystem, the limited knowledge of nutrient dynamics and, consequently, the problems in establishing causal links between increased nutrient loading and increased algal blooms. Most authors, therefore, made a plea for a precautionary approach to dealing with nutrients. The role of scientific research, if not a predictory one, would, according to Salomons et al. (1988) (see above), be to identify the most sensitive areas and processes. Another common element was that eutrophication and eutrophication problems in the North Sea were confined to the continental coastal waters and the Skagerrak-Kattegat area.

Another similarity is that most of the contributions described in this section underlined the lack of long-term data. For this reason, the existing time-series, established by the CPR, the Helgoland Reede measurements and the *Phaeocystis* Marsdiep, played the major role in the different analyses. This, despite the fact that neither of these three time-series can be regarded as ideal from a monitoring methodology point of view: the CPR data only provide indirect and rather coarse information on phytoplankton, and both the Helgoland Reede and the Marsdiep monitoring sites represent only one measuring location, irregularly influenced by different water masses.

There are, however, also some noteworthy differences and discrepancies, in particular between the North Sea and the Skagerrak-Kattegat contributions. The latter were very pertinent about nitrogen being the main factor limiting primary production. Moreover, clear proposals were given as to the desired policy direction, namely a 50% reduction in both nitrogen and phosphorus inputs, based upon a comparison with the situation in the 1950s and 1960s. There was, furthermore, a remarkable consensus between Danish and Swedish scientists about these conclusions and proposals. For the North Sea, on the other hand, the situation was less clear. The results of the German eutrophication project pointed to phosphorus as the main limiting factor. Also Dutch research, for example by De Jonge (1990) (see also 3.2.5), focused on the importance of phosphorus. Another difference with the Swedish-Danish approach was that no specific proposals for policies were given, but instead, rather general support for a precautionary approach. But the major difference between the North Sea and the Skagerrak-Kattegat situation was that for the North Sea there was no scientific consensus about increased nutrient inputs causing increasing algal blooms. Rather, it may be concluded that the scientific community supported the view that there had not been a consistent increase in the frequency and/or intensity of such blooms.

4.4.3 Science-policy interactions 1988–1990

In the previous sections scientific developments following the second North Sea Conference (INSC-2) were described. The current section is concerned with the question how the science-policy network dealt with the implementation of the eutrophication-related decisions of INSC-2. The focus of the analysis is on the relevance of science in the implementation work, and the extent to which new scientific insights were of relevance for the implementation process. The implementation tasks formulated by INSC-2 can be summarized as follows:

- To strengthen the scientific credibility of the 50% reduction decision;
- To find answers to the question in which parts of the North Sea nutrient inputs will cause pollution, including the question how to define pollution;
- To monitor nutrient concentrations and eutrophication effects.

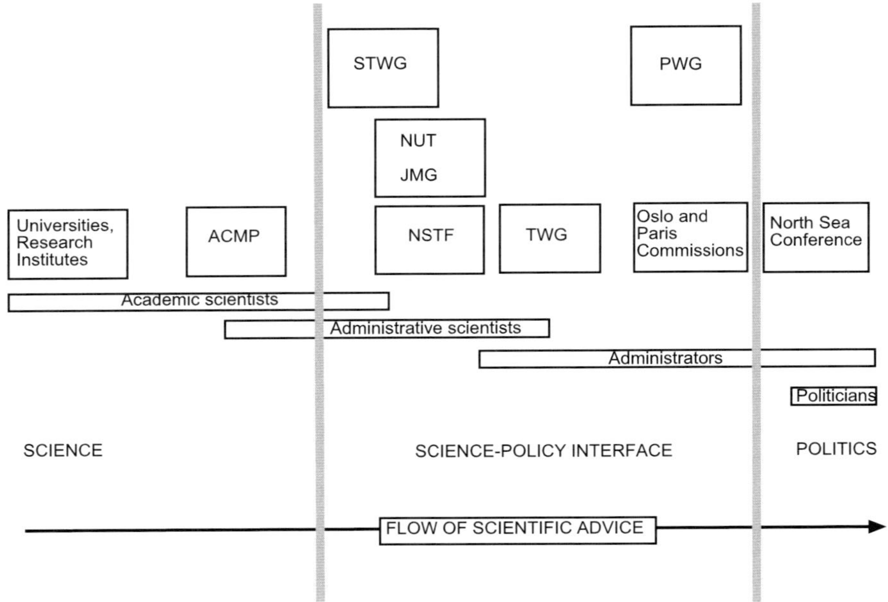

Fig. 4.4. The main elements of the science-policy network in the 2nd half of the 1980s. A schematic indication is given of the type of members of relevant working groups (Academic scientists, Administrative scientists, Administrators or Politicians), the main field covered (Science, Policy and Management or Politics) and the direction of the flow of scientific advice. Acronyms: see list of Acronyms

The main bodies responsible for the implementation were the nutrient working group (NUT), the North Sea Task Force (NSTF), established by INSC-2 (4.2.5), and relevant ICES working groups. The position of these bodies in the science-policy network, as at the beginning of 1988 and as far as relevant for the marine eutrophication case, is shown in Fig. 4.4, which is an updated version of Fig. 4.2. Figure 4.4 shows that I regard ICES to be the most scientific (and least political) body, while NSTF and NUT are considered intermediates between science and politics. An important argument for this qualification is the fact that the members of ACMP, acted – at least officially – as scientist and not as national representatives (compare 3.4.1)[6].

It should be noted that there are several overlaps between the activities of the different bodies, although each developed, as will be shown in the following, its own particular focal points. The main role of ICES was to provide scientific information and advice to NSTF and Osparcom, regarding the understanding of environmental processes and the scientific and methodological basis for monitoring

[6] In 1993 the ACMP was replaced by the ACME (Advisory Commission on the Marine Environment) with national representation

and assessment, and to manage the Osparcom database. As will be shown in this section, during the period 1988–1990 NUT activities concentrated on the development of criteria for eutrophication problems areas and on measures to reduce nutrient inputs. NSTF had been established as a joint body of ICES and Osparcom (see 4.2.5). Therefore, this body has a central position as an intermediate between science and policy. The NSTF activities started with an inaugural meeting in March 1988. The remit, as adopted by INSC-2, was specified further at this meeting and formally adopted by ICES and Osparcom later that year. The ICES Council underlined that the emphasis of NSTF should be on enhancing scientific knowledge. Osparcom acknowledged that many elements of the NSTF programme were already part of its working groups. NSTF should, therefore, have a mainly coordinating role. The tasks of NSTF can be summarized as to decide on monitoring requirements, to advise on tasks to be undertaken by relevant Osparcom and ICES groups, to advice on research, to coordinate the elaboration of an assessment report for the North Sea and to decide upon the final content of this report.

The first regular meeting of NSTF (The Hague, December 1988) was largely dedicated to discussions about the tasks of the Group and its position within the science-policy network. The meeting agreed that the objectives of NSTF were principally of a long-term character and would include:

- To provide an organizational framework for discussion between policy-makers and scientists;
- To screen and coordinate scientific work carried out within ICES and Osparcom groups;
- To produce a new assessment of the North Sea in 1993;
- To provide reports on selected subjects to Osparcom, ICES and INSC-3.

This section has been structured according to six categories of tasks, for which ICES, NSTF and NUT were responsible. It concerns:

1. Understanding;
2. Monitoring;
3. Assessment;
4. Structuring and categorizing;
5. Prediction;
6. Remedying.

The sequence in which these categories are presented here is intentional: Generally, understanding of the problem is a prerequisite for monitoring and assessment. The first three categories will be of major importance for the elaboration of the last three categories. However, also reverse interactions are possible: for example, the setting of quality objectives, which is part of category 4, may provide further guidance for the development of monitoring and assessment. For a better understanding of the temporal interactions within the science-policy network, the sequence of meetings of the various groups during the period 1988–1990 is outlined in Fig. 4.5.

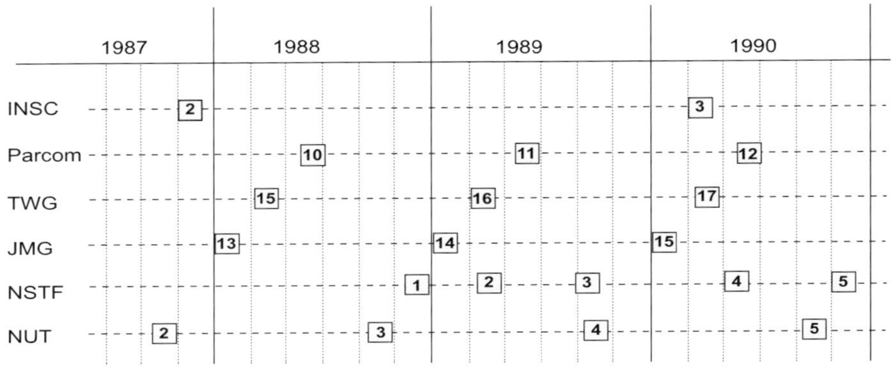

Fig. 4.5. Meetings of North Sea Conferences (INSC), the Paris Commission (Parcom), the Technical Working Group (TWG), the Joint Monitoring Group (JMG), the North Sea Task Force (NSTF) and the Nutrient Working Group (NUT)

1. Understanding

A better understanding of causes and effects of marine eutrophication is a prerequisite for the successful development and implementation of monitoring, assessment, structuring, prediction and measures. It was one of the main objectives of the specific research programmes, initiated after the oxygen depletion events in 1981 and the *Chrysochromulina* bloom of 1988. An overview of results of several scientific activities was given in 4.4.2.

Both NSTF and NUT had been commissioned with the tasks of exchanging research information and preparing advice on additional research. At the third meeting of NUT (The Hague, October 1988), information on ongoing and completed research projects in Germany, Portugal, the UK, The Netherlands, Sweden and Denmark was presented, according to a standard lay-out as decided at NUT-2 (4.2.4). However, no integrated analysis of the material had been carried out (NUT 1988). An overview of what was actually known about marine eutrophication was presented by Gerlach at NUT-3. His analysis had been carried out upon request of the Joint Monitoring Group (JMG). JMG needed this information in its assessment of the need for including nutrients in the Joint Monitoring Programme (JMP) (see further this section). According to Gerlach, blooms of species with deleterious effects were now more frequent than 20 years ago, but there was still insufficient proof of a cause and effect relationship between increased nutrient levels and phytoplankton blooms in general. The analysis furthermore showed what was not known about marine eutrophication, and this list was much longer than the summing-up of known facts (see further below). Also in NSTF, information on ongoing, completed and planned research projects was exchanged, and at NSTF-1 it was decided that the UK would establish a database on research projects (NSTF

1988). NSTF-2 (April 1989) decided, moreover, that NSTF would coordinate cruises of research vessels (NSTF 1989a).

Gaps in knowledge. An important aspect of understanding is the identification of gaps in knowledge, so as to be able to initiate research, necessary to fill these gaps. At the first meeting of NSTF, nine specific themes were identified for which it was considered necessary to fill gaps in knowledge. Three of these were directly or indirectly related to marine eutrophication and concerned:

- An improved understanding of nutrient dynamics and, in particular, their relation to occurrences of exceptional algal blooms;
- An assessment of critical loads of nutrients, metals and man-made substances;
- More knowledge of general ecosystem effects, on plankton, benthos, birds, fish and mammals, and especially on North Sea seal stocks.

NSTF-2 completed a comprehensive list of research areas to improve the understanding of nutrients and nutrient processes. A draft list had been prepared by Gerlach for the NUT-3 meeting, as a response to questions from the JMG (see above). The main research themes identified were:

1. Whether other areas than the eastern North Sea and the Skagerrak/Kattegat area were affected by increased nutrient concentrations and correlated phytoplankton blooms;
2. The role of hydrographic fronts, freshwater inputs and climatological factors in the triggering of algal blooms;
3. Components of the nitrogen pool, including distribution patterns;
4. Effects of imbalances in the ratios between nitrogen, phosphorus, silica and micro-nutrients in their role as growth limiting factors or triggers of algal blooms;
5. The effects of low concentrations of toxic pollutants on phytoplankton populations;
6. The effects of zooplankton grazing on phytoplankton dynamics;
7. The effects of algal blooms;
8. Inputs of nitrogen and phosphorous from land-based sources, including seasonal variation;
9. The dominance of individual algal species in phytoplankton blooms.

The need for research within these fields was motivated by the need for a better understanding of algal bloom ecology, which should lead to a better prediction and causal understanding of bloom events, facilitating proper targeting of remedial or preventive actions and the development of more realistic simulation models. All countries were requested to provide information on whether any of these issues was covered by national research projects. In its five-year plan, covering 1989–1993, NSTF announced that it would conduct research into a number of specific aspects of the North Sea environment, amongst which processes relevant to nutrient cycles, including biological aspects and means of characterization of key fluxes. Furthermore, targeted assessments would be carried out into specific topics, one of which the behaviour of nutrients.

It is obvious that the above research priorities had, to an important degree, been determined by the impact of the *Chrysochromulina* event. But not only NUT and NSTF had been influenced by this event. In 1988 the ICES Council decided upon the installation of a new working group, the Working Group on Phytoplankton Bloom Ecology. Already in 1984 a Working Group on Harmful Effects of Algal Blooms on Mariculture and Marine Fisheries had been installed (see 4.2.2), and the formation of the new group underlined the need for more basic knowledge concerning plankton blooms.

2. Nutrient Monitoring

As explained in 4.2.2, the ICES Advisory Committee of Marine Pollution (ACMP) considered it premature to start nutrient monitoring and was in favour of monitoring primary production. Upon request of Osparcom, ACMP was also working on an overview of trends in nutrients, but available data were generally not suited for trend analysis (see further this section). Also NUT had covered nutrient monitoring in its first two meetings, but it had not yet been possible to come to an agreement on common guidelines for nutrient monitoring (4.2.4). As a result of the decisions of INSC-2 to reduce nutrient inputs by 50% and the subsequent recommendation by Parcom to initiate nutrient reduction measures (see further 4.4.4), the question whether or not to include nutrients in the Osparcom Joint Monitoring Programme (JMP) had become more urgent. The JMP was managed by the Osparcom Joint Monitoring Group (JMG), but with the installation of NSTF there was now a third group interested in monitoring. In the following, the activities of these three groups, JMG, NSTF an NUT, relevant for nutrient monitoring, are described.

Mandatory nutrient monitoring. Shortly after INSC-2, in January 1988, mandatory nutrient monitoring was discussed by JMG. Before embarking upon the inclusion of nutrient parameters in the JMP, JMG wished to have an overview of what was known about the relationship between nutrients and eutrophication. Such an evaluation was carried out by Gerlach and discussed in the third meeting of NUT. Gerlach concluded that, although a clear causal relation between nutrients and plankton blooms could not be established, the reduction of nutrients was the only means of controlling phytoplankton numbers (see further 4.2.4). The NUT-3 meeting decided that the paper would be further improved and submitted to the 1989 JMG meeting. But JMG-14 could not come to a decision to advice to the Parcom Technical Working Group (TWG) that nutrients should be included as mandatory parameters in the JMP, as proposed by NUT-3. TWG-16 (March 1989) thereupon decided that it would be up to NUT to decide whether nutrient monitoring should be mandatory.

The Monitoring Master Plan. At the first meeting of NSTF (December 1988) the disadvantages of the JMP were discussed. This programme was restricted to estuaries and coastal zones and the information from the programme was rather heterogeneous, which would make it difficult to come to a comparative assessment. The meeting decided to develop a master plan for monitoring the North Sea, which would be "more comprehensive and disciplined" than at present (NSTF 1988). This plan would be based upon guidance from ICES. According to ICES, several questions would have to be answered in the process of developing a monitoring scheme, such as what to monitor, the reason for monitoring a particular variable, and how long the monitoring should be continued in order to meet the defined aim.

NSTF's critique on the JMP was discussed in JMG-14 (January 1989). The meeting supposed that there might be a feeling among policy-makers that the information produced by the JMP was not suitable for decision-making, and JMG decided that more should be done to bring forward the positive aspects of the programme. The meeting was against the establishment of a separate monitoring programme for the North Sea and was of the opinion that, in case additional monitoring was required, this should be done in the framework of the JMP (JMG 1989). NSTF-2 (April 1989) acknowledged that the existing JMP could be made more effective, if national implementation would be improved, but maintained its decision to propose to Osparcom to establish a Monitoring Master Plan (MMP) for the North Sea.

The main objective of the MMP was "In the longer term, to develop an adequate depth of coverage which will provide all the necessary information that is required to measure the condition of the North Sea, including investigations on trends in physical, chemical and biological parameters" (NSTF 1989a). One important element of the MMP was the mandatory monitoring of nutrients, to be carried out and evaluated by NUT. For the short term, an expanded programme of measurements would be carried out in 1990 and 1991, with the aim of obtaining data for the 1993 QSR. It should fill the gaps in knowledge regarding the spatial distribution of nutrients and contaminants. The MMP also contained recommendations on the improvement of the quality of monitoring data, including the adherence to quality assurance guidelines, sampling at one site by more than one party and a quality control by JMG and ICES.

The step-wise procedure. Also in NUT the discussion on nutrient monitoring continued. At NUT-3 a proposal by The Netherlands on a step-wise procedure for developing quality objectives was discussed (see further this section). One element of this procedure was the selection of sub-areas to be monitored, including monitoring parameters (NUT 1988). Sub-areas would be selected on the basis of eutrophication symptoms. Monitoring in the selected areas should start in 1990 and be continued for at least five years, in order to be able to establish a baseline for assessing temporal trends. NUT-3 furthermore decided that nutrient monitoring should be carried out in winter and that nutrient concentrations should be normalized for salinity. The nutrient aspects of the MMP were discussed at NUT-4 and the meeting decided that nutrient monitoring according to the step-wise procedure would be harmonized with the MMP (NUT 1989). NUT-4 recognised that there were principal differences between nutrient monitoring and the monitoring of other substances, mainly because nutrient monitoring involved a strong research element. The step-wise procedure therefore also contained research and field surveillance activities.

The main objectives of the mandatory nutrient monitoring programme established by NUT were:

1. To assess the scale, intensity and frequency of eutrophication problems in space and time (spatial trend monitoring);
2. To assess whether improvements would occur in the actual/potential problem areas (see also 5.4.3), following the introduction of reduction measures, and to assess whether the situation in non-affected areas would remain unchanged (temporal trend monitoring);
3. To further develop the understanding of causal relationships between inputs and effects;
4. To provide high quality harmonized data for the validation and clarification of predictive mathematical models;
5. To assist in the development of environmental quality objectives;
6. To assist in the development and fine-tuning of reduction measures.

With regard to the monitoring strategy and the tuning with NSTF activities, it was agreed to use as much as possible the data collected in the framework of the MMP, but to complement these with additional monitoring in problem areas (see further this section). Such complementary monitoring would not only concern additional monitoring stations, but also additional parameters. The regular MMP nutrient parameters were P and N compounds, chlorophyll *a* and silicate, together with salinity, suspended solids, temperature, dissolved oxygen and secchi-depth. These would have to be monitored in the winter period. The supplementary parameters were algal composition, primary production and special observation parameters, such as water colour, foam and mass mortality. The monitoring frequency in the problem areas would be at least every two months. The assessment of the results of nutrient monitoring would be done by NUT. At the 11th meeting of Osparcom (June 1989) both the MMP and the step-wise approach to nutrient monitoring, as proposed at NUT-3, were adopted.

Algal blooms early warning. A specific issue, directly related to the 1988 *Chrysochromulina* event, were early warning systems for algal blooms. The 1988 Osparcom meeting discussed possibilities for establishing an international early warning system, and also NUT-3 paid attention to the subject. The NUT-3 meeting agreed upon a system of aerial surveillance of algal blooms, the so-called ALGPOLREP programme, to be carried out as part of pollution surveillance flights in the framework of the Bonn Agreement[7]. The programme was adopted by Parcom in 1989. In addition to the international ALGPOLREP programme, several national algal warning programmes were in operation or being developed, again underlining the great impact of the *Chrysochromulina* event.

---

7 The Bonn Agreement is an international agreement by North Sea coastal states, together with the European Community, to:
-offer mutual assistance and co-operation in combating pollution;
-execute surveillance as an aid to detecting and combating pollution and to prevent violations of anti-pollution regulations.

3. Assessment

Where understanding (see above) mainly concerns knowledge of basic mechanisms and the development of research to cover gaps in this knowledge, assessment involves a validation of the overall situation, including the degree of human impact and proposals for priorities for actions to be taken. Assessment is, therefore, an activity in which "value aspects" (compare chapter 1) are involved. It is carried out by the science-policy interface and in many cases the result of what can be termed "negotiated science." One of the major tasks of NSTF was to produce a new North Sea QSR in 1993. Important data for the preparation of the QSR were to be derived from the MMP (see above). In addition, available data would have to be used, amongst others nutrient data. Upon request of Osparcom, ICES working groups were analysing nutrient data sets with the aim of establishing temporal trends for nutrients. Also some ad-hoc requests had been made to NSTF. The Preparatory Working Group (PWG), preparing the third North Sea Conference (INSC-3), had asked NSTF to produce first analyses of the 1988 *Chrysochromulina* and seal epidemic events for INSC-3.

Nutrient trend analysis. At NUT-3 (October 1988), ICES brought forward the problems encountered in establishing a report on trends in nutrients. Osparcom had requested ICES to analyse available data on nutrients for this purpose (see also 4.2.2). The 1988 ACMP meeting had concluded that a final report on nutrient trends could not be submitted to Osparcom before 1991 or 1992 because of problems with the quality of existing nutrient data (ICES 1989). These data originated from various sources, amongst others research cruises and fish monitoring programmes. ACMP therefore proposed to improve nutrient analysis by organizing an intercalibration exercise. Another problem was the insufficient submission of nutrient data. Osparcom had requested countries to submit such data to ICES on a voluntary basis, but only few countries had done so. The final results of the nutrient analysis were discussed in the 1990 ACMP report (ICES 1990). Because of the lack of data, the analysis focussed on Norwegian waters, the North Sea and the Baltic. The main conclusions were that:

- There had been an increase in the anthropogenic supply of nutrients (from land and air) into the Baltic Sea and the North Sea;
- Parts of the coastal North Sea and the whole Baltic Sea could be clearly identified as having increased winter nutrient levels;
- Changes over time could be identified for nutrient levels in Dutch estuaries, the German Bight, the Kattegat and, especially, the Baltic Sea.

With regard to the analysis, ACMP noted important differences between the North Sea and the Baltic Sea. Trends in the Baltic Sea were more apparent because of the better international data set and peculiarities of the Baltic, such as the relatively larger nutrient supply (per unit volume), the stratification and the longer flushing time. For the North Sea on the whole, a statistically sound statement regarding trends could not be made, due to the low quality of the data. Therefore, answers to questions about nutrient trends could not be given solely on the basis of statistical analyses. ACMP therefore concluded:

"both previous and contemporary nutrient data are unsuited to the identification of trends because data are too sparse, temporally and spatially. Equally, it is clear that nutrient introductions to the North Sea from anthropogenic sources have increased and that this has led to increased nutrient levels and biological production in some areas. Effort now needs to be applied to determining the best method of monitoring future nutrient changes in the area" (ICES, loc.cit.).

Interim Quality Status Report 1990. The catastrophic events of the year 1988 were of course of high political importance, and PWG, responsible for the preparation of INSC-3, had already at the first meeting of NSTF submitted a request for a progress report by NSTF, which should contain new scientific information on items on which ministers could take decisions (NSTF 1988). Items to be covered explicitly were the *Chrysochromulina* bloom and the seal epidemic. Reports on these issues were prepared within a very short period of time because of the limited time available until INSC-3. These reports were comprehensively discussed at the third NSTF meeting (September 1989) (NSTF 1989b). With regard to algal blooms, important input had been provided by an ICES workshop on the *Chrysochromulina* bloom, held early 1988 (see 4.4.1). In the report of this workshop it was, among others, concluded that a high N/P ratio of upwelling water had caused P limitation and *Chrysochromulina* becoming toxic (Skjoldal and Dundas 1991). Interestingly, only the main conclusions of the workshop report were taken over by NSTF and not the critical analysis ACMP had given in its 1989 report (ICES 1989). ACMP considered it unfortunate that the media attention for the *Chrysochromulina* bloom had created the popular impression that all algal blooms were noxious. Moreover, as noted by ACMP, the bloom was exceptional only in that *Chrysochromulina polylepis* had not previously been recorded as toxic over large areas. The species itself was a natural part of the algal population, and the biomass of the bloom in 1988 had not been particularly high. ACMP also put the bloom into economic perspective: From Danish fish farms no losses had been reported and the losses of Swedish and Norwegian farms were about 10 million Euros. The Norwegian loss of 800 tonnes of fish corresponded to 0.6% of the 1988 production (ICES, loc.cit.).

NSTF also presented information on other algal blooms. The report contained a list of 14 blooming events, caused by ten different species. The majority of these so-called "exceptional" blooms had been recorded in the Kattegat and continental coastal waters and in areas with reduced salinity due to freshwater inflow. Although algal blooms were a natural event, it was stated that there was evidence of recent, more frequent occurrences. It was noted, however, that also the increased observation due to greater public awareness and more extensive mariculture might have contributed to the increased number of incidents reported. Reference was also made to the results of the Continuous Plankton Recorder (CPR), which showed that changes in plankton composition had occurred over the entire Northeast Atlantic Ocean. Interestingly, this statement was followed by a reference to the recent GESAMP Report on Nutrients and Eutrophication in the Marine Environment, according to which there was clear evidence of an association between increases in nutrient inputs and/or changes in nutrient balance and enhanced frequency and/or persistence of troublesome algal blooms in waters with restricted circulation and exchange, and that such areas were encountered under certain hydrographic and climatological conditions along the coast of mainland Europe and in the Kattegat and inner Skagerrak.

NSTF had also prepared a number of recommendations to INSC-3 regarding further research into algal blooms. It concerned, among others, the life cycle of toxic algae, the role of nutrient ratios in algal species composition and the impact of a 50% reduction of nutrient inputs. The NSTF assessment of algal blooms, together with an assessment of the seal epidemic, were published as part of the so-called "Interim Report on the Quality Status of the North Sea," which was the scientific contribution to INSC-3. The Interim QSR, composed by the North Sea Conference Secretariat, also contained an update of data on inputs of contaminants to the North Sea (INSC 1990).

The 1993 North Sea QSR. The preparation of a new quality status report for the North Sea can be regarded as the main task of NSTF. At the first NSTF meeting, the ICES representative presented guidelines for the preparation of regional environmental assessments. According to ACMP, the primary purpose of a regional environmental assessment was "to provide an authoritative synthesis and evaluation of scientific information available." Such an assessment was "a product of rigorous review of data to determine the nature and severity of environmental disturbances and trend resulting from anthropogenic activity. The results could be used to determine the adequacy of existing environmental controls and the viability of its resources and amenities" (NSTF 1988). During the discussion it was stated that the primary aim of an assessment was to summarize current understanding of the effects of human activities, rather than directly addressing the necessity of protection measures. An assessment was, therefore, largely a scientific undertaking, although it would serve, indirectly, to demonstrate the need for and effectiveness of measures. Interestingly, from the Dutch side a paper was presented addressing the question how much effort to invest in improving the quality of information, considering the limited lifetime of political interest in policy issues. It was proposed to analyse the decision-making process, with the aim of developing criteria for judging the appropriateness of information, necessary for taking major policy decisions. Following this proposal, the suggestion was raised to convene policy-oriented meetings at regular times. It was, however, decided not to fix hard rules for doing so.

Upon the initiative of ICES, it was decided to follow a subregional approach with respect to the 1993 North Sea QSR. The main argument put forward by ICES was that the former two QSRs had covered the whole North Sea, concentrating on areas with high levels of pollution and, thus, presenting little information on less problematic areas. The North Sea was divided into 11 subregions, each with a lead country, responsible for the assessment. The subregional assessments, together with overall studies by ICES and Osparcom working groups and data from the MMP, would produce the main material for the so-called holistic assessment.

4. Defining and categorizing marine eutrophication

Quality Objectives and Quality Standards. An important task of NUT was to develop quality objectives for marine eutrophication (see 4.2.4). Because INSC-2 had decided on a 50% nutrient input reduction into areas where nutrients may cause pollution (4.2.5), an additional task had emerged, namely the designation of such areas. NUT-2 (1987) had already dealt with quality objectives and discussed the development of an international quality objective for eutrophication. But at NUT-3 (October 1988) it was generally felt that it would be premature to do so. Delegations pointed to the complexity of nutrient dynamics and the lack of proof of harmful effects, and questioned the usefulness of quality objectives in management. There was, however, the general feeling that the setting of quality objectives might be useful in the longer term (NUT 1988). It was within this perspective that the Dutch delegation presented its so-called "step-wise procedure for elaboration of quality objectives and standards." This procedure, which was adopted by NUT-3, consisted of the following six steps:

1. An inventory of negative eutrophication symptoms;
2. The selection by each country of one or two representative sub-areas (also called problem areas), in which relevant eutrophication symptoms should be monitored in the framework of the national monitoring programmes. It was underlined that the frequency of monitoring should be appropriate for recording the often rapid changes in eutrophication phenomena;
3. The selection of suitable parameters and methodologies;
4. The execution of monitoring in the selected areas, for the selected parameters;
5. Assessment of monitoring and research results;
6. The possible establishment of quality objectives and standards.

The consideration of the possible need for quality objectives was envisaged for 1995, in the framework of the discussion about the eventual need for further reduction measures. The step-wise procedure was endorsed by the 11th Parcom meeting (June 1989) and each contracting party was asked to follow the procedure as closely as possible. It is noted here that it concerns a request and not a mandatory activity and that, therefore, the step-wise procedure was implemented in different ways and to a different degree by the OSPAR countries.

Eutrophication problem areas. A second element of categorising marine eutrophication is the designation of eutrophication problem areas. INSC-2 had decided that nutrient inputs into areas "where these inputs are likely, directly or indirectly, to cause pollution" must be reduced by 50% between 1985 and 1995. The 1988 Parcom meeting had thereupon decided that NUT should prepare an overview of such regions. At NUT-3 this request was discussed and it was decided that The Netherlands would take the lead in preparing a series of maps, defining problem areas, including the criteria that had been applied. This task received a higher level of urgency by the decision of the 1989 Parcom meeting that a map with potential eutrophication problem areas would have to be finalized by NUT-4 and submitted to INSC-3. At NUT-4, therefore, a final draft version of a map was discussed, which already contained contributions by several countries. According to the introduction to the map, the drafters had taken into consideration major nutrient sources, giving rise to elevated winter concentrations and adverse eutrophication effects (or increased risks of such effects), resulting from excessive nutrient supply. Also secondary factors, such as climatological and hydrological conditions had been taken into consideration (NUT 1989). The map itself (figure 4.6) did not contain a clear-cut designation of potential problem areas, but four categories of parameters, which were regarded indicative of potential problem areas. It concerned:

- Elevated winter nutrient concentrations;
- Occurrence of exceptional algal blooms. For this category the observed occurrence of five species of toxic and nuisance algae was presented (see legend to figure 4.6);
- Oxygen deficiency, often related to excessive algal biomass;
- Reduced fauna, or even mortality of species, often related to toxic algal blooms and/or oxygen deficiency.

From the material presented it was concluded:

"On the basis of extensive and increased anthropogenic inputs of nutrients over the last few decades, and on the basis of the various adverse eutrophication effects, many coastal zones of the North Sea, including the Kattegat and the Skagerrak, have increased nutrient levels, and are therefore identified as potential eutrophication problem areas, whilst some others can be identified as problem areas."

The United Kingdom and France could not support this conclusion and had serious problems with the map, which they could not accept as a "statement of scientific fact." The main reservations of these countries were:

- Disagreement with the definition of eutrophication on which the map was based. It was stated that eutrophication was not just a consequence of nutrient enrichment from anthropogenic sources. France and the UK favoured the definition proposed by the Commission of the European Communities (CEC), according to which "eutrophication means the enrichment of water by nutrients, causing an accelerated growth of algae and higher forms of plant life to produce an undesirable disturbance to the balance of organisms and to the quality of the water concerned."
- Lacking evidence for elevated winter concentrations above background in coastal waters of the English Channel and the western North Sea.
- Confusion in the map about species of algae, which may be toxic and do not need increased levels of nutrients to grow. It concerned three of the five species shown in the map (*Gonyaulax*, *Alexandrium* and *Dinophysis*), which could cause Paralytic Shellfish Poisoning (PSP) or Diarrhetic Shellfish Poisoning (DSP) at low numbers and which did not need to bloom to show effects.
- The *Phaeocystis* blooms shown in the map had been taken from the publication by Lancelot et al. from 1987 (see 4.2.1), and in this publication the presence as species was listed and not the occurrence as exceptional bloom with secondary effects.

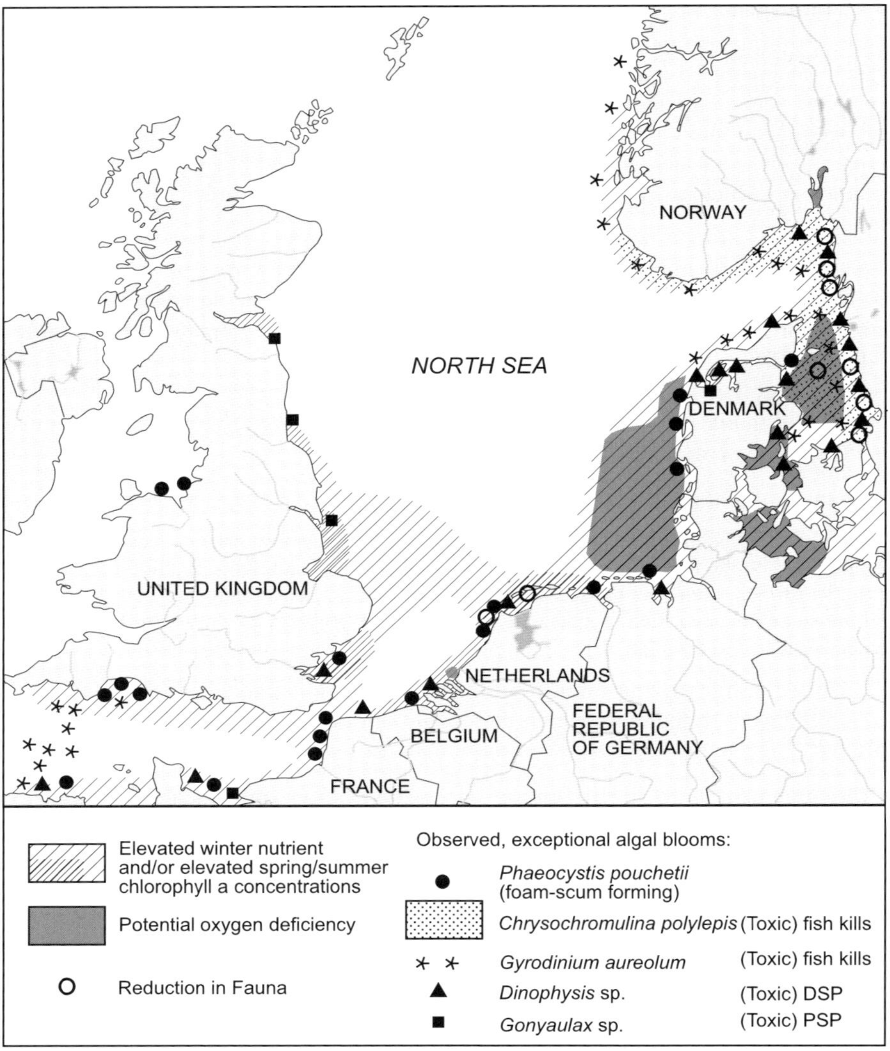

Fig. 4.6. Map of potential Eutrophication Problem Areas. Final Version 1989. Redrawn from NUT-4, Annex 6 (NUT 1989)

The majority of the meeting supported the map and it was agreed that it would be submitted to INSC-3, together with an attachment in which the reservations of France and the UK were listed.

5. Prediction

As indicated in several parts of the previous and current chapters, predicting the effects of human activities on the ecosystem was one of the major challenges to ecology, and mathematical models were regarded the most promising instrument for prediction (compare 2.3.1, 4.2.2 and 4.2.4). At its second meeting, NUT had identified the selection of suitable mathematical models as one of its future tasks. With the help of models it should be possible to quantify the distribution of nutrients and, in particular, the input of nutrients to problem areas (see above). With the aid of models it should also be possible to define any further input reductions, necessary to reach quality objectives for problem areas. For NUT this would imply working towards the adoption of models, which should include transport processes, interaction of nutrients with the sediment and ecological processes, such as algal growth (NUT 1987). At NUT-3 (1988) several national models were presented and it was agreed that it was necessary to evaluate the limitations and expectations of existing models. Belgium would take the lead in preparing such an evaluation. The evaluation should, in particular, address the ability of models to help answering the questions addressed in the so-called Gerlach study (see above). Furthermore, it should be defined in which context mathematical models could be used as management tools for decision-making (NUT 1988). At NUT-4 (September 1989) Belgium presented the first results of the evaluation. Also other model comparisons were presented by some delegations. An important point of discussion was whether models were already sufficiently sophisticated to allow a prediction of the results of a 50% reduction of nutrient inputs. Several delegations felt that this was not the case, mainly because of the poor understanding of nutrient dynamics. The meeting agreed to report to the Preparatory Working Group of INSC-3 that existing models were not yet sufficient to assess the contribution of different North Sea states to the nutrient loading of the North Sea, nor the results of reduction measures.

Also NSTF worked on the development of models for assessment and management purposes, which was a specific task given to NSTF by INSC-2. The use of models was addressed at the first meeting of NSTF in December 1988. Like NUT, NSTF concluded that an overview of available models should be made. NSTF considered a review of models necessary, preceding the preparation of the 1993 QSR. In the second NSTF meeting (April 1989) the future line of work regarding modelling was elaborated. The meeting differentiated between two aspects of modelling; first, the development of models, which was primarily a scientific and technical matter, and, second, the output of the models, which was something policy-makers had an interest in. The overall objective of the review of models was to prepare a specific chapter for the QSR on the role of modelling for assessment purposes. For the short term, NSTF considered it necessary to identify to what extent models could help increasing the understanding of physical, chemical and biological processes, and it was agreed that Belgium would coordinate such an inventory. It was also agreed to organize a workshop in which a comparison of existing models would be carried out.

6. Measures

One of the central tasks of NUT was to exchange information on measures to reduce nutrient inputs. A detailed account of national measures, reported in the first two NUT meetings, was presented in 4.2.4. INSC-2 and Parcom had decided upon catalogues of measures to be taken for the reduction of both phosphorus and nitrogen inputs. Parcom had, moreover, decided that NUT would have to prepare assessments of the national action plans. In the third and fourth meetings of NUT, first overviews were presented of the national implementation of the decisions of INSC-2 and the recommendations of Parcom (compare also De Jong 1990). The national action plans were most concrete with regard to the (further) installation of secondary and tertiary treatment stages of sewage treatment plants, and several countries, amongst which Germany, The Netherlands, Denmark and Sweden, presented figures of large investments which had been done for this purpose or were envisaged for the near future. Also additional regulations for waste water treatment had been or were being imposed upon industry. The reduction of inputs from agriculture, in particular nitrogen, was regarded as problematic by most participants. Although a substantial reduction might be expected from existing and planned measures, the 50% reduction goals would only be achieved with very stringent measures. The United Kingdom explicitly reserved its position with regard to reduction measures, with a view to the costs involved and the uncertainty that the measures would indeed be effective in preventing eutrophication problems. It was stated that there were no eutrophication problems in UK coastal waters. Generally, however, action was taken to reduce nutrient inputs to the aquatic environment.

Analysis of science-policy interactions

As a result of the establishment of NUT, the outcome of INSC-2 and the installation of NSTF, new political impetus had been given to marine pollution policies in general and policies for the reduction of nutrient inputs in particular. This development was even enhanced by the 1988 *Chrysochromulina* bloom. The new groups started to operate within the existing OSPAR system, which can be regarded as rather rigid. However, backed by politics, they were able to initiate considerable changes to the existing structures and to introduce new working methods. Both NUT and NSTF started with inventories and analyses of available data, monitoring programmes, research programmes and mathematical models. Generally, it was concluded that there was a great variety of material, but that only little was suitable for use in an overall assessment of the North Sea ecosystem or for the evaluation of policies. A second step was, therefore, to start international coordination of research, monitoring and modelling, aiming at filling gaps in knowledge and stimulating the development of useable models and research. One tangible result was the development of a Monitoring Master Plan (MMP) for the North Sea, which included mandatory nutrient monitoring. For most actions, however, the time span until INSC-3 was much too short to already deliver concrete results. But the problems with acquiring suitable results from research, monitoring and modelling were not only due to time constraints. The introduction of the MMP, together with activities such as improving data handling procedures, developing quality assurance procedures and the development of new guidelines, caused an increasing need for co-ordination and integration, as well as increasing pressure on national resources.

But not only problems of a logistic character had been introduced. Already before its start, the MMP, intended to provide important basic material for the 1993 QSR, was criticized for not being compatible with the JMP. The mandatory monitoring of nutrients, an important achievement of the MMP, was seriously criticized by ACMP, since it would not deliver the information it was supposed to do. ACMP stated in its 1990 report: "The temporal and spatial variability in the North Sea would confuse the interpretation of NSTF-MMP nutrient data to such an extent, that any change in the nutrient levels would not be demonstrated unequivocally" (ICES 1990). Instead, ACMP favoured measurements at carefully chosen representative stations, with a frequency of once per day or every second day, supplemented with synoptic measurements once or twice a year. Such problems, inherently related to the complexity of the marine ecosystem, also hampered the development of suitable models, environmental quality objectives and the definition of eutrophication problem areas.

I furthermore conclude that the science-policy interface, i.e. NUT and NSTF, did not critically discuss the need and relevance of the 50% reduction discussion on the basis of scientific reviews becoming available (see 4.4.2). One reason could be that the groups were too busy organising their implementation and preparation tasks, and that there was little time left for fundamental scientific discussions. This could have been done in the framework of the preparation of the interim QSR, but this document focused, as a result of political pressure, very much on the *Chrysochromulina* bloom.

Finally, since INSC-2 there had been very little time for the building up of critical scientific information. The critical information that was available was fully insufficient to counter the outcry following the *Chrysochromulina* event.

4.4.4 Political developments 1988–1990

The impact of the London Conference

The activities at the science-policy level, described in 4.4.3, were determined by three main factors, namely the outcome of INSC-2, the preparation of INSC-3 and the *Chrysochromulina* and seal epidemic events of 1988. But also a number of other political developments in 1988–1990 influenced the activities within the science-policy field. In June 1988 Parcom adopted Recommendation 88/2 "On the reduction of nutrients to the Paris Convention Area." The Recommendation called upon contracting parties to substantially reduce, between 1985 and 1995, inputs of nitrogen and phosphorus into areas where these inputs might cause pollution. It is clear that this Recommendation was strongly influenced by the decisions of INSC-2, which also follows from its formulation, which is almost identical to that of

INSC-2. As a rationale for the Recommendation the 1988 *Chrysochromulina* bloom was given, as well as several international political actions related to marine eutrophication, among which INSC-2, the European Community and the Helsinki Convention (see below). The Recommendation also asked for the preparation by NUT of an overview of regions "where inputs of nutrients are likely, directly or indirectly, to cause pollution." With this recommendation the marine eutrophication problem had been extended from the North Sea to the Northeast Atlantic Ocean. One year later, Parcom issued Recommendation 89/4 "On a coordinated programme for the reduction of nutrients." This Recommendation provided guidance on the implementation of Recommendation 88/2 by giving a comprehensive list of actions to be taken in order to reduce nutrient inputs. It concerned activities in the fields of agriculture, wastewater treatment, industry, aquaculture, nitrogen immission from combustion of fossil fuels and detergents.

Also the European Community increasingly addressed (marine) eutrophication issues. In June 1988 the Council of Ministers adopted a "Resolution on the protection of the North Sea and other waters in the Community." According to the Resolution, the Council "notes with concern the extensive growth of algae in certain areas of the North Sea and the Baltic, including the Skagerrak and the Kattegat, in May and June 1988, which is a symptom of a serious ecological imbalance." Other reasons for the Resolution were the 1988 seal epidemic and the "excessive fertilization and eutrophication of parts of the North Sea as well as the Baltic and other waters." In the Resolution, the Commission was invited to develop proposals concerning the reduction of nutrient inputs from diffuse sources, particularly from agriculture. Also proposals should be developed regarding the treatment of municipal sewage and industrial waste water. The Council considered such measures to be a contribution to the implementation of the decisions of INSC-2. Proposals for Directives on nitrates from diffuse sources and the treatment of urban and industrial waste water were already under development by the Commission.

Also in 1988, a ministerial meeting of the Helsinki Commission (Helcom) decided to reduce inputs of nutrients (but also hazardous substances) to the Baltic Sea by 50%, not later than 1995. This decision was without doubt inspired by the outcome of the London Conference, not the least because four members of the Helsinki Convention were also parties to the North Sea Conferences.

To underline the high political interest in marine pollution issues at the end of the 1980s, two additional political activities are mentioned. In 1988 the Karslkrona Conference on the Health of the Seas was held and in 1989 the Nordic Council organized an International Conference on the Pollution of the Seas.

The third North Sea Conference

The third International Conference on the Protection of the North Sea (INSC-3) was hosted by The Netherlands from 7–8 March 1990. This was less than two and a half years after INSC-2, and, as demonstrated in the foregoing, a time span much too short for substantial progress in the implementation of the INSC-2 agreements. As shown in 4.4.3, most of the actions agreed upon by INSC-2 were still in a preparatory state. Already in 1987 the timing of INSC-3 had been fixed by the Dutch

Minister Smit-Kroes, who was very interested in matters of North Sea pollution and who had a personal interest in chairing INSC-3. Ironically, the Dutch cabinet fell before the fixed date and INSC-3 was now chaired by a new Minister responsible for North Sea affairs, May-Weggen. It was mainly because of the early timing of INSC-3 that only few new political initiatives could be taken.

The main issues addressed at INSC-3 were inputs of hazardous substances, inputs of nutrients, dumping and incineration at sea, pollution from ships, pollution from offshore installations, protection of species and habitats, fisheries, and enhancement of scientific knowledge. In the following, the issues inputs of nutrients and enhancement of scientific knowledge will be addressed.

Inputs of nutrients. PWG had intended to come to a decision about the designation of specific areas, where nutrient inputs were likely to cause pollution. It had also attempted to present an overview to the Conference of the contribution of the different North Sea states to the nutrient loading of the North Sea. But, as explained in 4.4.3, both intentions could not be fulfilled because neither NUT nor NSTF had been able to carry out these requests, the main reasons being the lack of time and the lack of suitable data and mathematical models. These activities were, therefore, postponed by the Conference, at which it was agreed (§10) "To identify some coastal zones of the North Sea, including the Skagerrak, as being actual eutrophication problem areas and, in view of the increased inputs and levels of nutrients, some other coastal zones as being potential problem areas." It was furthermore agreed (§13) to establish common assessment and reporting procedures for the calculation of the reduction of nutrient inputs, and the determination of the sensitive areas from §10. Proposals for such procedures would be submitted to the fourth North Sea Conference, which was to be held in 1995.

The Conference furthermore decided upon a number of very specific measures for the reduction of nutrient inputs. However, as can be inferred from the full text below, these measures only [illegible] 'ied to the sensitive areas, for those cases for which it could not be proven [illegible] puts would not harm the marine environment. In §11 the North Sea Minister[illegible] ed

> "that for the North Sea catchmen[illegible], as a minimum level of treatment, urban areas (e.g. 5000 p.e. or more) and industries [illegible] a comparable waste water load, should be connected to sewage treatment plants wit[illegible] ondary (biological) or equally effective treatments, unless, on a case by case basis, c[illegible] hensive scientific studies demonstrate to the satisfaction of the competent internati[illegible] horities, that this discharge will not adversely affect the North Sea environment o[illegible] or regional level. In these cases primary treatment should at least be provided. F[illegible] nation should be provided in time for an assessment at the meeting of the Oslo and [illegible] mmissions at ministerial level in 1992."

In §12 of the Declaratio[illegible] specific requirements were listed for inputs to areas "where these inputs a[illegible] directly or indirectly, to cause pollution." For municipal treatment plants [illegible] pacity of more than 20,000 p.e., effluent concentrations of nitrogen wer[illegible] ess than 10–15 mg/l and of phosphorus at less than 1–2 mg/l. These valu[illegible] in line with the requirements of the proposed

EC Urban Wastewater Directive and thus, at least for the EC countries[8], no new political development. §12 also contained measures to be applied by industry and agriculture, again under the condition that it concerned inputs to areas where the inputs might cause pollution. For industry, Best Available Technology was required for treating industrial effluents. For agriculture, several practices were listed which should aim at achieving an "environmentally acceptable relationship between crop uptake and the amount of nutrients applied in manure and fertilizer."

Enhancement of scientific knowledge. As a basis for further measures, NSTF was invited to continue to implement its programme and, in particular, to assess research carried out on exceptional algal blooms and the seal epidemic. NSTF was furthermore asked to address in the 1993 QSR the overall ecological situation of the North Sea, including a number of so-called sensitive issues. Amongst these was the impact of fishing activities on the North Sea ecosystem. This was one of the new political issues of INSC-3. The protection of species and habitats was a second issue that had not been addressed by previous North Sea Conferences, which had, so far, been mainly dealing with pollution issues. NSTF was requested to co-ordinate relevant actions and measures with regard to the protection of species and habitats.

Two additional new tasks for NSTF are mentioned here. The first was to elaborate techniques for the development of ecological quality objectives, the second to consider possibilities for developing analytical tools to assess and compare the effects of policy decisions. These interrelated tasks underline the political desire for rational decision-making.

Future conferences. Denmark invited to the fourth North Sea Conference in 1995. It was also agreed to arrange a working group meeting at ministerial level to be held in 1993. At this meeting, the 1993 QSR would be discussed, as well as shipping issues and the problems caused by the inputs of pesticides and nutrient from agriculture. For the latter issue also the Ministers of Agriculture would be invited.

4.5 Summary and conclusions

In this chapter, the following questions have been investigated:

1. Which factors, among which science, have been relevant for the construction of the marine eutrophication problem?
2. To what extent were political decisions on marine eutrophication based upon science?
3. What did politics expect from science in the implementation of political decisions and has science been able to meet the expectations?

[8] United Kingdom, Denmark, The Netherlands, Belgium, France

4. How has the science-policy network, in particular the science-policy interface, developed as a result of political developments, and which role has it played in the use of science in political decision-making?

In the following sections the main findings with regard to these four questions are summarised.

4.5.1 The construction of the marine eutrophication problem

Whereas in the 1970s (chapter 3) marine eutrophication was predominantly an issue of scientific discussion, the period 1981 to 1990 (this chapter), showed a rapid development of political interest, first at the national but, as of 1985, also at the international level. In 1987 this resulted in the decision by INSC-2 to reduce inputs of nutrients to the North Sea by 50% between 1985 and 1995. Interestingly, it was not in the Baltic Sea that the political awareness of marine eutrophication received a strong impetus, but in the transition area between the Baltic Sea and the North Sea: the Kattegat and Belt Seas. In 1981 severe oxygen depletion and fish mortality occurred here, which caused strong public worries, especially in Denmark. In that same year also in German waters, both in the Baltic Sea and the North Sea, large areas with oxygen depletion were recorded. In the following years, oxygen depletion events were again recorded in these areas. The developments, leading from these events to the international recognition of the problem in 1985 and the formulation of concrete measures in 1987, have been placed in the framework of the social construction of environmental problems. In Sect. 4.3 it was shown that the main elements in the construction of an environmental problem, as formulated by Hannigan (1995), were clearly present in the marine eutrophication case. It concerns knowledge, timing, luck, disasters and entrepreneurs. The disasters, described above, coincided with an increasing political interest in marine pollution by the North Sea countries, caused to a large extent by the initiative of the Federal Republic of Germany to organize an international political conference on the protection of the North Sea in 1984. This Conference was followed by two additional conferences in 1987 and 1990. The central entrepreneurs were Danish scientists and Danish civil servants, who managed to use the North Sea Conferences, in particular INSC-2, as a vehicle to upgrade the marine eutrophication problem from a national to an international issue. The main rationale was the conviction of Danish scientists that international nutrient transport had been an important cause of the oxygen depletion events in Danish waters. But the decision of INSC-2 to reduce nutrient inputs must, first of all, be seen in the light of the political mood of the 1980s. This mood was, also in the framework of a precautionary approach, in favour of firm decision-making with regard to reducing inputs of polluting substances: In 1983 several European states decided to reduce SO_2 emissions by 30% (Wetstone 1987). In 1987 the Rhine Ministers Conference decided to reduce, by 50%, the inputs of several polluting substances into the Rhine river, and also in 1987 a global agreement was reached on a 50% reduction of CFCs (the Montreal protocol).

The political impact of INSC-2 was enhanced by two catastrophes, which occurred in 1988, namely the toxic bloom of *Chrysochromulina* in the Skagerrak and the epidemic of the harbour seal in several parts of the North Sea. INSC-2 and the 1988 catastrophes also had an impact on other political bodies, such as Parcom, Helcom and the European Commission, all of which introduced measures intended to reduce nutrient inputs to the marine environment.

The increased political interest also caused a strengthening of the international science-policy network for marine pollution and an intensification of activities within this network. Most important for the marine eutrophication case were the creation of the Parcom nutrient working group (NUT) in 1985 and the establishment of the North Sea Task Force (NSTF), as a joint body of Osparcom and ICES, in 1987. In addition to the push created by INSC-2, the preparation of the 3rd North Sea Conference (INSC-3), scheduled for March 1990, less than 2½ years after INSC-2, caused a pull on bodies working at the science-policy interface to produce scientific answers and instrument useable for policy makers.

4.5.2 Science and political decision-making

A central issue discussed in this study is whether objective scientific knowledge can be and should be the basis for political decision-making (compare chapter 1: rational decision-making). In the classical model of rational decision-making, as applied in this study, it is assumed that the discovery and subsequent agenda setting of an environmental problem, initiate a political need for scientific information. According to the concept of rational decision-making, political decisions should be based upon sound science. The required information should not only explain the causes of the problem (and in particular the role of human impacts), but also provide the scientific basis for developing political answers to solving it. If such information is not directly available, it should be developed by the initiation of targeted research. In the case of the 1981 oxygen depletion events, both the scientific "discovery" and "alarming," leading to political awareness of the problem, and the resulting political initiation of research, fit into this model.

Instant knowledge

The "discovery" of an environmental problem by the scientific community and the subsequent "alarming" of society, are accompanied by the delivery of what may be termed "instant" scientific information. Also in the phase directly following the political agenda setting, the information provided to politics will consist of an assessment of available data and knowledge, which is often limited, given the novelty and unexpectedness of the problem. This was also the case for the events covered in this chapter, oxygen depletion and the *Chrysochromulina* bloom, even though negative effects of marine eutrophication had already been observed in other marine waters. First results of (mainly literature) studies into the oxygen depletion events became available in the first half of the 1980s. The information was still of a rather general and coarse character, but, as shown in this chapter, had a substantial influence on the decisions taken at INSC-2. In Sect. 4.3 it was concluded that the knowledge used in the formulation of the 50% nutrient reduction decision was derived mainly from instantly available Danish research, which had not been subject to international review. Generally, there had not been an international scientific discussion about the need for, or the extent of nutrient reduction. The results of scientific studies, initiated to find answers to the question regarding the role of increased nutrient inputs in the oxygen depletion events, did reveal increased inputs of nitrogen and/or phosphorus substances, but, generally, failed to find causal relationships with changes in primary production. The main grounds for the problem of finding such causal relationships were the complexity and dynamic character of the marine ecosystem, which must be added to the already mentioned novelty and unexpectedness of the problem. Interestingly, the main differences in opinion within the scientific community were not about scientific facts, but about the interpretation of these facts, in particular the seriousness of marine eutrophication and, consequently, the need for reducing nutrient inputs. Generally, scientists from the European mainland countries supported the application of the Precautionary Principle in environmental policies, and were in favour of reduction measures. Scientist from the United Kingdom, on the other hand, did not regard it necessary to reduce nutrient inputs, at least not in UK waters, and stressed the need for more scientific proof of adverse eutrophication effects. The latter position was generally shared by the ICES Advisory Committee on Marine Pollution (ACMP), the main scientific advisory body within the science-policy network.

As discussed in 4.3, the premature introduction of knowledge for use in the policy process, i.e. before the knowledge has been accepted as credible by the international scientific community, may have negative consequences for the further process (Lambright 1995; Jasanoff 1990). One consequence may be that scientific controversies arise after new knowledge becomes available, which is not in support of the agreed policies. During the period 1988–1990 this had already happened, be it to a limited extent.

The decision to limit both phosphorus and nitrogen inputs, must, however, be valued as predominantly the result of the input of scientific information into the political decision-making process. The emphasis of policies for combating eutrophication had, until the mid of the 1980s, been on reducing phosphorus inputs to freshwater systems (compare chapter 3). The political decision to reduce nitrogen inputs as well, was taken, despite the fact that this was expected to be much harder than reducing phosphorus inputs, and mainly because there was increasing scientific evidence of nitrogen being the main limiting factor for primary production in marine systems.

But INSC-2 not only formulated clear reduction goal for nutrients. It also introduced a condition under which the reduction would be mandatory, namely only for those inputs that "were likely to cause, directly or indirectly, pollution." With this decision a need for additional knowledge had been generated, needed for the designation of areas where nutrient inputs would cause pollution, as well as for the development of criteria for "pollution."

4.5.3 The implementation of the North Sea Conference decisions

As a result of INSC-2, several political decisions and questions, waiting for scientifically-based solutions and answers, were on the agendas of working groups within the international marine pollution science-policy network, in particular NUT and NSTF. It concerned the definition of areas affected by increased nutrient inputs, the development of quality objectives for eutrophication, the development of predictive models, as well as finding the causes of the *Chrysochromulina* bloom. Following INSC-2, the working groups within the science-policy network had started with the collection and analysis of information, necessary for these implementation tasks. At the time the working groups started their activities, the level of understanding of marine eutrophication, in particular the knowledge of effects of increased nutrient loading on primary production, was limited, as can be inferred from the overview and analysis of the status of knowledge in the second half of the 1980s (section 4.4.2). There was, generally, consensus about the fact that the complexity of the marine ecosystem and the importance of other forcing factors, in particular weather and climate, made it very hard to link increased primary production, or changes in the composition of the phytoplankton, to changes in nutrient inputs. There were too few suitable long-term data series to allow proper statistical analyses, and also the increased observer effect, resulting from the increasing scientific interest in the issue, was acknowledged as a factor complicating the analyses of temporal developments. There was, furthermore, broad support for the fact that possible adverse effects of increased nutrient inputs were confined to the coastal zone of the mainland. It will not be surprising that the working groups at the interface of science and policy, in particular NUT and NSTF, were confronted with this same problem of insufficient understanding of processes relevant for managing marine eutrophication. This was the main reason why little progress was made with the tasks of developing suitable models for calculating national contributions of nutrient inputs and the designation of eutrophication problem areas. It had been the intention to provide INSC-3 with information on these tasks as a basis for supplementary decision-making.

But the complexity of the ecosystem and the lack of proper data were not the only reasons for the problems that had arisen in the process of finding suitable answers to the political requests. Additional (but interrelated) factors, identified in 4.4.5, were time constraints, problems of organization and lack of consensus. There were less than three years between INSC-2 and INSC-3, which was fully insufficient to provide answers to some of the main political questions emerging from INSC-2. What becomes obvious here, is that time is one of the principal incompatibility factors between science and politics: science, especially the study of large ecosystems, has a long-term perspective, whereas politics have a much narrower time horizon (compare Porritt 1993). Also the political controversy between the United Kingdom and France, and the other North Sea states, about the need for reducing nutrient inputs, has played an important role in blocking an agreement about a common map with eutrophication problem areas. The arguments raised against the draft map were scientific ones and must be regarded as valid. The background for using these arguments was, however, of a political nature, having a direct connection with the obligation to reduce nutrient inputs to such areas, as agreed at INSC-2.

4.5.4 Strengthening the science-policy network

The international marine pollution science-policy network, as at the beginning of the 1980s, was the result of the first wave of environmental awareness from the beginning of the 1970s (see 2.6). For the Northeast Atlantic Ocean, including the North Sea, this network consisted of ICES and Osparcom. In the course of the 1980s a second wave of environmentalism occurred in Northwestern Europe, which, for the marine environment, resulted in an extension of the marine science-policy network. This extension materialized in the form of additional ICES and Osparcom working groups and the addition of new elements (figures 4.2 and 4.4). With regard to the latter, especially the North Sea Conferences deserve attention. Also new ICES and Osparcom working groups were created. It concerned NSTF, NUT, the ICES Working Group on Harmful Effects of Algal Blooms on Mariculture and Marine Fisheries (1984) and the ICES Working Group on Phytoplankton Bloom Ecology (1988). The fact that three new working groups, specifically dealing with eutrophication and eutrophication effects, had been established, reflects the impact of the oxygen depletion events of the beginning of the 1980s and the 1988 toxic bloom of the alga *Chrysochromulina polylepis*. With the introduction of these new working groups, the "policy-research connectivity" had been strengthened and, thus, the potentials for the transfer of scientific information from science to politics and requests from politics to science. An important supplementary factor in the enhanced connectivity is the increasing role of so-called "administrative scientists." Because of the more intense political interest in marine pollution and the resulting increase in administrative efforts, especially at the interface of science and politics, the demand for scientifically skilled civil servants increased (compare Van der Windt 1992). In the course of the 1980s this became visible in the membership of, in particular, NUT and NSTF.

But with the strengthening of the science-policy interface, there has also been an increase in the passing on of responsibilities from politics to the science-policy level. The decision of INSC-2 to reduce nutrient inputs by 50% had been possible only because of the inclusion of the condition that such would only be necessary for those areas, where the inputs would cause pollution. With this decision a heavy burden was placed on the working groups at the science-policy interface, namely to develop a common definition of pollution caused by eutrophication. INSC-3 shifted even more responsibilities to the science-policy level by agreeing upon requirements for sewage discharge, which would be mandatory only if it could be scientifically proven that untreated discharges would not "adversely affect" the marine environment. In the light of the above described problems with the collection and application of scientific information, such a task seems hardly feasible. INSC-3 also commissioned Osparcom and NSTF with the development of techniques for the elaboration of environmental quality objectives and the development of analytical tools for assessing the effects of political decisions. The elaboration of these tasks must be regarded as problematic, given the complexity of the North Sea ecosystem, the many gaps in knowledge, the still insufficient research and monitoring infrastructure and, most important, the value-laden aspects of these tasks (compare 1.2.1).

5 The management of marine eutrophication

"The ultimate objective of all this hard work is a mechanistic understanding, based on scientific principles, from which management strategies can be designed to restore coastal ecosystem functions and biological communities that have been damaged by nutrient enrichment." (Cloern 2001)

In the previous chapter a description and analysis were given of the construction of the marine eutrophication problem and the formulation of political decisions to reduce nutrient inputs to the North Sea. The central issues addressed were whether and to what extent science had played a role in these processes. In Chap. 4 already a start was made with the analysis of the implementation of the decisions of the second North Sea Conference (INSC-2), but it was shown that the focus of activities was on the preparation of the third North Sea Conference (INSC-3) of 1990. The focus of the present chapter is on the management of the marine eutrophication problem, i.e. the last phase in the policy life-cycle (figure 1.1). Because in 1988 a Parcom recommendation on the reduction of nutrient inputs to the Northeast Atlantic Ocean was adopted (4.4.4), the area covered in this chapter comprises both the North Sea and the Northeast Atlantic Ocean.

The following five questions will be addressed:

1. How has the knowledge basis with regard to marine eutrophication developed after INSC-2 and INSC-3?
2. Has ecology been used as a means for justifying decisions?
3. Has ecology contributed to the fine-tuning of decisions and the elaboration of management instruments?
4. How has the science-policy network, in particular the science-policy interface, functioned with regard to the use of science in policy making?
5. To what extent has new knowledge influenced the political status quo?

Question 1 is addressed in Sect. 5.1, which contains a description of so-called "new knowledge." As comprehensively described in the previous chapter, scientific research into the causes of several phenomena, ascribed to increased nutrient inputs (increased phytoplankton growth, toxic blooms, oxygen depletion, increased secondary production), intensified in the 1980s. From analyses by Nixon (1995) and Vidal et al. (1999) it can be concluded that a further intensification of marine eutrophication research occurred in the 1990s. Of particular relevance for this study are analyses of long time-series of several factors relevant for primary production. The most important data series available are the Continuous Plankton Recorder (CPR), the Marsdiep series and the Helgoland Reede series. Specific attention will, furthermore, be given to the results of the 1991 ICES marine science symposium on hydrographic variability and the1995 ICES "Arhus revisited" symposium, dedicated to the analysis of long-term data related to primary production and development of fish stocks.

The question whether ecology has been used as a means for justifying decisions (question 2) is the subject of Sect. 5.1.3, in which an analysis of new knowledge is given. In Chap. 4 it was concluded that the 50% nutrient reduction was based upon instantly available knowledge about the causal relationships between increased nutrient loading and changes in primary production. There was no scientific consensus about the relevance of increased nutrient loading for an increase of intensity and frequency of algal blooms, nor had the scientific knowledge, used in the decision, been subject to international scientific discussion. The analysis of new knowledge in Sect. 5.1 will focus on the question whether new scientific information justified the political decisions taken at INSC-2 and INSC-3. Another element of the analysis will be whether, as assumed in the rational policy-making model, proposed by Winsemius (1986) (see chapter 1), the uncertainty about the relation between nutrients and eutrophication phenomena decreased as a result of new scientific findings.

In Sect. 5.2 it will be investigated whether new knowledge has contributed to the fine-tuning of decisions and the elaboration of management instruments, i.e. monitoring, prediction, assessment and validation (question 3). This section comprehensively deals with the implementation of the political decisions regarding marine eutrophication. Of particular relevance is the analysis of the use of new knowledge in the implementation of these decisions. Most studies about the role of scientific information in public policies deal with decision-making (compare chapters 1 and 4), but there is little theoretical material on the use of scientific knowledge in the implementation of political decisions. According to Hannigan (1995), science is most important in the assembly phase of the construction of an environmental problem, whereas politics are predominant in the contesting phase. Also Hischemöller et al. (1998) underlined the role of the natural sciences in the signalling of a problem, but concluded that the solution to the problem was usually done by technological disciplines. It is important to distinguish between the use of scientific information for solving the problem, in this case through the reduction of nutrient inputs, and the fine-tuning of political decisions. Wettestadt and Andresen (1990) have suggested that crude knowledge is sufficient in the phase of political negotiations, but that in the following phase of implementation and compliance, additional knowledge may be necessary for the fine-tuning of policies. In Sect. 5.2 it will be investigated whether additional knowledge has indeed contributed to the fine-tuning of political decisions.

In Sect. 5.2 also the question will be addressed, which role the science-policy interface has played in the use of science in policy-making (question 4). One of the factors, necessary for the successful construction of an environmental problem, is, according to Hannigan (1995), the emergence of an institutional sponsor, who can ensure both continuity and legitimacy, after political decisions have been taken. In Chap. 4 the central role of the nutrient working group (NUT) in the preparation of the nutrient reduction decisions of INSC-2 was elucidated. As a result of the decisions of INSC-2 and INSC-3, the science-policy interface had been further strengthened and its activities intensified. It will be investigated whether and how the science-policy interface succeeded in sustaining the political interest in the issue of marine eutrophication. In addition to the institutional role, it will be investigated how the science-policy interface has dealt with the mandates with a high value-laden content, transferred to it by politics (4.5.4). It concerns the development of criteria (quality objectives) to judge the seriousness of eutrophication and the designation of eutrophication problem areas.

Question 5, the extent to which new knowledge has influenced the political status quo, will be investigated in Sect. 5.3. In this section relevant political developments during the period 1991–2003 are described an analysed. It concerns the Intermediate Ministerial North Sea meeting of 1993, the fourth and fifth North Sea Conferences (1995 and 2002 respectively), the OSPAR Ministerial Meetings (1992, 1998 and 2002), as well as relevant developments within the framework of the European Community. The central question addressed in this section is whether new scientific findings and insights, as well as the activities of the science-policy interface, have led to the modification of earlier political decisions, assuming a feedback loop in the policy life-cycle.

The main findings with regard to the above questions will be summarised and discussed in the final Sect. 5.4.

5.1 New knowledge

As stated above, new knowledge is understood to be the knowledge becoming available after INSC-2 and INSC-3. Two main categories of research results will be addressed. The first category, covered in 5.1.1, is about whether and which nutrients are responsible for observed changes in phytoplankton growth. The outcome of such studies is of major relevance for policy and management, dealing with the reduction of nutrient inputs and the identification of areas sensitive to nutrient loading. The emphasis will be on research into *Phaeocystis* blooms in the western Dutch Wadden Sea and on analyses of the data collected at the Helgoland Reede. For both, long time-series were available. The second category (5.1.2) comprises the results of scientific studies into the effects of increased primary production and changing species composition. It concerns studies into toxic and nuisance algal blooms, changes in benthos and oxygen deficits, all of which are relevant for the justification of the introduction of eutrophication reduction efforts. The necessity of justification is illustrated by the criticism on some policy measures, expressed already in 1990 by Gray. In a commentary in the Marine Pollution Bulletin, Gray (1990) questioned the wisdom of a general reduction of nutrients by 50%. He argued that the Skagerrak and the North Sea (with the exception of some coastal regions) were not eutrophied and that the investments in sewage treatment could better be used for other purposes. According to Gray, the introduction of phosphorus and nitrogen removal along the south coast of Norway, not including the Skagerrak coast, would cost some 1300 million Norwegian Crones (approximately 220 million Euros). Gray therefore stated:

"So to save the Skagerrak from eutrophication, that to me has not been proven, will cost Norwegian taxpayers enormous sums. Some money clearly must be spent to reduce local problems but most could be better spent on more severe marine pollution problems of which there are many in Norway."

5.1.1 What causes phytoplankton blooms?

Nutrients

***Phaeocystis* in the Marsdiep.** In the previous chapter (section 4.2.1) an extensive account was given of observations of increasing blooms of *Phaeocystis* in continental North Sea coastal waters. Generally, the increase was related to an increase in anthropogenic nutrient loading. In the course of the 1990s, several scientific studies were carried out, aiming at resolving the complicated life history of *Phaeocystis* and the role of nutrients in the development of *Phaeocystis* blooms. The scientific interest was fuelled by the political relevance of the role of anthropogenic nutrient enrichment because *Phaeocystis* was considered a nuisance alga (compare Lancelot et al. 1987).

The longest time-series of *Phaeocystis* in coastal marine waters is from the Dutch Marsdiep area. In 1986 Cadée published the results of his monitoring scheme, showing a steep increase in both duration and intensity of *Phaeocystis* blooms since 1974 (Cadée 1986; figure 4.3). In 1991 Cadée and Hegeman placed these data in an historical perspective by comparing them with data from almost a century before (Cadée and Hegeman 1991). On the basis of this analysis it can be concluded that the number of days of intensive blooming (>1000 cells/ml) had increased by a factor of 1.5 to 3 between 1897 and 1998, which is substantially less than the increase by a factor of 5 to 8 that had occurred between 1974–1976 and 1978–1989 (figure 5.1).

Schaub and Gieskes (1991) found a high correlation between Rhine discharges and phytoplankton biomass in the Dutch North Sea coastal zone, but could not relate the changes in growth to either nitrogen or phosphorus. They concluded that the variable composition of various nutrients, including silicate, was important for phytoplankton species composition.

In 1992 a paper was published in the journal Marine Biology (Riegman et al. 1992) in which the steep increase in *Phaeocystis* blooms was related to a decrease in the N/P ratio during the same period. The authors explained the decreased ratio by hydrological changes that had occurred in Lake IJssel, an important source of nutrient loads to the Wadden Sea. The changes, which had been caused by the construction of the Houtribdijk in 1975, had resulted in an increase in the phosphorus flow to the Wadden Sea. Riegman et al. (1992) also presented experimental evidence that a low N/P ratio favoured the growth of *Phaeocystis*.

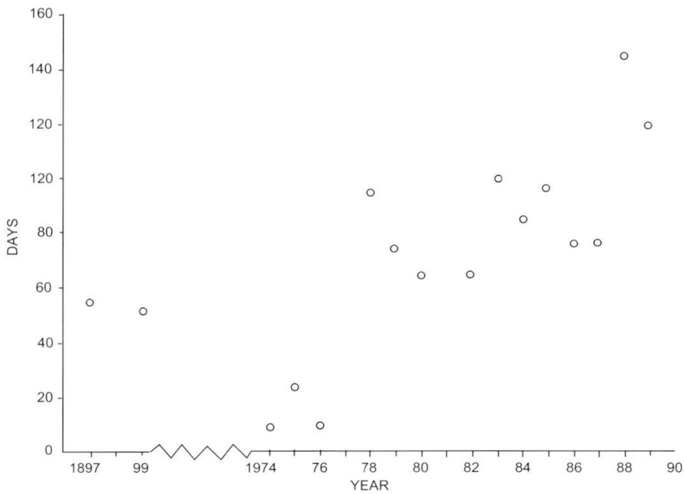

Fig. 5.1. *Phaeocystis* blooms in the Marsdiep in an historical perspective. Days with more than 1000 cells/ml. (Source: Cadée and Hegeman 1991)

Cadée and Hegeman (1993) pointed to the persisting high primary production in the Marsdiep, despite the fact that the phosphate load of the Rhine had almost continuously decreased since the beginning of the 1980s. This had resulted in a 40% decrease in phosphate concentrations in the Marsdiep, but *Phaeocystis* blooms had remained at a high level. Although Cadée had never related the high levels to either P or N, he now presented data in favour of N as the limiting factor.

De Jonge (1997) provided an alternative explanation by pointing to increased P loads from the English Channel. This hypothesis did not receive much support, and in the years to follow the emphasis in The Netherlands shifted from phosphorus to nitrogen as the factor considered most important for limiting primary production in the marine environment.

In a review article, Philippart et al. (2000) presented analyses of Marsdiep data for the period 1974 to 1994 and concluded that during this period several shifts in algal species composition had occurred. They related these to shifts in absolute and relative nutrient loads. The periods identified by Philippart et al. (loc.cit) were 1974–1976, 1978–1987 and 1988–1994.[9]

[9] Interestingly, these periods coincide with major shifts in the Rhine flow. From 1971–1978 the average annual flow at Lobith (Dutch-German border) was 55 km^3. The long-term average (1954–1995) is 72 km^3 per year. Starting in 1978, a period of above average flow began. During this period, which lasted until 1988, the average annual flow was 81.9 km^3. In 1989–1993 the Rhine flow was again below average, with an annual amount of 59.8 km^3

Again another causal relationship between *Phaeocystis* and nutrients was proposed by Van Beusekom et al. (2001). They found a very high negative correlation between NH_4 concentrations in the Rhine and *Phaeocystis* blooming. Since the mid 1970s, the ammonium concentrations in the Rhine had strongly decreased as a result of the introduction of biological sewage treatment. In this same period the duration of *Phaeocystis* blooms in the Marsdiep had increased. Already in 1985 Billen et al. had pointed to the importance of changes in ammonium concentrations and the undesired side-effects of the introduction of biological treatment (Billen et al. 1985).

What can be concluded from the above, is that in the course of time, the number of proposed causes of the increased *Phaeocystis* blooms in the Marsdiep increased. Moreover, not only man-induced changes in nutrient dynamics, but also natural factors were proposed. That this was not only the case for *Phaeocystis* in the Marsdiep, can be inferred from the following quote from a publication by Lancelot (1995):

> "Taken together, these data appear to be contradictory, suggesting a complex interaction between *Phaeocystis* blooms and natural changes and the influence of mankind. On the one hand, the recent dramatic upsurge of *Phaeocystis* colony blooms recorded in North Sea Dutch coastal waters since the early 1970s (Cadée and Hegeman 1991a, b) has been often associated with an increased eutrophication of anthropogenic origin. On the other hand, qualitative data available in the scientific literature [.....] and archival chronicles of the early part of this century give strong evidence of the presence of massive blooms of *Phaeocystis* colonies already at that time, suggesting that *Phaeocystis* blooms and the related undesirable foam forming are not a novel phenomenon recently induced by present-day eutrophication."

The Helgoland Reede series. Like the Marsdiep data, the Helgoland Reede time-series on nutrients, phytoplankton and several co-factors important for primary production, played an important role in the search for effects of increased nutrient loading. In the 1980s several analyses of the Helgoland data, collected since 1962, had been carried out in the framework of German research projects (sections 4.1.2 and 4.4.2). In the 1990s additional evaluations of time-series of three decades were carried out by Hickel and co-workers of the Helgoland marine research institute (Biologische Anstalt Helgoland, BAH), mainly commissioned by the German Environment Agency (UBA).

In the analyses by Hickel et al. (1993; 1997), the findings from earlier investigations regarding increases in phosphate and nitrate concentrations (compare 4.5.2) were largely confirmed. The observed increase of the phytoplankton stock at the Helgoland monitoring site was, however, attributed to a sudden increase in nanoflagellate biomass by the end of the 1970s. Nanoflagellate development could be correlated with the Elbe flow, but not with inorganic nutrients in the Elbe water (Hickel et al. 1997). In this respect, they also referred to publications about effects of large-scale meteorological changes on phytoplankton development. According to Hickel et al. (loc.cit.), neither mean nutrient levels in winter nor elevated nutrient loads of the Elbe had resulted in elevated phytoplankton stocks. They regarded it likely that eutrophication effects occurred in the outer German Bight, considering the oxygen deficiency in bottom water and large plankton blooms that had occurred here.

Radach (1998) presented an analysis and assessment of changes in the German Bight in 1962–1996, at the 1997 ICES phytoplankton variability symposium. Radach had quantified the ecological changes by means of an ecological development index (EDI), based upon a total of nine physical, chemical and biological parameters, measured at Helgoland. The parameter set included temperature, salinity, several nutrients and phytoplankton. On the basis of changes in the EDI, Radach (loc.cit.) distinguished four different periods within the investigated time frame, and related these different periods to changes in eutrophication of the German Bight. According to Radach (loc.cit.), "Climatic effects on variability, if present at all, are hidden by the much greater effects of river-induced eutrophication, expressed in phosphate and nitrate concentrations." Interestingly, the periods identified by Radach, showed a high similarity with the findings of Philippart et al. (2000) (see above), and with those of Reid and Lindeboom et al., presented later in this section, in particular the changes that had occurred around 1978 and 1985. Whereas Radach valued nutrient inputs as much more important than climatic changes, other researchers came to the opposite conclusion (see subsection "Climate and phytoplankton").

Danish coastal and marine waters. In 4.4.2 an account was given of the results of German and Swedish research projects, initiated after the oxygen depletion events of the beginning of the 1980s. The probably most comprehensive marine eutrophication research programme was carried out in Denmark during the period 1988–1994. The results were, among others, published in a scientific volume, edited by Jørgenson and Richardson (1996a), and a report of the Danish Environmental Protection Agency (Christensen 1996), of which an English translation was published in 1998 (Christensen 1998). The most important issue, addressed in the Christensen Report, was the effect of the Danish National Action Plan on the aquatic environment. The main goals of this Plan were a reduction of 80% P and 50% N inputs to the marine environment between 1987 and 1993 (see also 4.2.4). It was concluded that the main human contribution to the nutrient loading of Danish estuarine fjords derived from Danish arable land, and that the effects were greatest in the innermost parts of the fjords. It was also in these parts that effects of the National Action Plan could be documented, i.e. a reduction in phosphorus loading, a reduction of nitrogen point source loading and a reduction in phytoplankton biomass which, according to Christensen et al. (loc.cit.), "appears to be due to the reduced inputs of phosphorus." Also in coastal waters a reduction of phosphorus could be documented, but this had not led to a general improvement in the environment. In the open parts of Danish marine waters effects of the Action Plan had not yet been found.

Jørgenson and Richardson (1996b) presented a systematic analysis of observed long-term changes in the Kattegat and Belt seas, with specific emphasis on the oxygen situation and nutrient loading, and possible causal relationships between the two. They first of all argued that it was most likely that the observed general decrease of the oxygen content in the bottom water of the Kattegat (from 4 mg/l in

1965 to 3 mg/l in 1990) had been the result of an increase in pelagic primary production by a factor of two in the same period. They then investigated the question whether the increased primary production had been caused by increased nutrient availability, or whether other factors could be made responsible. Starting with the latter, they investigated the relevance of climatic factors for the upwelling of nutrients. For the cubed wind speed, used as a proxy for wind generated turbulence, data were presented, showing an increase in autumn and winter values as of 1960. For the summer and spring period, the increase was much less pronounced, which was the reason why Jørgenson and Richardson (loc.cit.) concluded that it was unlikely that changes in turbulence could explain the changes in primary production. They then carried out a comprehensive analysis of long-term changes in nutrient transport to the Kattegat, on the basis of which the following was concluded:

"Thus, it seems unlikely that changes in inorganic nutrient input capable of supporting the observed increase in primary production have occurred during recent decades at the borders of the Kattegat and its surrounding seas. It is unclear for both the Skagerrak and the Baltic borders whether or not changes have occurred in the transport of organic nutrient material. However, even if they have occurred, it seems unlikely that their magnitude would have been sufficient to have caused the recorded changes in primary production" (Jørgenson and Richardson, loc.cit.).

With regard to nutrient loading of the Kattegat, it was also concluded that nutrient pulses from the German Bight only played a limited role. The only changes that had, according to Jørgenson and Richardson (loc.cit.), been documented for the period under investigation, were increased nitrogen runoff from land and increased atmospheric nitrogen deposition, both of which were, to a large extent, the result of the intensification of Danish agriculture.

The relevance of the conclusions of both Christensen et al. and Jørgenson and Richardson lies in their political content. The Danish political initiatives with regard to marine eutrophication were to an important extent focussed on the international dimension, namely the contribution of nutrients from non-Danish sources to eutrophication in Danish waters. Also in the public debate the importance of pollution from outside Denmark was an issue. Both in 1993 and 1995, for example, warnings had been given by scientists for possible oxygen depletion as a result of large amounts of nutrients, discharged by continental rivers. On 24 March 1995 the Danish newspaper "Politiken" published a satellite picture of nutrient rich water in the German Bight that would "hit" Denmark in the weeks to follow, and that would cause large plankton blooms and subsequent oxygen depletion in Danish waters. However, neither in 1993 nor in 1995 such extraordinary blooms were observed.

How politically sensitive the unwelcome conclusions of the Danish researchers were, is underlined by what happened in the years 2002 and 2003. In 2002, oxygen depletion in Danish coastal waters initiated a public debate about the causes, in particular whether nitrogen inputs from outside Danish waters could be blamed or whether it was mainly an internal Danish problem. The scientific credibility of researchers from the National Environment Research Institute (NERI), who had concluded that inputs from Danish agriculture were the main cause of the problems (Aertebjerg et al. 2003), was questioned and a decision was taken in the Danish parliament that the NERI research would have to be evaluated by an independent scientific review panel. In the spring of 2003 the evaluation panel, consisting of John Gray from Norway and Patricia Glibert, Robert Diaz and Nancy Rabalais from the USA, reviewed the NERI research and held interviews with NERI critics (NERI 2003). A major issue of criticism against NERI was that it had not paid sufficient attention to the relevance of the Jutland current, transporting nutrients from the North Sea into the Kattegat (compare 3.2.6 and 4.1.1). The panel concluded in its final report, after having heard the opinion of several experts, that this current was an episodic event, which, in the worst case, might lead to an additional transport of 17,000 tonnes of nitrogen into the deep waters of the Kattegat.[10] Generally, the panel was positive about the scientific quality of the NERI work and supported the view of NERI that inputs from land were the largest contributor to eutrophication and oxygen depletion in Danish coastal waters.

The North Sea Project. Not only in the continental North Sea states targeted research into marine eutrophication had been initiated. In 1987 the UK had launched a major research programme, which lasted until 1992. The main aim of the so-called North Sea Project (NSP) was the development of environmental water quality models with a prognostic capacity for determining the fate of pollutants (Simpson 1994). The field data for validating the models were collected through cruises, carried out in the southern North Sea from 1988 to 1989. During this period, the same track was repeated 15 times on a monthly basis and, thus, data for a whole seasonal cycle collected. On the basis of the results of the project, some interesting conclusions with regard to eutrophication were drawn, putting the relevance of anthropogenic nutrient inputs for increased primary production into perspective.

Howarth et al. (1994) concluded that regeneration of organic material was the major cause of changes in dissolved nitrate in winter, rather than the supply of new nutrients from rivers, the atmosphere and the ocean. According to Howarth et al. (loc.cit.), this implied that "productivity of the North Sea may be primarily influenced by the flushing characteristics, which retains nutrients within the region, rather than by supply from the rivers, as has been previously postulated."

Based upon synoptic chlorophyll data, Tett et al. (1994) showed that phytoplankton concentrations were greatest near the continental coast. The mean summer levels were, however, highest at sites of intermediate mixing and not at sites of greatest nutrient availability. Moreover, the large spring bloom in these waters occurred only under favourable illumination conditions in May and June. In a more recent publication, Hydes et al. (1999) confirmed these findings. Based upon numerical modelling, using data from the North Sea Project, supplemented with newer data, it was concluded that the high productivity of the North Sea was maintained by both the total amount of nitrate supplied to the system and recycling in shallow waters. In the coastal waters of Germany and The Netherlands, the degree of recycling was five, whereas it was two off the UK coast. Hydes et al.

[10] According to Rasmussen and Andersen (2003) an annual amount of 293,000 tonnes of nitrogen is transported into the Kattegat and Belt seas, of which 165,000 tonnes are transported out of the area into the Skagerrak.

(loc.cit.) also underlined the relevance of light conditions for production, which was also one of the reasons for the lower productivity in UK coastal waters.

The research findings presented in this section have a clear political relevance. First, the hypothesis that increased nutrient inputs would cause more production was put into perspective: the relatively high productivity in continental coastal waters was, according to Hydes et al. (loc.cit.), apparently caused by more favourable physical features (recycling, light), rather than higher nutrient inputs. Hydes et al. (loc.cit.) even postulated: "Although total production has been enhanced by increased inputs of nutrients, the evidence from the winter distribution of phosphates is that there is no statistically significant evidence for a net accumulation of phosphate and organic carbon in the system." Second, evidence was presented that production in UK coastal waters was relatively low, supporting the claim by the UK that there were no eutrophication problems in these waters (compare 4.2.5). The modelling results, presented by Hydes et al. (loc.cit.), furthermore showed that the N/P ratio in many areas of the southern North Sea was lower than in any of the sources of the waters. It was postulated that the apparent N deficit was caused by denitrification. Also this research outcome is politically relevant because evidence was provided that (part of the) nitrogen is removed from the system in a natural way.

Light limitation

It goes for itself that light is just as important for phytoplankton growth as are nutrients. Still, as put forward by Colijn and Cadée (2003), research into phytoplankton growth had developed a narrow focus on nutrients. According to Colijn and Cadée (loc.cit.) eutrophication has been one of the main reasons for intensive studies of nutrient-phytoplankton relationships in the Wadden Sea, but that there has generally been little emphasis on irradiance as a limiting factor. They explained this development by the political interest in anthropogenic nutrients, which can, contrary to the light regime, be influenced by policies. The high relevance of irradiance, especially in shallow coastal waters such as the Wadden Sea, was clearly demonstrated by Colijn and Cadée (loc.cit.), who concluded on the basis of their analysis that in many cases both spatial and temporal light limitation far exceeded nutrient limitation.

Zooplankton grazing and phytoplankton blooms

In the foregoing sections, the emphasis was on the possible relationship between increased nutrient supply and the increased occurrence and intensity of phytoplankton blooms. But, as shown in several studies, zooplankton grazing can play a principal role in the control of phytoplankton blooms. In a review paper on eutrophication, Brockman et al. (1988) (see also 4.4.2) concluded that zooplankton grazing matched phytoplankton production only during the summer months, and that spring and autumn primary production was not kept under control by herbivores. In a comprehensive review of eutrophication in the Dutch coastal zone, Klein and Van Buuren (1992) discussed in more detail the possible role of zooplankton grazing in the control of phytoplankton blooms. They concluded that microzooplankton "can exert an enormous impact on the phytoplankton standing stock." They also stressed the limited data available on microzooplankton and pleaded for more research.

In 1994 Scholten et al. (1994) published the results of experimental studies in freshwater systems, showing the impact of toxic substances on the grazing ability of zooplankton. On the basis of their findings they concluded: "Disfunctioning of water fleas and other zooplankton seems to be a more important factor than fertilisation in the transformation of aquatic ecosystems into eutrophicational states." They also stated that "the concept of the environmental hazard of phosphate for aquatic ecosystems should, therefore, be reconsidered." The publication caused considerable irritation amongst responsible authorities, who regarded it an attack on established practices (i.e. the use of phosphate free detergent and phosphate removal in sewage treatment) for combating freshwater eutrophication problems. In press articles also the fact was mentioned that the studies had been financed by the phosphate industry. Questions were asked in the Dutch Parliament whether the practice of phosphate removal could still be considered effective. In several replies to the conclusions of Scholten et al. the argument was used that, so far, a relationship between phytoplankton blooms and zooplankton density had not been established. However, what Scholten et al. were pointing at was not zooplankton biomass but their ability or capacity to graze. This ability could be seriously reduced in the presence of low concentrations of toxic substances. Although originally applied to freshwater systems, the studies raised the interest of authorities responsible for marine waters. Therefore, the relevance of toxic substances for zooplankton grazing in coastal waters was investigated in the framework of the Dutch BEON programme for policy relevant ecological studies in the North Sea. On the basis of field studies, mesocosm experiments and model studies, it was concluded that the grazing capacity of zooplankton, especially copepods, was sensitive to low concentrations of toxic substances, most notably the pesticides lindane and mevinfos and that, generally, the effects of pesticides on marine zooplankton and, consequently, on the development of phytoplankton blooms, had been underestimated (Jak and Scholten 1994; Jak and Michelsen,1996).

Climate and phytoplankton

Already in the previous decades British scientists had stressed the possible impact of climatological changes on changes in phytoplankton stocks, on the basis of data from the Continuous Plankton Recorder (CPR) (compare 2.5.2 and 3.2.5). In the course of the 1990s more information from the CPR became available, supporting the relevance of climatic changes for plankton development. The CPR team presented time-series of North Atlantic sea-surface temperature and zooplankton and phytoplankton abundance at the 1991 ICES hydrographic variability symposium, showing highly coherent patterns (CPR Survey Team 1992). Interestingly, the same patterns were found for other trophic levels, i.e. herring numbers and bird parameters, all showing a sudden change around 1980 (Aebischer et al. 1990). At the same symposium Colijn (1992) presented a comprehensive review of the causes of changes in plankton communities. In his paper he also discussed the CPR data and pointed to some methodological weaknesses of the CPR, such as the inability to determine small phytoplankton and the fact that the CPR does not monitor nearshore areas. With regard to plankton blooms in coastal waters, Colijn stated that these were under the predominant influence of eutrophication (Colijn, loc.cit.).

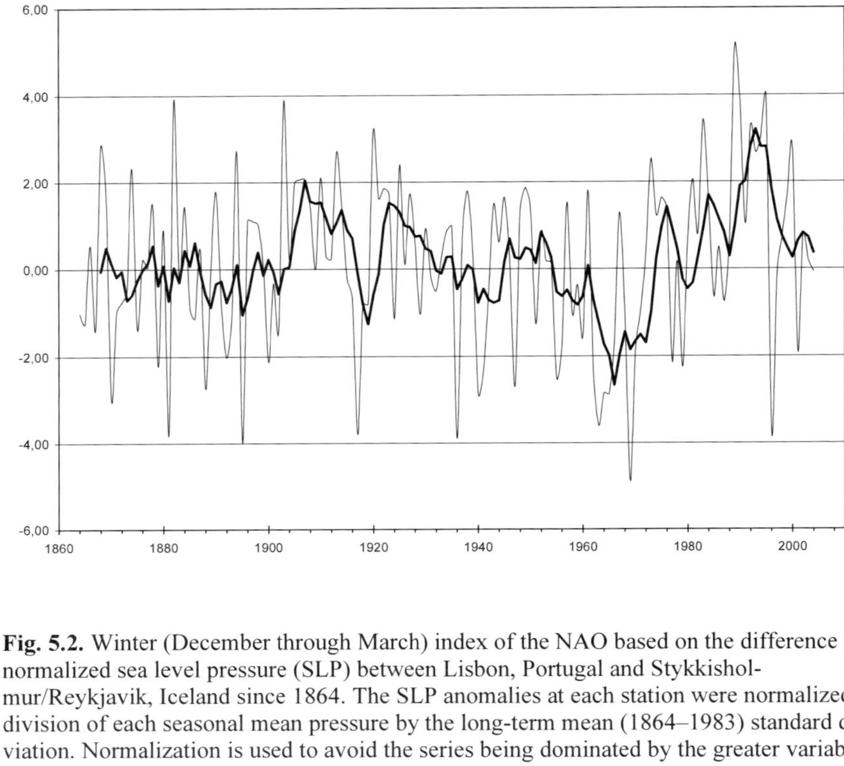

Fig. 5.2. Winter (December through March) index of the NAO based on the difference of normalized sea level pressure (SLP) between Lisbon, Portugal and Stykkishol-mur/Reykjavik, Iceland since 1864. The SLP anomalies at each station were normalized by division of each seasonal mean pressure by the long-term mean (1864–1983) standard deviation. Normalization is used to avoid the series being dominated by the greater variability of the northern station. Positive values of the index indicate stronger-than-average westerlies over the middle latitudes. Bold line: five-year running mean. Source: www.cgd.ucar.edu

At the 1993 scientific symposium on the Quality Status Report, Van Beusekom and Diel-Christiansen (1996) concluded that the North Sea could be divided into two regions, of which the northern one was influenced by climate, whereas the southern region was mainly eutrophication influenced. However, Lindeboom et al. (1996), at the same symposium, put the data from the coastal waters of the southern North Sea into a quite different perspective. They showed that the changes in *Phaeocystis* blooms in the Marsdiep area could also be explained as a sudden shift from a low to a high level, which had occurred around 1978, instead of a steady increase since 1973, as presented by Cadée in several publications (figure 4.3; figure 5.1). Lindeboom et al. (loc.cit.) also found this shift around 1978 for other parameters, among others macrozoobenthos biomass, eiderduck fledglings and the increase in nanoflagellates at Helgoland (see above). As possible triggers for this "system shift" the authors suggested climatological factors and a salinity anomaly which had entered the North Sea in 1978.

In the second half of the 1990s and the beginning of the 2000s the impact of climatic changes on phytoplankton, in particular related to changes in the North Atlantic Oscillation (NAO, figure 5.2), was demonstrated in several papers, among which Reid et al. (1998), Planque and Taylor (1998), Lindahl et al. (1998) and Reid and Edwards (2001). The latter distinguished two contrasting periods in the North Sea, a relatively cold one between 1978 and 1982 and a warmer one after 1987. The changes were, according to Reid and Edwards (loc.cit.), the result of a combination of local, regional and far field hydrometeorological forcing, of which the most important the variability in oceanic inflow. The changes that occurred around 1987 were large enough to be called a "regime shift," and Reid and Edwards (loc.cit.) pointed to some coinciding events from 1988, namely the seal epidemic and the *Chrysochromulina* bloom (see also 5.4.2).

5.1.2 The impacts of marine eutrophication

Where the foregoing section was concerned with the question what causes phytoplankton blooms in general, this section will in particular address studies into the impacts of increased primary production and changes in the species composition of the phytoplankton stock. Knowledge of causes of and changes in impacts of eutrophication is very relevant for the justification of measures, such as the introduction of sewage treatment, including nitrogen and phosphorus removal and regulations for the reduction of nutrient discharges from agriculture. The negative effects of such changes have been discussed in detail in the previous chapters and concern the increased occurrence and intensity of toxic and nuisance phytoplankton blooms, oxygen deficiency and concomitant kills of fish and benthos and increased growth of macroalgae.

Another possible effect of increased primary production is increased secondary production. In the 1980s, several observations of increased biomass of benthos had been published. Generally, these increases were attributed to increased nutrient input and in most cased regarded as an adverse effect because of the concomitant shift towards short-lived opportunistic species. However, also positive effects of increased primary production have been documented. It concerns the increase of secondary production, resulting in an increase in harvestable products (fish, shellfish).

Toxic and nuisance blooms

The 1988 *Chrysochromulina* bloom (see 4.4.1) had an enormous impact on science and politics in the North Sea countries. It had reinforced the discussion on the possible causal relationship between increased and/or changed nutrient inputs into the marine environment and increases in the frequency and intensity of such blooms (compare 3.2.3 and 4.1.4). As described in 4.1.4, there was little scientific evidence for the latter, although this was mainly so because there was no proper data base for evaluating and assessing temporal trends. New fuel to the toxic bloom debate was provided in 1990 by a paper by Smayda that carried the title "Novel and nuisance phytoplankton blooms in the sea: evidence for a global epidemic" (Smayda 1990). Smayda (loc.cit.) presented "evidence" from several coastal waters around the world, on the basis of which he concluded that "a long-term trend in increased frequency and dynamics of novel phytoplankton blooms of indigenous species, both benign and harmful ones, has accompanied nutrient enrichment of coastal waters and inland seas on a global scale." Most of the cases presented were from the Skagerrak, the Kattegat, the Wadden Sea and the North Sea. In the following, it will be investigated whether new research and analyses from these areas were in support of Smayda's sweeping statements.

A second factor in the discussion about toxic blooms was the increasing N/P ratio, observed in the North Sea and the Skagerrak since the beginning of the 1990s. This increased ratio was the direct result of the success of phosphorus reduction policies and the continued high nitrogen inputs. Smayda's article was used to underline the possible adverse consequences of the increasing N/P ratio, in particular an increase of toxic and nuisance blooms (Zevenboom and De Vries 1996).

***Phaeocystis* a nuisance alga?** In 1987 Lancelot et al. (1987) had called attention for what they described as a "disquieting increase" in *Phaeocystis* blooms in the North Sea (see 4.2.1). Moreover, the foam produced by *Phaeocystis* could cause "great nuisance" for recreational activities. According to Lancelot et al. (loc.cit.) *Phaeocystis* blooms might also have a negative effect on the atmosphere because of the production of dimethylsulphide (DMS), which might contribute to the acidity of rain water. In 1995 Lancelot published a review article in which it was stated that the *Phaeocystis* foam caused "serious environmental as well as economic problems" without, however, providing data on either of the two (Lancelot 1995).

In the magazine of the Dutch Institute for Coastal and Marine Management (RIKZ) "Zoutkrant," Peperzak and Nieuwerburgh (1998) stated that *Phaeocystis* had a negative effect on fish and shellfish and might cause oxygen deficient water. According to model calculations, the damage caused by *Phaeocystis* was estimated at Hfl. 20 million. Also in this article no facts were presented on the nature and extent of the damage, neither were time-series of foaming events given.

In the framework of the development of eutrophication criteria for the OSPAR Common Procedure for the Identification of the Eutrophication Status of the Marine Environment, a Belgian project made an attempt to come to grips with the term "undesirable disturbance," which is the central element in the OSPAR definition of a Eutrophication Problem Area (see 5.2.3). Of course, *Phaeocystis* was one of the prime parameters to be investigated and the results were quite interesting. A questionnaire was circulated among beach tourists and fishermen with the question which factors they regarded causing most nuisance, respectively which factors were economically most detrimental. For both parameters, *Phaeocystis* had such a low scoring (9% nuisance for tourist and probably no economic losses for fishermen) that it was concluded that "undesirable disturbance" was not a suitable criterion for identifying the eutrophication status of Belgian coastal waters (Rousseau et al. 2004).

Also Cadée and Hegeman (2002) questioned the harmful nature that had, for many years, been associated with *Phaeocystis* blooms. According to these authors, "The wane of these blooms may produce considerable amounts of foam on the beaches, but there is no reason to see *Phaeocystis* as harmful."

It was not only damage caused by *Phaeocystis* for which evidence was lacking. Also its "disquieting increase" could, with the exception of the Marsdiep, not be substantiated. In several publications that had appeared in the 1990s, the increase of *Phaeocystis* in Dutch coastal waters was mentioned, but, in fact, reference was only made to the Marsdiep situation. For example, the following statement from the above mentioned article by Smayda (1990) exclusively refers to publications by Cadée about developments in the Marsdiep: "Thus, between 1973–1985 an epidemic of *Phaeocystis pouchetii* blooms has characterised Dutch coastal waters." Also Lancelot (1995) referred to Cadée only with her statement: "accurate quantitative data on *Phaeocystis* blooms in the continental coastal waters trace back only to the early 1970s."

Zevenboom and de Vries (1996), while discussing *Phaeocystis* blooms in the Dutch coastal zone, Marsdiep area and Wadden Sea, concluded: "Its relatively high abundance, increased concentration, frequency and duration of blooms are strongly linked to increased eutrophication," thereby referring to publications by Cadée, Lancelot et al. (1987) and the OSPAR 1992 Report on nutrients in the Convention area. The 1987 Lancelot article was addressed in 4.2.1 and does not provide data on increases in *Phaeocystis,* other than from the Marsdiep. The OSPAR Report (for further discussion see section 5.2.3) contains literally the same conclusion, however without any reference to a scientific publication or to monitoring data. In the list of references on which the report is based, the above mentioned publications by Lancelot et al. (1987), Smayda (1990) and Cadée are included, though.

In the National evaluation report of the joint assessment and monitoring programme 1995 of the Netherlands (Akkerman 1997), it was concluded that "In Dutch coastal waters a sharp increase (by a factor of 2–3) in annual averaged chlorophyll *a* and primary production took place in the late 1970s as a result of increased nutrient enrichment." This conclusion was based on two references, namely Cadée and Hegeman (1993) and Colijn (1992). In the latter reference, data for chlorophyll *a* in the Dutch coastal zone (period 1975–1990) were analysed, and the only conclusion given was that the data showed "a large interannual variation in both the timing and the frequency of blooms over the years" (Colijn. loc.cit.).

Whereas in the above examples the Marsdiep data were extrapolated to Dutch coastal waters and even to continental coastal North Sea waters, Anderson et al. (2002), in a global review of harmful algal blooms and eutrophication, generalised the Marsdiep series for the whole North Sea by stating that "Mass occurrences of this species began in 1977 in the North Sea (Cadée and Hegeman 1986) and increased in cell abundance and bloom duration through 1985."

My conclusion is that there has never been a firm scientific basis for using *Phaeocystis* as an indicator of negative eutrophication effects. Neither could a general proliferation be made plausible. In fact, there is only one time-series substantiating an increase since the mid 1970s, namely the Marsdiep time-series.

The 1988 *Chrysochromulina* bloom revisited. The 1988 *Chrysochromulina* bloom, which had started along the Swedish west coast (see 4.5.1), was generally attributed, at least partly, to human induced changes in nutrient inputs. Reid (1997), however, postulated a surprising alternative. He pointed to the fact that the bloom had been preceded by five consecutive months of very high runoff of the Göta Alv River. According to Reid (loc. cit.), this had caused shallow and stable stratification and, thus, ideal conditions for phytoplankton growth. Increased nutrient inputs were not considered a causative factor since the Göta Alv was, according to Reid (loc.cit.), low in nitrogen because it had a forested catchment. A possible stimulating factor, mentioned by Reid, were dissolved organic substances.

A comprehensive evaluation of the causes and impacts of the 1988 *Chrysochromulina* bloom was given in a review paper in the Marine Ecology Progress Series (Gjøsaeter et al. 2000). According to Gjøsaeter et al. (loc.cit.) benthic communities had recovered surprisingly fast. The original idea that elevated surface nitrate concentrations had triggered the bloom, was considered unlikely because large influxes of nitrogen were a recurrent phenomenon. The publication carried the subtitle "a catastrophe or an innocent incident?" Although this question was not answered with a clear yes or no, the authors placed the catastrophic label attached to the 1988 bloom into a long-term perspective. According to Gjøsaeter et al. (loc.cit.), "The 1988 bloom may properly be seen as an ecological perturbation triggered by peculiar – but not very atypical – hydrological and meteorological conditions." Because of the fast recovery of the system, they regarded the Norwegian Skagerrak system as having high resilience and stability, and concluded that blooms like the one from 1988 might reoccur, but were unlikely to have long-lasting effects.

Eutrophication and the proliferation of toxic blooms. Above, two specific cases of nuisance and toxic blooms were discussed. An important question in the 1990s was whether a general increase in toxic and nuisance blooms had occurred, as a result of increased nutrient inputs to the marine environment and/or through changes in the nitrogen, phosphorus and silicon ratios. Since the 1988 *Chrysochromulina* bloom, the search for toxic algae had increased, among others through the introduction of national warning programmes, intended to provide an early warning for fishermen and beach tourists. In a critical review of phytoplankton bloom variability, presented at the 1991 ICES Hydrographic variability symposium, Colijn (1992) pointed to the "striking" correlation between areas where blooms had been reported and the vicinity of these areas to marine research stations. The area covered in Colijn's analysis was the North Sea, the Skagerrak-

Kattegat and the Baltic. Colijn (loc.cit.) therefore concluded that "observer effort, mentioned as one of the possible causes of the increase in toxic algal blooms (Smayda 1990) is really one of the factors to be dealt with." With regard to blooms in general, Colijn (loc.cit.) concluded: "despite the huge amount of information, no clear evidence for an increase of bloom event in the ICES areas in general can be given. Locally, areas can be found where changes have occurred. Some of these changes were temporary, others seem fairly permanent."

In a more recent review by Anderson et al. (2002), presented at the Symposium "Nutrient over-enrichment in coastal waters: Global patterns of cause and effect," it was concluded that there was little question that nutrient loading fuelled high biomass algal blooms, and that there was clear evidence for direct stimulation of some harmful algal blooms (HABs) by nutrient over-enrichment. At the same time these authors stated that the linkages between other harmful algal blooms and eutrophication were more complex. They therefore stressed that "It is important to avoid ascribing the apparent global increase in HABs solely to pollution or eutrophication, although the public and the press often assume this linkage. There are many causes for the expansion and eutrophication is but one of these mechanisms" (Anderson et al., loc.cit.).

Oxygen depletion

As described in 4.1, the political interest in marine eutrophication started with oxygen depletion events in the German Bight and the Kattegat and Belt seas, in the beginning of the 1980s. Oxygen depletion and concomitant fish and benthos kills had, therefore, a very important justification function for the introduction of nutrient reduction measures. In Danish enclosed coastal waters in the Kattegat and Belt area, oxygen depletion events were a recurrent event in the 1980s and 1990s, but in offshore waters of the Kattegat serious oxygen depletion events did not occur after 1983. In the German Bight serious oxygen depletion events did not occur after 1983 either, and during the period 1984–1987 a rapid recovery of the benthos was observed (Niermann et al. 1990). According to Van Beusekom et al. (2003) low oxygen concentrations had been measured during irregular research cruises in the German Bight in the years 1989, 1994 and 2000. These observations did, however, not raise worries like those in the beginning of the 1980s. This was maybe the reason that in some analyses a quite different example of the adverse impact of eutrophication-induced oxygen depletion was put forward, namely the so-called black area event in the Friesian Wadden Sea in 1996 (BLMP 2000; Van Beusekom et al. 2003).

Since 1988 indications of oxygen depletion had been observed in the sediments of the East-Friesian Wadden Sea. Anoxic parts of the surface sediment of the littoral, so-called "black spots," had been registered by workers of the research station of Norderney (Höpner and Michaelis 1994). In the spring of 1996, sudden large anoxic sediment areas were observed in the East Friesian part of the Niedersachsen Wadden Sea. By the end of May, 24 km^2 of the littoral had become anoxic and on June 12th, the anoxic area had reached a coverage of 36 km^2 (Michaelis 1997). The anoxic conditions in the sediment surface were accompanied by a massive mortality of the benthos. The event caused considerable public and political uproar in Germany and was generally linked to excess nutrient inputs to the sea. Politicians and environmental groups demanded tighter measures, especially for traffic and agriculture, to reduce such inputs. However, also alternative causes were put forward and, in order to create scientific consensus, the German Environment Ministry organised a scientific workshop in July (Henke 1997). At this workshop, several causes of the event were discussed, including eutrophication, specific weather conditions and the import of organic material from the North Sea. At the end of the day the general picture had emerged that the high black spot incidence had been caused by the input of organic material from the German Bight, produced by a large bloom of the diatom *Coscinodiscus concinnus*. Under certain conditions, this species produces oil with n-C_{14} and n-C_{16} fatty acids. An oil slick had been observed by satellite north of the East Friesian islands on May 22nd.

Although it was rather clear that there was no link to eutrophication, the representative of the German Environment Ministry presented a quite different picture in his summary of the workshop, illustrating the political need for linking an environmental disaster to excess nutrient inputs, so as to justify measures to reduce such inputs:

"The entire discussion, which today was focused on black spots, gives me the impression that the Wadden Sea – maybe we can say, the East Friesian Wadden Sea in general – has become more prone to disturbances. The particular situation leading to the appearance of the black spots in the Wadden Sea may have resulted from a coincidal collection of particularly unhappy circumstances. But I do believe that we have a fundamental underlying disturbance, as opposed to you, Dr. Bakker, who says that this is just the normal situation, which is only disguised at times. This fundamental disturbance is probably based on 'overfeeding', to put it mildly" (Henke, loc.cit.).

Interestingly, several historical records of comparable events exist, in which a high salinity and a low turbidity in the coastal waters, caused by an exceptionally low river runoff in the preceding winter, are described. Michaelis (1977) recorded a massive mortality of cockles and other invertebrates in the summer of 1976 in the same area. The event was accompanied by the presence of a *Coscinodiscus concinnus* bloom and anoxic areas. Delafontaine and Flemming (1997) referred to a *Coscinodiscus* oil slick in the central North Sea, reported by Grøntved in 1952. Van Bennekom et al. (1975) mentioned a "massive flowering" of the diatom *Coscinodiscus concinnus* in 1849 in Dutch coastal waters, after which, according to these authors, such blooms had not been reported again until the mid 1960s. Roskam (1970) gave an extensive description of a bloom of *Coscinodiscus concinnus* off the Dutch coast in May–June 1964. As a result of long-lasting easterly winds and an offshore surface current, the diatoms were transported into the Wadden Sea, where they caused anaerobic conditions, resulting in the formation of H_2S and benthos mortality. Gieskes (1973) reported on a massive dying-off of *Coscinodiscus concinnus* in 1972 off the Dutch North Sea coast. Beukema (1986) judged these last two events as the result of very special meteorological conditions. On both occasions they coincided with a high salinity and a low turbidity in the coastal waters, caused by an exceptionally low river run-off in the preceding winter.

Increased secondary production

In the 1980s several cases were presented of increases in benthic biomass, which were attributed to increased primary production. Two often referenced cases of anthropogenically induced increased secondary production are studies by Pearson, Rosenberg and Josefson in the Kattegat and eastern Skagerrak (Pearson et al. 1985; Josefson 1990), and those of Beukema and Cadée (1986) in the western Dutch Wadden Sea. On the basis of a comparison of benthic data from the Kattegat from 1913 and the year 1984, Pearson et al. (1985) were able to demonstrate an enormous increase in benthic biomass, which they mainly attributed to enrichment. Josefson (1990) presented the results of monitoring biomass and abundance of macrobenthos at 14 stations in the Skagerrak-Kattegat area. The total biomasses at the 14 stations showed a linear increase by a median factor of 1.8, primarily from 1981 to 1988. Josefson (loc.cit.) concluded that the most likely cause of the increase were increased human-generated nutrient inputs.

Beukema and Cadée (1986) found a doubling of macrozoobenthos biomass in the western Dutch Wadden Sea between 1970 and 1985 (see also 4.2.1), for which they considered increased primary production the most plausible explanation. In the following years, several publications by Beukema confirmed the findings from 1986.

Tunberg and Nelson (1998) put the macrobenthos biomass changes in the Skagerrak-Kattegat in the perspective of climatic changes. They found distinct cyclical patterns of seven to eight years in the biomass of macrobenthos along the Swedish west coast, which correlated with changes in the North Atlantic Oscillation Index. The NAO Index was negatively correlated with river flow from western Sweden, while this flow was positively correlated with macrobenthic biomass. Tunberg and Nelson (loc. cit.) concluded that "climatic variability in the region may be a more basic causative factor for benthic disturbance than eutrophication and other possible factors which have previously been proposed." Also Richardson and Cedhagen (2001) discussed the relevance of changes in the NAO index for changes in the benthic community. As one possible causative factor, they mentioned the fact that positive NAO indices in the Skagerrak are marked by westerly winds, which normally prevent the exchange of bottom water in the fjords, and, thus, the supply of oxygen to the benthos. As also put forward by Reid and Edwards (2001) (see above), NAO values had changed around 1978 from mainly negative to predominantly positive values (Richardson and Cedhagen 2001).

The earlier assumed link between eutrophication and increased macrozoobenthos biomass in the western Dutch Wadden Sea (Balgzand area) was placed in a broader perspective in more recent publications. In 1997 Beukema and Cadée reevaluated several possible causes of the observed increase in macrozoobenthos biomass, amongst which enrichment and changes in climatic factors (Beukema and Cadée 1997). They concluded that eutrophication was indeed the most plausible cause of the increase, but that the enrichment had only been effective in those parts of the Balgzand where environmental conditions were favourable. In areas with harsh conditions, food was not the limiting factor and eutrophication thus not of much influence (Beukema and Cadée, loc.cit.).

Essink et al. (1998) made a comparison of macrozoobenthos development at different locations in the entire Wadden Sea, including the Balgzand area. They concluded that the severity of the winter was an important synchronising factor for macrozoobenthos, a severe winter usually being followed by increased growth. The cause-effect relationship between macrozoobenthos biomass and factors such as primary production or eutrophic state was, with the exception of the Balgzand, less clear (Essink et al., loc.cit.).

Other examples of changes in North Sea benthos, first attributed primarily to enrichment and recently reassessed as primarily being caused by climatic changes, come from the Dogger Bank and the coastal waters off Norderney (Kröncke et al. 1998; Wieking and Kröncke 2001; Kröncke and Wieking 2003).

Eutrophication: "A blessing in disguise"?

Whereas in the 1970s increased nutrient supply was generally regarded as a principally positive thing (more nutrients, more primary production, more fish), the emphasis in the 1980s was on negative aspects of eutrophication. This change in perception becomes also clear from the definition of eutrophication, which changed from nutrient enrichment sensu strictu to nutrient enrichment causing negative effects on the ecosystem, the latter used in the EU Nitrate Directive (see 5.3.3). In 1991, however, Boddeke and Hagel of the Dutch Institute for Fisheries Research (RIVO) published a paper in which they attacked this attitude. In their paper "Eutrophication of the North Sea Continental Zone, a blessing in disguise," Boddeke and Hagel (1991) stated that the continuing decrease of phosphorus inputs to the southern North Sea "is likely to have negative effects on the production of fish and shellfish in the Southeastern North Sea." This postulation was questioned in several scientific papers, among others by Cadée and Hegeman (1993), who had found no decrease in primary production in the Marsdiep, despite decreasing P concentrations. Boddeke, however, continued his crusade for not continuing P reductions and for even supplementing the sea with phosphorus. By this, he received much press attention in the mid 1990s. From the side of green organisations and responsible authorities in the Netherlands the proposals of Boddeke evoked strong reactions, among others because his ideas were considered "dangerous," since they interfered with established nutrient reduction policies. Probably because of the substantial scientific counterarguments against their phosphate theory, Boddeke and Hagel had, in the meantime, adjusted their ideas in such a way that they now pointed to changes in N/P ratio as the main cause of reduced fish stocks. In 1997 this hypothesis was confirmed by a study of the University of Groningen, commissioned by the Dutch Fisheries Producers' Organisation (Nanninga 1997). However, in a subsequent workshop attended by many Dutch scientists, Nanninga's conclusions could not be held upright (Wolff, pers.comm.).

The Boddeke and Hagel action was not an isolated case, nor was it something new. Proposals to use excess nutrients for fertilising the sea had already been proposed in the past, and kept popping up at irregular intervals. In the 1970s, for example, the Dutch Ministry of Transport had asked for scientific advice regarding plans to dump manure into the North Sea (Gieskes, pers.comm.). Such an idea was announced again in 1998 as being the ideal solution to the never ending Dutch manure problem. Lindeboom et al. (1988) discussed several options for relocating nutrient discharge outlets. One suggestion was to discharge part of the Rhine load via a pipeline 30 to 60 km into the North Sea. In this way the coastal waters would be relieved and primary production in offshore waters would increase. How this would influence secondary production was, however, unclear. They referred in this respect to a large scale fertilisation experiment that had been carried out in Scotland. The enrichment with phosphate had resulted in increased growth of mainly unwanted macrophytes. Lindeboom et al. (loc.cit.) suggested to carry out large-scale experiments, in order to determine whether enrichment of certain marine areas could become a future policy option.

In 1996 an EU funded research programme into the effects of fertilisation with phosphorus of a Norwegian fjord was announced. This so-called "Maricult" plan met with heavy criticism. A commentary in the New Scientist carried the title "Norway's fish plan 'a recipe for disaster'" (MacKenzie 1996). Also in Dutch and German newspapers the plan was heavily criticised. At the 1996 OSPAR joint Commissions meeting, a paper was presented by the World Wide Fund for Nature (WWF), in which the experiment was called a "warning precedent," and in which the question was raised how the experiment, which was to be carried out within the OSPAR convention waters, could be compatible with the provisions of the 1992 OSPAR Convention regarding the 50% reduction of nutrient inputs (see also 5.3.2.) (WWF 1996b).

In their analysis of the impacts of increased nutrient supply on secondary production (see above), Beukema and Cadée (1997) used the term "mild eutrophication" to describe the situation in the Wadden Sea. This mild eutrophication was valued as positive. In a global review of nutrient enrichment and secondary production, Nixon and Buckley (2002) presented several examples of positive effects of enrichment. They concluded: "Despite a recent review concluding that there is little or no reason to expect that the production of fish and other animals will increase with nutrient enrichment or eutrophication, there is a variety of evidence that anthropogenic nutrients can stimulate secondary production in marine ecosystems." Nixon and Buckley (loc.cit.) also stated: "Concerns over the growing nutrient (especially N) enrichment of coastal marine waters are clearly valid and deserve the attention of scientists and managers, but the recent demonising of N ignores the fact that nutrients are a fundamental requirement for producing biomass."

5.1.3 New knowledge: an analysis

As shown in the foregoing, several (review) articles about the relationship between eutrophication and observed biological changes had been published in the course of the 1990s. An important feature of these publications was that they could draw upon time-series, which were at least a decade longer than analyses from the 1980s. The following analysis focuses on the questions whether the newly gained knowledge was in support of the political agreements for the reduction of nutrient inputs, and whether is has led to a reduction of scientific uncertainty, which is essential for elaborating management instruments.

Justification

The justification of agreements is related to two aspects, namely the seriousness of eutrophication effects and the relevance of nutrient reduction for reducing these effects. It is in this respect important to note that serious and large-scale events, generally linked with increased nutrient inputs, had happened in the beginning of the 1980s and in 1988, but that in the years thereafter no such incidents had occurred. The importance of large-scale events for politics was comprehensively discussed in Chap. 4. The non-occurrence of new events was most probably the reason why the 1996 black spot event was embraced by politics, even though there was ample scientific material to demonstrate that it had not been caused by excess nutrient inputs. Moreover, already in the following year the ecosystem recovered, and signs of the so-called "black spot disease" were not or hardly observed after 1996. Also the impacts of oxygen depletion on the benthos in the German Bight disappeared surprisingly fast. New knowledge also put the 1988 *Chrysochromulina* bloom into a different perspective. A rapid recovery of the benthos was observed and the event was assessed as a natural perturbation of the ecosystem that could be expected to happen again. One of the classic examples of negative eutrophication impact, the increase in *Phaeocystis* blooms in the Marsdiep, remained the only case of such a consistent increase in the North Sea. However, the nuisance aspect of the blooms, i.e. the negative effects of the foam produced, could not be substantiated and appeared to be exaggerated. The increase of toxic blooms in general, as a result of increased and/or changed nutrient inputs, as prophesied at the beginning of the 1990s, did not manifest itself. My conclusion is that the impact of and damage caused by most of the described eutrophication events, turned out to be less severe than originally thought.

Another element of the seriousness of eutrophication induced events is their scale. As already stated above, large-scale incidents did not happen in the 1990s and the real impact of nutrient loading appeared to be limited to the inner parts of fjords and coastal areas subject to stratification. The second aspect, relevant for justification, is the relation between increased nutrient inputs and eutrophication effects. As comprehensively described in Chaps. 3 and 4, changes observed in the marine environment, such as oxygen depletion, increased primary and secondary production, as well as changes in phytoplankton and benthic stock species composition, were linked to increased nutrient inputs. The scientific discussion focused on which of the nutrients nitrogen and phosphorus was the limiting factor for primary production. In the course of the 1990s, however, this emphasis shifted in several respects. Firstly, the nitrogen or phosphorus discussion became more differentiated with new knowledge becoming available about the relevance of changes in the ratio of different nutrients for shifts in algal species composition. Also the importance of growth factors other than nitrogen and phosphorus, for example organic substances, was underlined in several publications. Secondly, new life was injected into the discussion on the role of zooplankton grazing for the control of plankton blooms, with publications on the impact of pesticides on the grazing ability of zooplankton. Third, and most important, were the many scientific publications about the relevance of climate forcing for explaining a broad spectrum of observed changes. For almost all these changes, correlations were found with changes in the NAO Index. Interestingly, there was a high level of agreement between the different publications on the timing of major shifts, occurring in the investigated parameters. During the period under investigation two such shifts were observed, one around 1978 and one around 1988, both coinciding with substantial changes in the NAO Index. As causal explanations for these correlations, among others, changes in precipitation and changes in wind energy and direction were proposed. Changes in precipitation exerted influence on coastal waters through changes in the flow of rivers, which, in turn, could influence stratification and/or the input of nutrients and plant growth factors. Changes in wind energy and direction are relevant for turbulence and stratification and, thus, for the oxygen situation in the bottom water.

My conclusion is that new knowledge, produced in the 1990s and early 2000s, was generally not in support of the assumed seriousness of marine eutrophication impacts, and, thus, did not provide a contribution to the justification of eutrophication reduction policies. On the contrary, new knowledge was published in which the positive aspects of nutrients, i.e. increased production of shellfish and fish, were underlined. Also new knowledge on the relevance of nutrients for controlling eutrophication effects proved not to be in support of nutrient reduction policies. It had not been possible to link a specific compound to changes in phytoplankton, nor had it been possible to substantiate a specific nutrient reduction percentage. Generally, the relevance of increased anthropogenic nutrient inputs was put into perspective by new knowledge about the predominant role of climatic factors.

Uncertainty

An important assumption underlying the model of rational management is that, generally, more research will lead to less uncertainty (compare chapter 1). Reducing uncertainty may increase scientific consensus, which is considered important for the backing of policies. Reducing uncertainty is also relevant for the fine-tuning or amendment of agreements and the development of management instruments. The main uncertainties that existed in the second half of the 1980s, the period in which the political decisions on nutrient reduction measures were agreed upon, were which nutrient, N or P, was the limiting factor for primary production, what reduction percentage was needed to restore the marine environment and to prevent negative eutrophication effects and, finally, which areas were most susceptible to enhanced nutrient supply. Moreover, there was the general question how to differentiate between man-induced and natural causes of eutrophication effects.

The question, whether the uncertainty regarding these questions was reduced by new knowledge, cannot be answered in an unequivocal way. On the one hand, the relevance of both N and P for phytoplankton blooms was confirmed in several studies. At the same time, however, the importance of several other nutrient related parameters was underlined, in particular nutrient ratios and growth substances other than N or P. The relevance of this new knowledge for policy and management is illustrated by the following conclusion by Philippart and Cadée (2000), in a discussion about the relevance of nitrogen for primary production:

"For example, primary production in shallow marine waters may also be related to additional nutrient sources (e.g. inputs from the open sea), ambient concentrations of other nutrients (P, Si), co-limitation effects between nutrients and light, and rates of biochemical processes that affect nutrient loadings (e.g. denitrification rates). In addition, total primary production will be governed by the species composition, life-history characteristics and stoichiometry of the autotrophic components present. If these factors are not taken into account, management regulations that aim to diminish the effects of eutrophication hold the risk of seriously under- or overestimating the nutrient reductions which are thought necessary to obtain their goals."

Even more important was the relevance of factors other than nutrients, such as zooplankton grazing and, most notably, climatic forcing. With the introduction of these alternative causes of eutrophication effects, the level of uncertainty in fact increased.

With regard to the question about marine areas susceptible to eutrophication, the new knowledge that appeared in the 1990s and early 2000s showed a quite consistent picture, namely that eutrophication effects were confined to bays and fjords and coastal areas prone to stratification.

The general question, whether new knowledge has reduced uncertainty regarding the differentiation between man-induced and nature-induced eutrophication effects, cannot be answered in a straightforward way either. My conclusion is that new knowledge has certainly provided ample evidence of the importance of natural factors for eutrophication effects to occur or to be aggravated. It has, however, generally not been possible to assign specific events to specific natural and/or man-induced causes only, or to indicate the relative importance of either of the two for the investigated case. One of the most important remaining uncertainties in this respect is related to the occurrence of toxic and nuisance phytoplankton blooms.

5.2 New knowledge and the science-policy interface

The central question, addressed in the remainder of this chapter is whether the newly gained knowledge, which, as demonstrated above, was certainly not in support of nutrient reduction policies set in motion in the second half of the 1980, would indeed cause a change in these policies. A second central question is whether and to what extent the new knowledge was usable in the implementation of the 50% nutrient reduction policies. In accordance with the agreements of INSC-2 and INSC-3, the implementation process focused on:

- The development of appropriate measures to achieve nutrient reductions;

- The development of a monitoring programme for nutrients and eutrophication phenomena;
- The assessment of the quality status of the marine environment;
- The designation of areas sensitive to eutrophication;
- The development of quality objectives for nutrients and marine eutrophication phenomena;
- The prediction of the results of nutrient reduction.

The development of measures to reduce nutrient inputs is directly concerned with solving the problem. It is to a large extent connected with the development and introduction of technical solutions in sewage treatment, agriculture and traffic. From the perspective of the application of marine ecological knowledge, it is hardly relevant, and will therefore not be addressed in this section. The other tasks can be regarded as a contribution to the fine-tuning of the very general 50% reduction decision. They have a close relation with marine eutrophication knowledge and will be the focus of the analysis in this section. A wish, explicitly expressed by politicians, was that in elaborating these tasks the relevant working groups should use sound knowledge about the marine ecosystem. What is relevant in this respect, is to realise that the above tasks have different qualities, ranging from mainly factual analysis, i.e. monitoring nutrient concentrations, via assessing ecosystem quality, in which an evaluation of factual data is carried out, to mainly value-laden issues, i.e. developing criteria to distinguish between eutrophication problem and non-problem areas and elaborating quality objectives. In the analysis in this section these different qualities must be taken into consideration, in particular the use of science to solve value-laden questions (compare 1.2.1).

A second focal point of the analysis in this section is the development of the role of the science-policy interface for marine eutrophication. The implementation of the North Sea Conference decisions had, to a large extent, been transferred to the Oslo and Paris Commissions (Osparcom) and its working groups. In other words: the "ownership" of the problem was no longer a political one, but in the hands of the administration. Together with the ownership, mandates and, thus, discretionary powers with regard to value-laden issues, had been transferred (4.5.4). The most important questions addressed in the analysis of the science-policy interface are how it has dealt with the mandates from politics with regard to value-laden issues, and which role it has played in sustaining the interest in the marine eutrophication issue. Particular attention will furthermore be given to the role of ICES. For most activities in the OSPAR working groups, scientific advice by ICES was asked for by the Commission. ICES, in particular the ICES Advisory Committee on Marine Pollution (ACMP), can, therefore, be regarded as the main body responsible for integrating and transferring results of academic science into the OSPAR working groups.

Section 5.2 is structured in accordance with the implementation tasks listed above. The development of a nutrient monitoring programme is addressed in Sect. 5.2.1. In 5.2.2 the assessment of the quality status of the marine environment is covered. The designation of nutrient sensitive areas and the development of quality objectives for marine eutrophication are described and analysed in 5.2.3, under the header "categorizing marine eutrophication." Section 5.2.4 covers the prediction of the possible effects of nutrient reduction measures.

During the period, covered in this chapter, the OSPAR hierarchy has changed several times. An overview of the (changing) OSPAR structure is in Fig. 5.3.

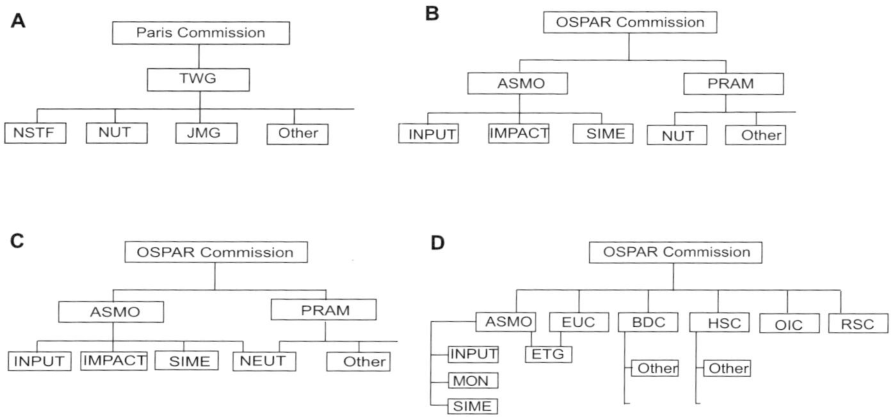

Fig. 5.3. The changing structure of OSPAR. Only the working groups addressed in this study have been listed.
A. Situation until 1992. TWG = Technical Working Group; NSTF = North Sea Task force. NUT = Nutrient Working Group; JMG = Joint Monitoring Group.
B. Situation 1992-1998. Second tier level: ASMO = Assessment and Monitoring Committee. PRAM = Programmes and Measures Committee. Third tier level: INPUT = Working Group on Inputs to the Marine Environment; SIME = Working Group on Concentrations, Trends and Effects of Substances in the Marine Environment; IMPACT = Working Group on impacts on the marine environment.
C. Situation 1998-2002. NEUT = Working Group on Nutrients and Eutrophication.
D. Situation as of 2002. Second tier level: EUC = Eutrophication Committee; BDC = Biodiversity Committee; HSC = Hazardous Substances Committee; OIC = Offshore Industry Committee; RSC = Radioactive Substances Committee. Third tier level: MON = Working Group on Monitoring; ETG = Eutrophication Task Group.

Note the changing role of the working groups dealing with eutrophication: NUT was responsible for both the impacts of eutrophication and the nutrient reduction aspects. In 1992 these tasks were split up between ASMO and PRAM after which NUT only dealt with measures for the reduction of nutrient inputs. In 1998 the NEUT group again covered both aspects. With the new structure of 2002 all issues on the OSPAR agenda were addressed by Committees responsible for both impact and management aspects.

5.2.1 A mandatory programme for nutrient monitoring

The North Sea Monitoring Master Plan

Since the acknowledgement of marine eutrophication as an international issue in 1985 (chapter 4), there had been a discussion on the need for a mandatory international monitoring programme for nutrients. Within the framework of the monitoring programme of the Oslo and Paris Conventions, the Joint Monitoring Programme (JMP), it had not been possible to come to an agreement about the inclusion of nutrients (4.4.3). Therefore, in 1989 Osparcom had decided that the Nutrient Working Group (NUT) would be responsible for all matters regarding nutrient monitoring and should elaborate the principles of a co-ordinated programme. However, in 1989 the North Sea Task Force (NSTF) had decided to establish a so-called Monitoring master Plan (MMP) for the North Sea, which should deliver data for the North Sea Quality Status Report (QSR), which was to be published in 1993 (see further 5.2.2). The MMP, which was mandatory for the North Sea states, also contained parameters relevant for eutrophication, namely P and N compounds, chlorophyll *a* and silicate, together with salinity, suspended solids, temperature, dissolved oxygen and secchi-depth. Criticism was expressed by ACMP, which considered the MMP insufficient for acquiring good quality data for future temporal trend purposes. According to ACMP

"The temporal and spatial variability in the North Sea would confuse the interpretation of NSTF-MMP data to such an extent that any change in the nutrient levels would not be demonstrated unequivocally. Measurements every day, or every second day, at carefully chosen representative stations, combined with synoptic measurements once or twice per year on a very dense network of stations over the entire area, would constitute a better approach" (ICES 1990).

It had always been ACMP's strategy to focus on the relevance of nutrients for primary production, instead of simply measuring nutrient concentrations in seawater (compare 4.2.2). However, the first international mandatory nutrient monitoring to be carried out now in the framework of the MMP, focused on geographical coverage, instead of providing the basis for temporal trend monitoring.

The actual measurements in the framework of the MMP were carried out in 1990/1991, so as to deliver data for the 1993 QSR in time. With regard to the success of the programme, the QSR concluded:

"Both the quantity of data and the coverage achieved represent a considerable advance over previous studies [....] However, despite the initial undertakings as to what was to be done and how, the data still suffer to some extent from a lack of comparability and insufficient information on quality assurance procedures" (NSTF 1993).

ACME[11] was harsher in its criticism, as worded in the 1993 ACME report:

---

[11] In 1993 the Advisory Committee on Marine Pollution (ACMP) was replaced by the Advisory Committee on the Marine Environment (ACME). The purpose of establishing the new group was to place the environmental issues within ICES in a broader perspective. An interesting difference with ACMP is that the ACME membership is according to national representation.

"A strong point of the monitoring programme is that data were collected in areas far enough from local pollution sources to be used as reference areas. However, the established off-shore sampling stations did not always give adequate results for a full assessment, or gave useful information for certain areas only. On the whole, there were insufficient data to allow an adequate assessment of the parameters covered by the programme. One of the major reasons for this insufficiency was that the agreed monitoring plan often was not carried out completely and sometimes monitoring guidelines were not followed. This meant that mandatory determinants were not measured and/or too few stations were sampled" (ICES 1994).

Although the MMP was intended to serve as data input to the QSR, this had apparently not been the case for nutrients. The spatial distribution of nutrients, presented in Sect. 3.4 of the QSR by means of isolines of nutrient concentrations, concerned data from the ICES oceanographic databank, the latest of which were from 1989.

The OSPAR Joint Assessment and Monitoring Programme

The above criticism by ACMP on the MMP was part of a more general evaluation of international monitoring programmes in the OSPAR and Helcom areas. In 1992, with the signing of the new OSPAR Convention (see 5.3.2), it had been decided that a new joint monitoring programme for the maritime area would be developed, to update and take over from the JMP and the MMP. In the 1992 ACMP report an analysis was given of existing monitoring programmes, for which two major types of deficiencies were found (ICES 1992). These were the inadequate translation of management requirements into the conceptual scientific design of monitoring programmes, and the lack of adherence to agreed protocols for sampling, analytical and data handling procedures. According to ACMP, the design of monitoring programmes should in all cases recognise the particular anthropogenic, environmental and oceanographic characteristics of the marine sector to be studied. In the 1993 report several recommendations for improving monitoring programmes were given. First, the purpose of the programme should be clearly defined; second, the programme should provide sufficient information to enable an adequate assessment and, third, the choice of determinants and locations of sampling should be related to the subjects under consideration. This could mean that a programme would not be uniform for the entire area (ICES 1994). ACME also underlined the importance of quality assurance, data handling, overall co-ordination and appropriate financing of international monitoring. Furthermore, it was stressed that all monitoring and assessment activities of a convention should be conducted under one umbrella.

After preparatory work of the Joint Monitoring Group (JMG) and the new Assessment and Monitoring Committee (ASMO), which had succeeded JMG following the establishment of the joint OSPAR Convention in 1992 (see further 5.3.2), a draft Joint Assessment and Monitoring Programme (JAMP) was adopted by the 1995 Osparcom meeting. The draft programme reflected several of the above ICES recommendations: It was based upon a series of questions relevant for management, so-called hypotheses, and also contained provisions for quality assurance and data handling.

The draft JAMP was structured according to the following seven categories:

1. Contaminants;
2. Nutrients and eutrophication;
3. Physical impact;
4. Litter;
5. Microbiological pollution;
6. Fisheries and mariculture;
7. Habitats and health of ecosystems.

This catalogue clearly shows how the scope of the work of the Oslo and Paris Commissions had broadened since the establishment of the Conventions in the mid 1970s. For each category, one or more hypotheses and related monitoring and assessment requirements were proposed. In total, the draft contained 70 hypotheses. For the category eutrophication, seven issues had been elaborated. These were:

2.1 Phytoplankton blooms;
2.2 Influence of eutrophication on community structures and higher trophic levels;
2.3 Oxygen depletion;
2.4 Seagrass;
2.5 Macroalgae;
2.6 Eutrophication and contaminants;
2.7 Effectiveness of measures.

The aim of the JAMP was to deliver the necessary data for the preparation of a Convention-wide Quality Status Report, which was to be published in the year 2000. For a good understanding, it must be noted that only part of the parameters of the JAMP were to be monitored in a joint programme. An important part of the data would have to be delivered by research and modelling. The term "Joint Assessment and Monitoring Programme" is thus somewhat misleading.

The adoption of a monitoring programme and its practical implementation are, as will be shown below, two very different things. Several basic conditions, necessary for a monitoring programme, had not yet been elaborated. It concerned, in particular, common guidelines for sampling and analysis and the handling of the data. In the following section, the implementation of the nutrient monitoring programme will be described. This process can be regarded as representative for the implementation of the whole programme.

The Nutrient Monitoring Programme

As stated above, in 1989 NUT had been made responsible for nutrient monitoring. This group had, however, postponed action on developing the principles for a common programme, pending the execution of the MMP. After it had become clear that the MMP would not be continued, NUT proposed in 1992 that nutrient monitoring should be mandatory for the Convention area. This proposal was adopted by the 1993 OSPAR meeting, provided that a satisfactory programme could be developed. A draft programme was elaborated by NUT and submitted to the first meeting of ASMO in 1994. With the establishment of ASMO, NUT was no longer responsible for monitoring because ASMO was the overall co-ordinating body for monitoring and assessment (see also figure 5.3). NUT also asked ASMO to provide further guidance on the monitoring of fluxes, as proposed by ICES (see also 4.1.4). In 1995 the nutrient monitoring programme was adopted by Osparcom, together with the full JAMP. The nutrient monitoring programme is presented in Table 5.1.

Table 5.1. The Nutrient Monitoring Programme. (Source: ASMO 1995)

	Non-problem areas	**Potential problem areas**	**Problem areas**
Nutrient enrichment			
NH_4-N	+	+	+
NO_2-N	+	+	+
NO_3-N	+	+	+
PO_4-P	+	+	+
SiO_4-Si	-	+	+
Salinity	+	+	+
Temperature	+	+	+
Frequency	About every three years during winter	Annually during winter and during direct and indirect effects monitoring	
Direct and indirect eutrophication effects			
Phytoplankton chlorophyll	-	+	+
Phytoplankton species composition	-	Composition: genera and nuisance/potentially toxic species	Composition: genera and nuisance/potentially toxic species. TOC and POC [a]
Macrophytes	-	Biomass	Biomass. Species composition and reduced depth distribution
O_2 (incl. % saturation)	-	+	+
Benthic communities	-	Biomass and species composition (if time series already exist)	Biomass and species composition
Frequency	-	Annually at times of maximum growth/ activity	

a Total Organic Carbon and Particular Organic Carbon.

An important aspect of the programme is the differentiation between mandatory monitoring requirements for Eutrophication Problem Areas, Non Problem Areas and Potential Problem Areas, which will be comprehensively addressed in Sect. 5.2.3. The adoption of the nutrient monitoring programme did not mean that the execution indeed started in that same year. It took another two years for guidelines to be developed and, in the meantime, nutrient monitoring was continued on a national basis by those countries that had initiated such programmes. In 1997 guidelines were in place, be it for nutrients only and not for the other parameters, and Osparcom decided that the programme should start in the winter of 1997/98. Unfortunately, this delay had made it impossible to use data from the programme for the 2000 QSR.

After the adoption of the JAMP, SIME (figure 5.3), the working group responsible for monitoring and assessment of substances, had been investigating to what extent national monitoring programmes complied with the OSPAR obligations. In 2001 SIME reported to ASMO that there was no precise OSPAR guidance about the frequency of temporal monitoring and the stations to be used in spatial monitoring, meaning that there were substantial differences between national programmes. ASMO 2001 requested SIME to elaborate such guidance, and also the Eutrophication Committee (EUC), the successor of the NEUT group (figure 5.3), made such a request to SIME, with particular reference to the nutrient monitoring programme. In the following year, SIME reported back to ASMO that the question of the frequency of monitoring and the location of monitoring stations depended on the management goals of the programme and that, thus, the first step should be to clearly define such goals (ASMO 2002). It was, however, unclear to SIME how to proceed because, after the completion of the 2000 QSR, the current JAMP was being revised. Moreover, also the recently adopted EU Water Framework Directive (WFD, see further 5.3.3) would have to be taken into consideration. One year later, in 2003, not much progress had been made by SIME. EUC therefore developed a proposal how to proceed with guidelines for the monitoring of nutrients and eutrophication effect parameters. EUC proposed to evaluate the level of confidence and precision of temporal trend monitoring of national data sets, used in the assessment of the eutrophication status in the framework of the Common Procedure (see further 5.2.3). ASMO 2003 agreed that the work on the nutrient and eutrophication guidelines would be continued with high priority, and a so-called intercessional correspondence group (ICG) was installed with the task of evaluating national data sets for their spatial and temporal resolution. On the basis of this evaluation, the guidelines for monitoring nutrients and eutrophication effects would be further refined. By 2005, this work had not yet been completed, due to problems with the timely delivery of national case studies, but also because no specific conclusions could be reached in the relevant technical working groups (EUC 2005).

Data Handling

An issue not yet addressed, but very relevant for the success of any monitoring programme, is the handling of the data collected. As will be shown in the following sections, there were recurrent requests from OSPAR groups for data to be used for assessment, evaluation and classification purposes, and also for the validation of models. Data handling includes the regular collection of data from different sources, quality assessment, i.e. checking whether data comply with agreed standards, such as availability of relevant co-variables, and the storage of the approved data in a data bank in such a way that easy retrieval for further analysis is possible. Already since 1984 the data from the JMP had been handled by ICES and stored in the ICES data bank. In 1986 Parcom had asked ICES to prepare an overview of trends in nutrient concentrations. However, nutrients were not part of the JMP and there were only limited nutrient data available in the ICES data bank, which were, moreover, derived from various sources and of poor quality (compare 4.2.2 and 4.4.3). Parcom had, therefore, requested countries to submit data to ICES on a voluntary basis, but, as reported by ICES in 1988, data submission had been insufficient, which made a trend analysis problematic.

The poor submission of data to ICES continued to be a problem in the 1990s. In its 1997 Report, ACME "noted with concern the rapidly deteriorating position with regard to the delivery of nutrient data." It was, however, not only the amount of data, but also the deterioration of the quality of the data, that worried ACME. The latter was thought to be the result of inadequate resources at the level of the institutes, delivering the data (ICES 1997). As will be shown in the following sections of this chapter, the decrease of both the quantity and the quality of nutrient data coincided with an increase in the need for such data by the working groups of OSPAR. In the 1998 report, ACME presented a table showing the decline in the delivery of data collected in the North Sea. For the other OSPAR regions the situation was even worse (ICES 1999). In 1999 ACME again addressed the nutrient data submission problem: there had been a reduction in the number of data delivered by 50–75% since the early 90s, and it was stated that for the coastal zone this reduction was even greater. According to ACME, "Various attempts to stimulate nutrient submissions via national data centres, relevant ICES working groups and OSPAR delegations have so far yielded little reaction" (ICES 2000). The data submission problem was discussed in the 1999 meeting of the OSPAR NEUT (Nutrients and Eutrophication) working group, apparently without much success: in the 2002 ACME report it was stated: "an increasing number of relevant data sets from almost all OSPAR countries are still not available" (ICES 2002b).

Conclusions

By 2005, two decades after the 1985 consultation meeting on eutrophication (4.1.5), there was still no really effective, co-ordinated and harmonised international monitoring programme for nutrients and eutrophication effects, and, consequently, no sound instrument for systematically tracing the effects of the implementation of the 50% reduction measures on a North Sea wide or Convention-wide level. This goes for the monitoring of nutrient reduction, as well as the possible effects of such a reduction. The causes of this failure are problems with incorporating scientific demands into the programme, logistic problems with the handling of data, problems with overcoming differences in national methodologies, resource problems and developments in the political field.

The scientific problems relate to the aims of the programme, the choice of parameters, the frequency of sampling, the location of monitoring stations, quality assurance and data handling. Already in 1986, at the first meeting of NUT, the Danish delegation had underlined that the frequency of measuring nutrients should be once every fortnight at least, which was related to the high variability of the concentrations of substances in seawater. It must be stressed that the problems, described here, mainly relate to the monitoring of nutrients. For the other parameters of the programme (see table 5.1), most of which will have to be monitored in eutrophication problem areas, comparable problems must be expected. Issues that have apparently disappeared from the discussions, but which are, according to recurrent comments by ICES, very relevant for the assessment of the impact of nutrients, are fluxes and primary production measurements.

The development of an international monitoring programme is apparently a time-consuming process. What is needed for the understanding of large-scale dynamic ecosystems, such as the North Sea, are long-time data series, meaning that an established programme should be in operation for at least ten years to be of any use in environmental management. What happened in practice was that even before the programme was in full operation, new political developments interfered with the cumbersome process of their scientifically sound design and execution: The MMP had been in operation for one season only, and was then overtaken by the political decision to develop the JAMP. The JAMP, as set out above, had not even come to its full functioning, before a process of revision began. The most recent political developments are the Water Framework Directive (WFD) and the EU Marine Strategy Directive (see further 5.3.3). It must be expected that the JAMP will have to adapted to the requirements of these Directives.

With regard to the monitoring of nutrient inputs, the picture is much more positive. In the course of the 1990s, international OSPAR programmes have been developed for the monitoring of inputs via rivers and the atmosphere to the OSPAR area. Since the beginning of 1990s, the so-called RID (Riverine Inputs and Discharges) Programme and the CAMP (Comprehensive Atmospheric Monitoring Programme), have delivered regular reports, integrating and assessing national riverine and atmospheric input data. Like the nutrient monitoring programme, both the RID and CAMP have been, and still are, faced with methodological, logistic and support problems, but these have, apparently, not been as big as for the monitoring of nutrient concentrations in seawater. The results of both programmes, in particular the RID, have been used as input to the 1993 and the 2000 QSRs and the progress reports to the North Sea conferences, and have been the most important source of information for evaluating the success of nutrient reduction policies.

5.2.2 Assessment: The 1993 and 2000 Quality Status Reports

In the 1980s two quality status reports (QSRs) for the North Sea had been produced, the first in 1984 (published in 1986), shortly before INSC-1 (1984), the second in 1987, shortly before INSC-2 (1987) (see 4.1.4 and 4.2.2). The assessment of marine eutrophication in both QSRs was based upon limited knowledge because the problem was relatively new and results of targeted research not yet available.

Also in the 1990s two QSRs were published. The 1993 QSR was developed by NSTF and submitted to the Intermediate Ministerial North Sea Meeting (IMM) of 1993 (see further 5.3.1). The latest QSR was issued in 2000 and developed under the responsibility of OSPAR. An important difference between the 1980 and 1990 QSRs is the fact that the latter two assessments could make use of an increasing amount of results of research into marine eutrophication. The aim of this section will be to compare the assessment of marine eutrophication from the 1993 and 2000 QSRs with the outcome of new scientific findings, as described in 5.1.

The 1993 North Sea Quality Status Report

The development of a new QSR for the North Sea had been commissioned to NSTF by INSC-2 (4.2.5). In 1990 an interim QSR had been submitted to INSC-3 (4.5.3). The procedure for developing the 1993 QSR was different from the way the 1986 and 1987 reports had been prepared. Upon initiative of ICES it was agreed to base the new QSR upon 11 subregional assessments, overall studies by relevant OSPAR and ICES working groups and data from the MMP. The subregional assessments would provide a better overview of the quality status of the whole North Sea because the previous QSRs had mainly focussed on problem regions with regard to pollution. The preparation of the subregional reports was done under the responsibility of lead countries, while the elaboration of the so-called holistic report was the direct responsibility of ICES and NSTF.

In the fifth meeting of NSTF (November 1990) it was agreed that input to the QSR would also be delivered by OSPAR and ICES working groups. NUT thereupon established an ad-hoc expert group to elaborate the QSR part on marine eutrophication. The remit of the group was:

- To produce a brief overview of nutrient inputs to the North Sea;
- To produce a short description of problem areas showing adverse eutrophication symptoms;
- To assess relevant nutrient data from the ICES data bank;
- To produce a contribution to the 1993 QSR based upon the outcome of the first three tasks.

The nutrient expert group, consisting of Colijn, Dooley, Owens and Skjoldal, elaborated a brief report (Colijn et al. 1992), the conclusions of which are reproduced in full below because they provide an excellent overview of the situation with regard to the scientific knowledge on marine eutrophication at the beginning of the 1990s:

1. "Nutrient levels in the North Sea are greatest in the coastal zones. Highest concentrations are consistently found in the southern North Sea. These result from a combination of the hydrography of the North Sea and the distribution of the riverine inputs.

2. Phosphate concentrations have increased in the coastal zones (approximating to the area bounded by the 33 salinity contour) from the levels measured in 1935/36. No increase in phosphate over this period is evident offshore from this area. No difference is apparent between this recent analysis and a similar analysis carried out for the period up to 1978.
3. The area of the North Sea showing increased phosphate concentrations in approximately 60,000 km^2 (approximately 10% of she surface area). The volume influenced is less than 1%.
4. There is a marked inter-annual variability in the area exhibiting elevated phosphate concentrations.
5. The historical data on nutrients levels in rivers and coastal waters are extremely limited in both the amount and quality. This makes it difficult to draw firm conclusions as to the exact natural background levels in the pristine state prior to human development in the watersheds and coastal regions. However, considerable quantities of the nutrients in the freshwater runoff to the North Sea result from man's activities, reflecting urbanisation, industrialisation and agricultural development. It is therefore also likely that a considerable proportion of the elevated nutrient levels in she coastal zone are the result of anthropogenic influences.
6. Contemporary data are also limiting in several respects. First, there is not a consistent monitoring of nutrient levels, oxygen concentrations, algal biomass and production over relevant space and time scales. Second, even if available, there is not a consistent policy of submitting data to the ICES data bank.
7. The nitrate to phosphate ratio of many of the inputs is significantly different from 16; this is reflected in certain areas of the North Sea. The deviation from 16 is most notable in the southern North Sea, and under certain hydrographic conditions can extend into the Skagerrak and Kattegat.
8. There are only limited data on the effects of these nutrient inputs.
9. In several parts of the North Sea there is evidence for an increase in algal biomass, change in species composition (from diatoms to flagellates), and an increase in nuisance blooms in recent decades. However, direct links with nutrients are not always clear-cut. There is some evidence for an increase in algal biomass, change in species composition, and increase in nuisance blooms in recent decades, although the links with nutrients are not clear-cut.
10. Although the frequency of reporting of toxic blooms has increased in recent years there are insufficient data to establish trends, and the links between these blooms and nutrients are equivocal. Changes in the ratio of available nutrients may be as important as absolute concentrations.
11. There are considerable gaps in knowledge of the subject of nutrients in the North Sea, which should be addressed by the initiation of a coordinated approach. These include:
(i) Lack of a suitable historical record of all relevant variables. This results in a lack of understanding of natural variability, which leads to an inability to distinguish this from man's influence. While nothing can be done to overcome the lack of historical data, it is doubtful that the current effort is rectifying the problem of understanding natural variability.
(ii) Lack of understanding of how a complex array of environmental variables (e.g. nutrients, light, species) interact in situ.
(iii) Lack of an understanding of toxicity in algae.
(iv) Insufficient decadal markers of change in the sediment record which might give an insight into past (and continuing) change in the North Sea.
(v) Lack of information on the processes leading to the accumulation of nitrite in the water column.
(vi) Are pelagic and benthic fish production related to eutrophication."

Some of these conclusions were criticised by national delegates in NSTF, indicating the political unease with the modest scientific support for the marine eutrophication problem. Especially Denmark raised comprehensive and severe criticism on the work of the expert group. Many of the conclusions of the expert group were, in the view of the Danish delegation, "misleading" because "Doubts and uncertainties of causative effects have been overemphasised and scientific papers documenting such effects have to a large extent been neglected" (NSTF 1992). Conclusion number 3 was called "nonsensical and misleading." In the view of the Danish delegation, it could have been equally correct to state that more than 50% of the length of the coastal waters of the North Sea and adjacent seas was affected by increased nutrient concentrations. In the introduction to the third draft the authors remarked that comments had been provided to the second draft, but that the group had not been able to fully respond to them. One reason for this were time constraints, the other that the group "wished its views to remain unchanged" (Colijn et al. 1992).

There are some striking differences between the experts' text and the contents of the 1993 QSR. In Sect. 5.2 of the QSR, which deals with nutrients and eutrophication, the subsection on eutrophication effects started with the factors that must be borne in mind when evaluating the impacts of eutrophication (NSTF 1993). It concerned the lack of reliable data, tidal mixing and the limits to knowledge, in all, a good reflection of the experts' text. In the subsection dealing with phytoplankton species composition and harmful algal blooms, the differences became more evident. In the expert report it was stated that the frequency of blooms was strongly influenced by the observation effort, but in the QSR algal bloom frequency was not put into this perspective. In conclusion number 9 from the expert report it was stated two times that there was no clear-cut relationship between nutrients and frequency and composition of algal blooms. In conclusion number 10, links between blooms and nutrients were judged to be "equivocal." In the QSR it was stated: "The many blooms of phytoplankton algae – including toxic species – reported for the North Sea, cannot be attributed to eutrophication effects alone." Also in the subsection "Oxygen deficiency" there are some obvious differences with the expert report. In the latter it was stated that reductions in oxygen concentrations were not linked with eutrophication per se, but that the extent and severity of the depletion could be enhanced by increased organic matter (resulting from nutrient inputs). It was concluded: "It is difficult to establish whether bottom water oxygen levels have fallen or increased in the North Sea as a result of eutrophication, due to the lack of historical data." In the QSR the following conclusion was drawn with respect to oxygen deficiency, for which a trend of decreasing concentrations was postulated: "It is likely that this trend is related to increased sedimentation and decomposition of organic material caused by eutrophication."

The differences between the QSR and the expert report are even more pronounced in Chap. 6 of the QSR, the "Overall scientific assessment." In Sect. 6.7, dealing with contaminants, concentrations, inputs and effects, the subsection on nutrients and eutrophication concluded, among others:

"The increased input of nutrients combined with the resulting change in nutrient ratios has altered phytoplankton community structure and succession in some regions of the North

Sea. These alterations have led to changed patterns in the flow of energy in the food chain, as evidenced by increased production and biomass of phytoplankton, changes in planktonic species composition including the occurrence of harmful algae, changes in benthic algae and animal communities, and increased consumption of oxygen in water and sediments, leading to reduced concentrations of oxygen and thereby mass mortalities of benthic organisms and fish" (NSTF 1993).

An interesting difference between the 1993 QSR and the 1987 QSR is the very high emphasis in the 1993 QSR on the differences in N and P inputs and the resulting increase in the N/P ratio. Time and again the possible relevance of deviations from the Redfield ratio, which was termed the "normal" ratio, was underlined, especially for the occurrence of toxic blooms (compare 3.2.2).

Something becoming very clear from both the QSR and the expert report is that there were, in fact, only three data series relevant for the analysis, namely the Helgoland Reede series, the Marsdiep series and the CPR measurements. Results of regular monitoring programmes had not been used or referred to (compare also 5.2.1). Especially when it comes to finding evidence for a relation between enhanced algal growth and increased nutrient inputs, the only "firm" data presented were the Marsdiep *Phaeocystis* and the Helgoland Reede data.

The 2000 North Sea Quality Status Report

Compared with the 1987 QSR, the 1993 QSR could draw upon the results of "only" five years of additional research results. As shown in Sect. 5.2.1, results of international monitoring had hardly played a role in the evaluation. The QSR published in 2000 could make use of more than a decade of targeted research into the impact of increased anthropogenic nutrients inputs into the North Sea and adjacent waters. As pointed out in Sect. 5.2.1, also for this QSR monitoring results were too late to be of use in the assessment. What is of particular interest, is to what extent the eutrophication assessment in the 2000 QSR reflected the increasing evidence from scientific research, analysed in detail in 5.1, that the effects of anthropogenic nutrient inputs to the North Sea appeared to be limited in space and severity. The eutrophication effects addressed were oxygen depletion, harmful algal blooms and effects on fauna. With regard to oxygen depletion, the QSR signalised "Trends of decreasing oxygen concentrations, which may be due to eutrophication, enhanced sedimentation and organic matter decomposition." These trends had been documented for deep waters in the Kattegat and basin water of Swedish and Norwegian fjords (OSPAR 2000). However, also "some improvements" were noted with regard to nuisance algal blooms, oxygen depletion and kills of benthos and fish in many parts of the North Sea. Also the Wadden Sea "black spot" incident of 1996 (see 5.1.2) was covered as an example of anoxic conditions in an intertidal area. The information provided in the QSR was in good agreement with the scientific analyses of the "black spot" case.

The section on harmful algal blooms started with the conclusion that there was no evidence of an increased incidence in the last 5–10 years. In this same section it was stated that noxious blooms of *Phaeocystis* and *Coscinodiscus* recurred on the south-eastern and eastern coasts of the North Sea, without, however, providing a reference. Interestingly, the possible impact of changes in the N/P ratio, postulated in the 1993 QSR, was put into perspective in the 2000 assessment. According to the report there was "some evidence" that changes in the N/P ratio could affect species composition and food web structure, but "additional evidence is required to demonstrate that changing rates of nitrogen and phosphorus input have had an effect on the North Sea ecosystem" (OSPAR 2000). Finally, the supposed eutrophication impacts on the benthos of the Dogger Bank, first addressed in the 1993 QSR, were put in a different perspective. According to the report, eutrophication could be a factor in the changes observed in the macrofaunal communities between 1987 and 1996, but also changes in hydrography and the cold winter of 1995/6 should be considered.

Negotiated science

My conclusion is that the 2000 QSR fairly well reflects the changes in the overall scientific "mood" with regard to the possible effects of increased anthropogenic nutrient inputs into the North Sea, as set out in Sect. 5.1 of this study. It is, therefore, the more striking that in Chap. 6 of the QSR, in which an overall assessment of the conclusions presented in Chaps. 1 to 5 is given, nutrient input has been placed into the highest impact category. A "hierarchical scheme of criteria" taking into account severity, scale and recovery aspects, had been used for the assignment of several human impacts to four impact classes (OSPAR 2000). Unfortunately, the scheme itself was not presented, nor the way of weighing, but if severity, scale and recovery aspects are applied to nutrients and eutrophication effects, none of these would have a high scoring. As has become clear from scientific evidence and the conclusions of the QSR itself, eutrophication effects are, generally, confined to a small part of the North Sea. There are only few known examples of severe impacts, among which the 1981 oxygen depletions, the 1988 *Chrysochromulina* bloom and the 1996 black spot events, the latter two probably not even having a direct relationship with increased nutrient inputs. Moreover, in all three cases a rapid recovery occurred. The obvious discrepancy between scientific facts presented and the overall conclusions drawn is most probably the result of the process of "negotiated science," in which uncertain scientific facts are mixed with political interests. This process was also visible in the 1993 QSR, but the discrepancies between facts and conclusions are much more pronounced in the 2000 QSR. Apparently, the bodies at the science-policy interface have used their discretion in the process of translating science into policy-useful science in such a way that the political status quo with regard to the seriousness of marine eutrophication and the need for nutrient reduction, was not threatened. What is striking, is that to my current knowledge, there has not been any reaction within the scientific community about these discrepancies. Maybe the scientific community has lost interest in the case. It is also possible that there is still broad consent with nutrient reduction measures for precautionary reasons. It is also possible that scientists, engaged in marine eutrophication research, are reluctant to criticize current policies because they are dependent upon official grants to do research. In Sect. 5.3 it will be investigated how politics responded to the messages of the 1993 and 2000 Quality Status Reports.

5.2.3 Categorizing marine eutrophication

The need to categorize the marine environment into areas more or less sensitive to the effects of enhanced nutrient loading, has its origin in the decision taken at INSC-2 (London, 1987) that a 50% nutrient reduction would only be necessary for discharges to areas, where these discharges might cause pollution (4.2.5). The implementation of this decision involves (integrated) knowledge about various aspects of the marine eutrophication process, among which transport and distribution of nutrients in the marine environment, the role of nutrients in the development of algal blooms and possible effects of increased primary production. What is furthermore needed, are criteria to differentiate between "safe" and "unsafe" nutrient loads, in other words, what constitutes pollution by nutrients? The scientific requirements, the management relevance and the value-laden content of the task of categorizing marine eutrophication, make this issue of central relevance in the analysis of the role of marine ecological knowledge in policy making.

Eutrophication problem areas

Following the 50% nutrient reduction decision of INSC-2, NUT had, under the lead of The Netherlands, been working on the preparation of a map of eutrophication problem areas in the North Sea (4.4.3; figure 4.6). However, the outcome of the work of NUT was criticised by France and the United Kingdom, and it had not been possible to come to an agreement on the designation of eutrophication problem areas during the preparation of INSC-3. Therefore, at INSC-3 (The Hague, 1990), it was agreed "to identify some coastal zones of the North Sea, including the Skagerrak, as being actual eutrophication problem areas and, in view of the increased inputs and levels of nutrients, some other coastal zones as being potential problem areas." Parcom was requested to identify such areas and NUT continued its work after INSC-3. In 1992 NUT finalised the so-called "Eutrophication Symptoms and Problem Areas" report, which was presented to the 1992 OSPAR Ministerial Meeting (see further 5.3.2). This meeting formally "took note" of the report, meaning that, at least for the time being, no political commitment was attached to it.

The contents of the report only slightly differed from the 1989 version (4.4.3). The report covered the issues nutrient enrichment, phytoplankton blooms, oxygen deficiency and increased growth of macroalgae, and contained various maps related to these issues. The first conclusion drawn on the basis of the information presented, was that seven areas had been identified as eutrophication problem areas (OSPAR 1993). These were presented in a so-called integrated administrative map (figure 5.4). It concerned sites on the northern French coast, Belgian and Dutch coastal waters, the German Bight, Danish coastal waters and Danish fjords, the Norwegian Skagerrak coast including the Oslo Fjord, the Swedish Skagerrak and Kattegat coastal waters and, finally, the Ythan estuary and Langstone Harbour areas in the UK.

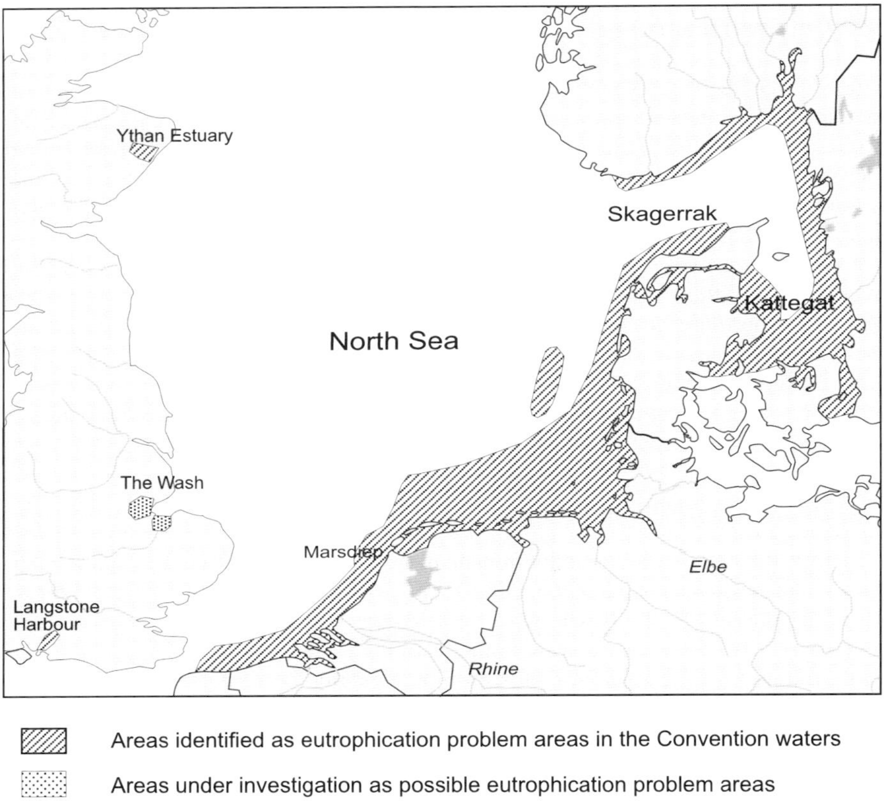

Fig. 5.4. Administrative map of eutrophication problem areas. Redrawn from Ospar, 1993

A second conclusion was, that in all these areas long-term increased anthropogenic inputs had resulted in eutrophication problems. Of these problems, especially the increase of phytoplankton blooms was underlined. According to the report, total algal biomass in the continental coastal waters was, on average, two to three times higher than in 1962, with a sharp increase of flagellates. The increase was related to the increase of nutrient inputs. Also in the North Sea, the Skagerrak and the Kattegat, the frequency, duration and intensity of blooms had increased, which might be a consequence of long-term increases in coastal nutrient levels. The information on which the maps and the conclusions were based, had been derived from 118 publications, which were also listed in the report. Unfortunately, the references were not directly connected to the maps or the descriptions of the various eutrophication parameters. It is, therefore, hard to assess the scientific backing of the data presented and the conclusions drawn.

When comparing the conclusions with scientific findings from the end of the 1980s and the beginning of the 1990s (see 4.4.2 and 5.1.3), it becomes very clear, however, that the effects of increased nutrient inputs, especially the supposed increase in algal blooms, had been overemphasised. Although also other possible causes, such as observed observer effort and meteorological and hydrographical factors, were mentioned as possible causes, the emphasis was clearly on increased nutrients being the main factor. The integrated map mainly reflected the areas with increased nutrient concentrations, although it covered a much larger area than the salinity-33 contour as identified by the nutrient expert group, which had prepared an assessment for the 1993 QSR (see 5.2.2).

One of the major problems in the analysis of the effects of increased nutrients on phytoplankton growth, as put forward in several scientific reviews, for example by Reid et al. (1990) and Colijn (1992), being the lack of long-term data, was not mentioned in the report. Also the poor spatial coverage of the North Sea with regard to long term monitoring sites was ignored. A clear example is the supposed increase in the occurrence of *Phaeocystis* along the continental North Sea coast although, in fact, as extensively argued in 5.1.2, there was only one long-term data set in the Marsdiep area.

The ICES working group on shelf seas oceanography had reviewed the report and concluded: "the document is not a scientific paper, rather it is a synthesis of many reports and discussion documents, many unpublished and not subject to peer review." The group furthermore "felt that in many aspects the figures and text are not consistent with current scientific understanding" (NUT 1992).

The Intermediate Ministerial North Sea Meeting (IMM). In 1993 the report was submitted to the Intermediate Ministerial North Sea Meeting (IMM, Copenhagen 1993, see further 5.3.1) but, as had been the case at INSC-3, it was again not possible to come to political agreement about the identified problem areas. In §17.1 of the Ministerial Declaration the areas were listed, which had been identified in the integrative map as being eutrophication problem or potential problem areas (see above) and, thus, subject to the requirement of a 50% reduction of nutrient inputs. It was, however, explicitly stated that the identification was based upon national criteria. In §17.2 Parcom was, therefore, requested to develop a common procedure for the identification of actual and potential eutrophication problem areas. A second request by IMM to Parcom was to develop further the strategy to combat eutrophication, i.e. the catalogue of decisions from INSC-3. The relevance of this strategy was worded in §36 of the IMM Declaration, in which the Ministers invited Parcom to "consider the size and nature for further reduction targets for nutrients in light of the strategy developed, the QSR and additional scientific knowledge." Progress in the above tasks was to be reported to the fourth North Sea Conference (INSC-4), scheduled for 1995 (see further 5.3.1). It is clear from the above that much (political) hope was put into finding scientific answers to questions regarding the strategic and management aspects of combating eutrophication.

Following the establishment of a joint OSPAR Convention at the 1992 OSPAR Ministerial Meeting (5.3.2), the work of the OSPAR working groups was divided into the categories assessment and monitoring, and programmes and measures. Consequently, the development of the Strategy to Combat Eutrophication was coordinated by the PRAM (Programmes and Measures) Committee, while all questions of a (marine) scientific nature were coordinated by the ASMO (Environmental Assessment and Monitoring) Committee (figure 5.3). The latter included the development of a common identification procedure for eutrophication problem areas, but also the scientific elements of the Strategy, being nutrient input levels, nitrogen to phosphorus ratio, ecological and quality objectives, regional differences and seasonal variations.

The remainder of this section is structured as follows:
First, the development of the Common Identification Procedure or Common Procedure, which took place during the period 1993–1998, is described and analysed. The emphasis is on the elaboration of science-based criteria to distinguish between eutrophication problem and non-problem areas. Next, the implementation of the OSPAR Strategy to Combat Eutrophication (the Strategy), after its adoption in 1998, is addressed. The first part of the implementation concerns the implementation of the Common Procedure, the second part the development of ecological quality objectives for eutrophication.

The development of the Common Procedure

At the first meeting of ASMO (March 1994), it was decided to install an ad hoc expert group, which should establish common criteria and a procedure for the identification of eutrophication problem areas. All countries were asked to submit their national criteria. The group should, moreover, make a start with updating the report on eutrophication symptoms and problem areas. It was of course ASMO's intention to come to a proposal for a common procedure and a commonly supported map in due time for the fourth North Sea Conference (INSC-4) in 1995. What followed, however, was a four year period of tough and cumbersome negotiations which, in 1997, resulted in the adoption of a common identification procedure, but not yet in the actual identification itself. In December 1994 the outcome of the work of the ad hoc working group was presented to the second ASMO meeting (ASMO-2), and at this meeting a draft of a "Common Procedure for the Identification of the Eutrophication Status of the Maritime Area of the Oslo and Paris Conventions" (the Common Procedure) was adopted, be it with study reservations by France, Spain and the United Kingdom (ASMO 1995). The draft procedure aimed at classifying the eutrophication status of the maritime area into three categories. These were:

- Eutrophication Problem Areas: areas for which there is evidence of an undesirable disturbance to the marine ecosystem due to enrichment by nutrients;
- Potential Eutrophication Problem Areas: areas of unknown status with regard to eutrophication. These are areas for which there is insufficient information available for classification as either a problem area or non-problem area with

regard to eutrophication or for which there are reasonable grounds for concern that elevated levels, trends and/or fluxes in nutrients may lead to an undesirable disturbance to the marine ecosystem;
- Eutrophication Non-Problem Areas; areas for which there are no grounds for concern that anthropogenic enrichment by nutrients has disturbed the marine ecosystem.

The draft Common Procedure contained a long list of parameters regarded necessary for discriminating between these categories (see Annex 2, § 4.2.1).

Fundamental questions. What became clear at ASMO-2 were the many unresolved questions regarding the proposed criteria and their application. These questions were not only regarding scientific, methodological and data requirement aspects, but also of a normative and, thus, political nature. Critical and fundamental national comments about the definition of eutrophication, the quantification of reference values for nutrients, the deviation from the reference value on the basis of which an area could be regarded as a problem area, and the weighing of the different criteria in the assessment procedure, had already been raised during the period preceding ASMO-2. The definition of eutrophication, proposed in early versions of the draft Common Procedure was "nutrient enrichment of the marine environment resulting from natural and/or anthropogenic influences." The UK and the European Commission were in favour of the definition given in the EC Nitrates Directive, according to which eutrophication was "the enrichment of waters by nutrients, especially compounds of nitrogen and/or phosphorus, causing an accelerated growth of algae and higher forms of plant life to produce an undesirable disturbance to the balance of organisms present in the water and to the quality of the water concerned." It is obvious that with the latter definition the presence of enhanced nutrient concentrations alone would not be sufficient to give an area the label of a problem area. The UK also pointed to the definition in the 1993 QSR, according to which not only nutrient enrichment, but also the increased primary production resulting from this enrichment was contained. Therefore, the draft presented to ASMO-2 contained not one definition, but took account of the various versions mentioned above.

The UK regarded the setting of general concentration standards as a means of distinguishing between problem and non-problem areas to be unrealistic, considering the high natural variability of the marine environment and the lack of understanding of the eutrophication process. Reference was in this respect made to the 1993 ACME report (ICES 1994) and the 1993 QSR (5.2.2). The UK also stressed the importance of differences in regional meteorological and hydrological factors for the development of eutrophication related phenomena. The available scientific knowledge of marine eutrophication and its effects were clearly in favour of the UK point of view (compare 5.1.3), and also other countries had underlined the need for a more flexible approach, for example with regard to reference levels for nutrient concentrations. ASMO-2, therefore, agreed that the assessment criteria would have to be region-specific. But even with this clear mitigation of the procedure, differences were too big to be solved by ASMO-2, and it was decided that the work be continued in the SIME (Substances in the Marine Environment) working group, one of the three so-called third-tier working groups falling under ASMO's responsibility (see figure 5.3). In April 1995 the results of SIME's deliberation were presented to ASMO. Being one level lower in the OSPAR hierarchy, this group had of course not been able to solve the principle problems inherent to the Common Procedure. What was returned to ASMO were, therefore, the same principal questions related to the relative priorities of the various assessment criteria, the region-specific quantification of the criteria and the integrated weighing of the criteria. At the 1995 ASMO meeting an attempt was made to quantify region-specific assessment criteria, but this failed and it was realised that much more time would be needed to finalise the Common Procedure. Future work on the Common Procedure should, according to ASMO-1995, focus more on a dynamic approach, i.e. on fluxes and processes, instead of on concentrations. The reason for this decision can be traced back to ACME which had, in its 1993 Report, criticised the monitoring of nutrient concentrations as a means of finding answers to effects of increased nutrient inputs (ICES 1994) (see also 5.2.1). Although France and Spain lifted their study reservations, those of the UK remained and Ireland raised new reservations. It was decided to propose to Osparcom that the work would be continued in an ad-hoc group and to ask the Commission to provide guidance for this work.

Guidance by Osparcom. The Osparcom 1995 meeting agreed to such an ad-hoc group and, moreover, that the further work on the assessment criteria should start with the parameters of the newly established nutrient monitoring programme, adopted by the same Osparcom 1995 meeting (see 5.2.1). ICES was requested to provide input to the work, by summarising new information on the relationship between inputs, concentrations and effects, and to provide information on monitoring the effects of anthropogenic nutrient inputs, as well as on possible effects of changing N/P ratios. Moreover, statistical analyses of data series of nutrient concentrations were to be carried out as a basis for the establishment of reference concentrations.

At Osparcom 1995 also the nature of the reservations of the UK and Ireland was elucidated, so as to provide additional guidance to the follow-up work. The reservations by Ireland related to the classification of non-problem areas, which was considered superfluous. Because of limited resources, the emphasis should be on identifying problem areas and potential problem areas. Ireland was, moreover, against the classification of areas of unknown status as potential-problem areas, and referred in this respect to the waters along its west coast, for which few scientific data were available, but for which there were no "reasonable grounds for concern that a potential problem may exist." Whereas Ireland's reservations were very much of a pragmatic nature, the UK objections were much more principal. According to this country, there were no common criteria for the application of the procedure. Furthermore, natural factors that might lead to eutrophication were given insufficient weight, and there was no guidance on how they should be taken into account. Moreover, fluxes and nutrient cycles were poorly quantified and understood. Generally, there were too many open ends in the draft and the UK did see no reason to give it greater weight than other working documents (OSPAR 1995).

The Drafting Panel on Eutrophication. The meeting of the ad-hoc working group, the so-called Drafting Panel on Eutrophication (DPEUT), was held in December 1995. According to the chairman, the work of DPEUT should have a stronger scientific element than had previously been the case. In order to examine links between nutrient concentrations, fluxes and effects, it was necessary that countries would make available all data sets. It would be the particular task of DPEUT to concentrate on elaborating guidance on the application of the assessment criteria. Following the decisions of ASMO-95, a more dynamic, process-oriented approach was to be followed, in which nutrient inputs and fluxes were coupled to plankton dynamics. Also some of the key biological effect criteria were to be quantified. Interestingly, Germany, supported by the Netherlands and Belgium, stated that nutrient enrichment was the basic parameter to be controlled, and that a reduction of inputs towards specified reference levels should be the ultimate goal for eliminating direct and indirect effects of eutrophication. This may be seen as an attempt to shift the emphasis of the procedure towards a source-oriented approach, which was more in line with Germany's national policies, and which might have been a way-out of the so far unsuccessful attempts to establish science-based criteria. Apparently, the other countries were in favour of using effect criteria as well, since the work on these criteria was continued in the years to come.

Towards a new draft. At the DPEUT meeting the fundamental problems, described above, could not be solved, and the meeting only resulted in some small amendments to the draft Common Procedure. Germany was requested to act as lead country for the collection of national quantified assessment criteria. At the 1996 ASMO meeting (March 1996), Germany presented a first synthesis of quantified assessment criteria. However, this synthesis only contained data from The Netherlands, Belgium and Germany, and the other countries were once more requested to submit their national data. ASMO-96 decided that several additional working group meetings would be necessary, if the goal of the adoption of a Common Procedure in 1997 was to be achieved. Between ASMO-96 and ASMO-97 a new draft of the Common Procedure was elaborated in three working group meetings.[12] The outcome, presented to ASMO-97, was an amended version, containing "solutions" to some of the fundamental disagreements. It is important to note that between the second and third working group meetings, a so-called Heads of Delegations meeting took place in which much of the finalisation of the draft Common Procedure was done, in order to be able to submit a draft in time for adoption by the 1997 Osparcom Meeting.

[12] The three meetings of the so-called EUT group were held in September and November 1996 and January 1997.

The Screening Procedure. The first problem discussed in the working group, concerned the reservations by Ireland regarding the wise allocation of limited resources in the application of the Common Procedure. In order not to burden countries with unnecessary monitoring and assessment activities, a so-called Screening Procedure, preceding the actual identification procedure, was proposed. The Screening Procedure was intended as a first "broad-brush" exercise for those areas where there was insufficient information available for an assessment, but which were likely to be non-problem areas. The aim of the Screening Procedure was to identify those areas, to which the actual identification procedure was to be applied. The Common Procedure now consisted of a Screening Procedure and a so-called Comprehensive Procedure, the latter the actual identification procedure.

The identification process. The second problem, addressed in the working group meetings, was the way in which the assessment criteria should be applied in the identification process. Although the working group recognised that guidance on the application of the criteria would have to be part of the Common Procedure, the outcome of the work was rather poor. The United Kingdom had submitted a discussion paper, containing a proposal for a comprehensive scoring process for the various criteria. However, this proposal was hardly discussed, among others because of time constraints. The actual guidance, contained in the new draft of the Common Procedure, was still very general, and would have to be specified and elaborated in the future.

Quantification. Lastly, there was the problem of quantifying the criteria. Lead country Germany presented a synthesis of national proposals for the quantification of assessment criteria at the third working group meeting. As was the case with the UK scoring proposal, there was insufficient time to discuss the synthesis. Moreover, the third working group meeting was dealing with a final draft of the Common Procedure, agreed upon by the Heads of Delegation meeting and, consequently, there was very limited room for amendments. Thus, also the quantification of criteria was postponed.

Analysis. Despite the still remaining deficiencies, the Common Procedure was adopted by the Osparcom Meeting of 1997. The full text is in Annex 2. Ten years after the decision of INSC-2 to reduce nutrient inputs to areas where they may cause pollution, a procedure for identifying such areas had been agreed upon. The actual designation of such areas had, however, still to begin. The work on the identification of eutrophication problem areas, carried out between 1987 and 1997, was characterised by ongoing attempts to find scientific answers to problems with a heavy political overlay. This is very well reflected in the following quote from a paper by the World Wide Fund for Nature (WWF), submitted to the first EUT meeting in September 1996:

"WWF wishes to express its concern to those progressive countries who apparently believe that it will be possible, by force of scientific argument, to arrive at a Common Procedure for the identification of the eutrophication status of marine areas. WWF understands the frustration that comes from the absence of a Common Procedure. But where opposition is politically driven, and proof is unattainable, it will be possible to extend the technical debate indefinitely" (WWF 1996a).

However, one year later, at the 1997 Osparcom meeting, the Common Procedure was indeed adopted. When comparing the official version with the first drafts from the beginning of the 1990s, it is obvious that there are hardly differences with regard to the assessment criteria, despite the fact that these had been criticised for their poor scientific and indicative quality. The main differences with the first drafts were that the criteria had to be applied region-specific, and that first a screening procedure would be have to be applied. Important gaps still existed in the quantification of the assessment criteria and in the weighing and integration procedure, on the basis of which the actual identification would have to be based.

The Strategy to Combat Eutrophication: Implementing the Common Procedure

The implementation of the Common Procedure is officially part of the OSPAR Strategy to Combat Eutrophication, elaborated under the responsibility of the PRAM Committee, parallel to the work on the Common Procedure. The Strategy was adopted at the 1998 OSPAR Ministerial Meeting and is reproduced in Annex 3. During the drafting process, the geographical scope of the Strategy was broadened from the North Sea only to the whole OSPAR Convention area. The aim of the Strategy is worded as follows: "The Commission will implement this strategy progressively by making every effort to combat eutrophication in the maritime area, in order to achieve, by the year 2010, a healthy marine environment where eutrophication does not occur." The Common Procedure (see above) is the core of the Strategy because the actions, prescribed in the Strategy, must be carried out for eutrophication problem areas and potential problem areas only. Therefore, the first step in the Strategy was to implement the Common Procedure.

The implementation of the Common Procedure. According to the time frame of the Strategy, the Screening Procedure would have to be finalised by the year 2000 and the Comprehensive Procedure by the year 2002. The reason for this tight schedule was the fact that a Ministerial Meeting of the OSPAR Convention was scheduled for 2003, and the results of the Common Procedure were to be presented at this meeting (compare 5.3.2). The Screening Procedure was indeed finalised in 2000. It was not carried out by Germany, Denmark, Sweden, Belgium and The Netherlands. These countries had sufficient information to decide that for their marine areas the Comprehensive Procedure should be applied (EUT 1997). Remarkably, also the Comprehensive Procedure was finalised within the set time frame. To guide OSPAR contracting parties in the application of the Comprehensive Procedure, the Osparcom 2002 meeting had formally adopted assessment levels for several of the harmonised assessment parameters (Table 5.2; compare Annex 2, §4.2.1), as well as a common methodology to apply and weigh these criteria. The scheme for integrating and scoring the assessed parameters is in Table 5.3. A flow chart, in which the interrelationships of the assessment parameters are shown, is in Fig. 5.5. National reports on the first application of the Comprehensive Procedure were finalised in 2002. These national reports were the basis for an integrated report on the eutrophication status of the OSPAR area, which was adopted by the 2003 OSPAR Ministerial Meeting (OSPAR, 2003a).

Table 5.2. Harmonised assessment parameters and their area-specific assessment levels. Source: OSPAR (2003a)

Assessment Parameter and Respective Assessment Levels		
Category I	**Degree of Nutrient Enrichment**	
	1	Riverine total N and total P inputs and direct discharges (RID) Elevated inputs and/or increased trends (compared with previous years)
	2	Winter DIN and/or DIP concentrations Elevated level(s) (defined as concentration > 50 % above salinity related and/or region specific background concentration) [a]
	3	Increased winter N/P ratio (Redfield N/P = 16) Elevated cf. Redfield (> 25)
Category II	**Direct Effects of Nutrient Enrichment (during growing season)**	
	1	Maximum and mean chlorophyll *a* concentration Elevated level (defined as concentration > 50 % above spatial (offshore) / historical background concentrations) [a]
	2	Region/area specific phytoplankton indicator species Elevated levels (and increased duration)
	3	Macrophytes including macroalgae (region specific) Shift from long-lived to short-lived nuisance species (e.g. *Ulva*)
Category III	**Indirect Effects of Nutrient Enrichment (during growing season)**	
	1	Degree of oxygen deficiency Decreased levels (< 2 mg/l: acute toxicity; 2–6 mg/l: deficiency)
	2	Changes/kills in zoobenthos and fish kills Kills (in relation to oxygen deficiency and/or toxic algae) Long term changes in zoobenthos biomass and species composition
	3	Organic Carbon/Organic Matter Elevated levels (in relation to Category III.1) (relevant in sedimentation areas)
Category IV	**Other Possible Effects of Nutrient Enrichment (during growing season)**	
	1	Algal toxins (DSP/PSP mussel infection events) Incidence (related to Category II.2)

a Other values less than 50 % can be used if justified.

When comparing the assessment levels and the scoring chart with the heavy scientific requirements from the period 1992–1997, it is clear that the approach chosen is a very pragmatic one. Important parameters in the scheme are winter dissolved inorganic nitrogen (DIN) and dissolved inorganic phosphorous (DIP) concentrations and chlorophyll *a* concentrations (table 5.2). For each of these parameters, elevated levels are defined as 50% above region-specific background concentrations. Five years earlier, such an approach would have been strongly criticised on scientific grounds. First, deriving background levels was considered problematic, due to the lack of long term data. Second, the indicative value of nutrient concentrations for eutrophication phenomena was doubted and the use of fluxes had been recommended several times. Generally, an exceedance level of 50% was applied, for which no scientific backing was given and which can only be regarded as a political choice. Also the overall scoring chart (table 5.3) and the flow scheme (figure 5.5) have little bearing on the results of scientific research as described in 5.1.

Table 5.3. Integration of categorised assessment parameters. Source: OSPAR (2003a)

	Category I Degree of nutrient enrichment Nutrient inputs Winter DIN and DIP Winter N/P ratio	**Category II** Direct effects Chlorophyll *a* Phytoplankton indicator species Macrophytes	**Categories III and IV** Indirect effects/other possible effects Oxygen deficiency Changes/kills zoobenthos, fish kills Organic carbon/matter Algal toxins	**Initial Classification**
A	+	+	+	problem area
A	+	+	-	problem area
A	+	-	+	problem area
B	-	+	+	problem area
B	-	+	-	problem area
B	-	-	+	problem area
C	+	-	-	potential problem area
C	+	?	?	potential problem area
C	+	?	-	potential problem area
C	+	-	?	potential problem area
D	-	-	-	non-problem area

(+) = Increased trends, elevated levels, shifts or changes in the respective assessment parameters
(-) = Neither increased trends nor elevated levels nor shifts nor changes in the respective assessment parameters
? = Not enough data to perform an assessment or the data available is not fit for the purpose
Note: Categories I, II and/or III/IV are scored '+' in cases where one or more of its respective assessment parameters is showing an increased trend, elevated level, shift or change.

In the flow chart a positive relationship is given between nutrient concentrations and the concentrations of nuisance and toxic species. In 5.1.2 it was shown that both for *Phaeocystis* and *Chrysochromulina* such a relationship is not as straightforward as presented in the chart. According to the flow scheme, the final result of nutrient enrichment is a negative impact on the structure of the ecosystem, the latter a poorly defined ecosystem property.

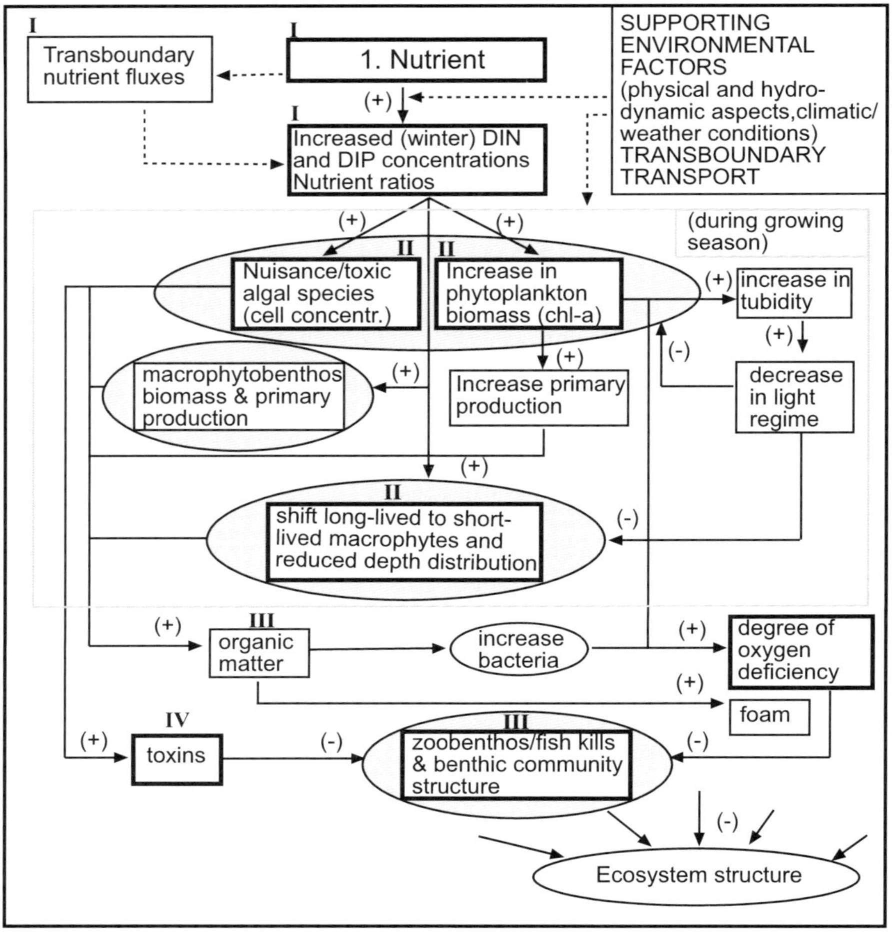

Fig. 5.5. Flow scheme of interrelations between the assessment parameters of the Common Procedure. Source: OSPAR (2003a). Main Interrelationships between the Assessment Parameters (in bold) of the Comprehensive Procedure. Parameters for which Assessment Criteria and their assessment levels are identified are shown in boxes with bold lines. Biological elements are shaded. Continuous arrow lines with (+) and (-) indicate "having stimulating effect upon", and "having inhibiting effect upon", respectively. Dashed arrow lines indicate 'having influence upon'.
Key:
I = Category I. Degree of Nutrient Enrichment (Causative factors)
II = Category II. Direct Effects of Nutrient Enrichment
III = Category III. Indirect Effects of Nutrient Enrichment
IV = Category IV. Other Possible Effects of Nutrient Enrichment.

Some principal problems of the Comprehensive Procedure became clear in its application to the Wadden Sea, an area for which, compared to most other parts of the North Sea, a very comprehensive data base exists. According to Van Beusekom et al. (2001), in which the results of this activity are comprehensively described, an integrated assessment should, ideally, be based upon causal links between causative factors and direct and indirect effects, taking into account the supporting environmental factors. However, whereas eutrophication in a strict sense, i.e. nutrient enrichment, can be described in an objective way, the definition of problems is to a large extent subjective. Moreover, it had been demonstrated that only few direct links between nutrient enrichment (riverine input) and effects could be established. Although for most phenomena, such as increased macrobenthic biomass, increased anoxia in sediments ("black spots"), increased macroalgal cover or decreased eelgrass stands, a certain role of eutrophication could be observed, a direct link that quantifies the effect of eutrophication on the undesired effects, could not be established. Many other factors, in particular climate and weather related parameters, influenced the occurrence or even triggered the outbreak of certain phenomena.

The report on the application of the Comprehensive Procedure also contained an elaborate chapter on the experiences of several countries, which generally underlined the above described problems (OSPAR 2003a). France had not used winter DIP and DIN concentrations because it regarded the link between nutrient concentrations and eutrophication too complex to define assessment criteria based upon nutrient concentrations. For the N/P ratio a reference value of 16 was fixed in the Procedure, but The Netherlands applied a value of 33 for its estuaries and coastal waters. Ireland had identified an average N/P value of 75 for its estuaries, due to the freshwater influence. The assessment parameter "elevated levels of phytoplankton indicator species" had not been applied by the UK because this country was still studying the use of phytoplankton as indicators of eutrophication.

According to the Comprehensive Procedure, the area classification should be done as a three-step process. The first step concerned the assessment levels of the agreed harmonised assessment criteria. The second step was to provide a score for the harmonised criteria, in accordance with the assessment levels, leading to an initial classification according to Table 5.3. The overall conclusion was that all parties had carried out this step in a harmonised way. The third step was to carry out an appraisal of all parameters in relation to the supporting environmental factors, i.e. region-specific characteristics in hydrodynamics, weather, climate and physical conditions. With regard to this step, it was concluded that the appraisal differed between parties "leading to a non harmonised final classification." On the basis of these overall conclusions, it was recommended to improve the assessment tools to allow for further harmonisation. These improvements related to the derivation of background values for specific parameters, the nature of the classification process and to research needs. It was also recommended to further develop tools, including numerical models, for the elaboration of nutrient budgets for specific areas, so as to account for transboundary transport of nutrients. Finally, the need for improving monitoring and developing guidance on monitoring frequency and coverage was underlined.

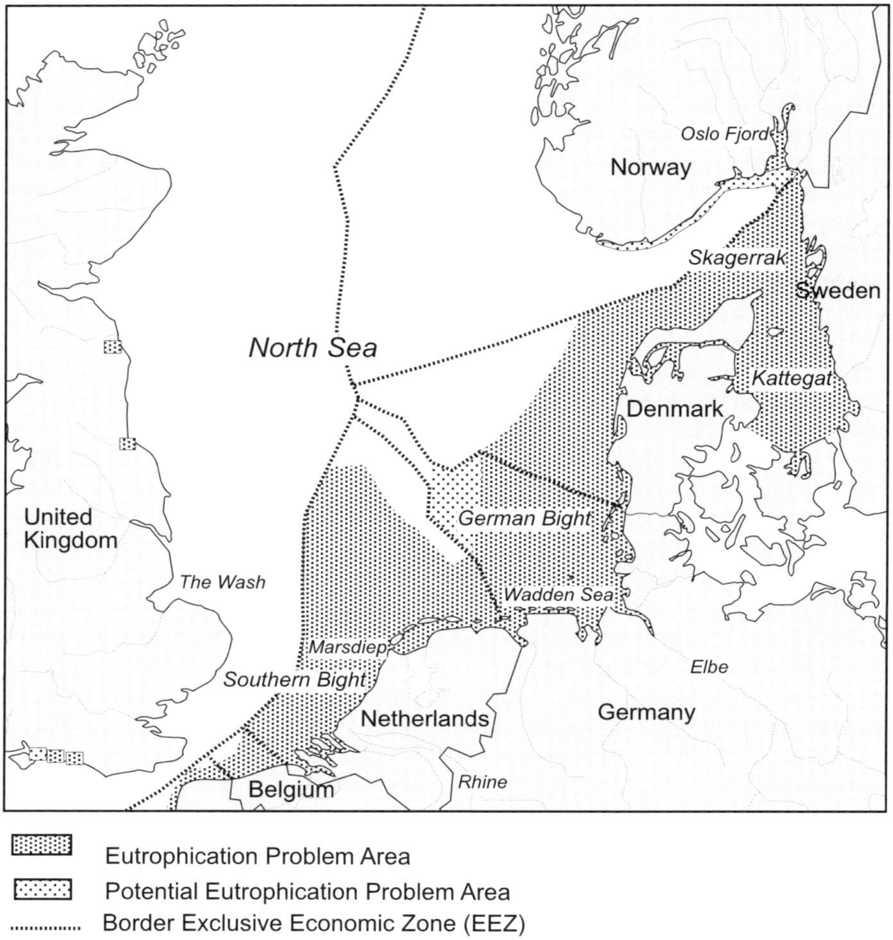

Fig. 5.6. Outcome of the first application of the Common Procedure in the North Sea. Based upon OSPAR (2003a). (Note: In the original maps from OSPAR (2003a) the part of the problem area in the Danish EEZ, bordering the central North Sea and the Skagerrak, has a different type of shading, for which no explanation is given in the document)

Analysis. Considering the fact that more than a decade had elapsed in which much effort had been given to elaborating a really harmonised, science-based identification system, it is remarkable how much the above conclusions and recommendations matched the situation of the end of the 1980s, beginning of the 1990s. Also the flow scheme and the scoring procedure are to a high extent comparable to the way in which during the period 1990–1992 the identification was carried out. Still, there are some substantial differences between the so-called "integrative administrative map" of 1992 (figure 5.4) and the outcome of the Comprehensive Procedure for the North Sea, which is shown in Fig. 5.6. The most striking one is the fact that almost the whole Dutch and a substantial part of the Danish Exclusive Economic Zones (EEZ) have been designated as eutrophication problem areas. This outcome is in strong contradiction with the scientific knowledge that has become available in the course of the 1990s (compare 5.1.3 and 5.2.2). It had taken fifteen years to agree upon a selection of areas to which the decision of INSC-2 to reduce nutrient inputs by 50%, would apply. Despite many efforts it had, however, not been possible to base this selection upon common harmonised scientific criteria, as is obvious in Fig. 5.6. The delimitation of eutrophication problem areas in the North Sea is one according to the national borders of the EEZ, rather than based upon common criteria derived from monitoring data and scientific research findings.

During the period 2003–2005, discussions on several principal and practical questions with regard to the Common Procedure continued, resulting in the adoption of an updated Common Procedure in 2005 (OSPAR 2005). The updated Common Procedure does not principally differ from the first version. It contains the quantified assessment criteria (table 5.2), already adopted by Osparcom in 2002, and applied in the first application (see above). In addition, there is more specific guidance on the integration of parameters and the classification of areas, as well as a procedure for dealing with transboundary nutrient transport. The update also contains a chapter on synergies between the Common Procedure and the classification of EU coastal waters under the Water Framework Directive (see further 5.3.3). The next classification of the maritime area is due in 2008.

The Strategy to combat eutrophication: ecological quality objectives

In accordance with the original decision of the Intermediate Ministerial North Sea Meeting (IMM 93, see 5.3.1), the Strategy consists of two categories of actions, the first related to the development of appropriate nutrient reduction measures (source-oriented actions), the second to evaluating the eutrophication status of the sea (target-oriented actions). According to the Strategy, the main elements of the target-orientated approach are an evaluation of the situation in the maritime area that is expected following the implementation of agreed measures, and the development of ecological quality objectives (the full text is in Annex 2, § 3.4). The source oriented actions, which must be carried out by the member countries of the OSPAR Convention, are listed in Annex 2, § 3.5.

In the remainder of this section, the target-oriented actions will be discussed in more detail. It concerns the development of ecological quality objectives and the activities necessary to evaluate the status of the maritime area after completion of the 50% nutrient reduction.

Ecological Quality Objectives. In the Strategy, the following "working definition" of Ecological Quality Objective is given:

1. "Ecological quality" is an expression of the structure and function of the ecological system, taking into account natural physiographic, geographic and cli-

matic factors, as well as biological, physical and chemical conditions, including those from human activities;
2. "Ecological quality reference level" is the level of ecological quality where the anthropogenic influence on the ecological system is minimal;
3. "Ecological quality objective" is the desired level of ecological quality relative to the reference level.

This set of definitions can be traced back to activities, following the decision of INSC-3 (The Hague, 1990) to request NSTF "to elaborate techniques for the development of ecological objectives and its coastal waters." In the Ministerial Declaration of INSC-3, this decision was part of the category "Enhancement of Scientific Knowledge," underlining that the development of ecological goals was regarded a predominantly scientific undertaking (4.4.4). Until 1995, three international workshops were held, at which the above mentioned common definitions and a conceptual framework were developed (Nordic Council 1999). The emphasis in the approach taken was on deriving parameters from basic ecosystem processes and their interaction with human activities. Due to the abstract, process-oriented approach, little progress was made in the development of concrete proposals, and the only common result was the above set of definitions.

In 1995 INSC-4 requested OSPAR to develop ecological goals for species and habitats and, by this, the North Sea states put more pressure on the process. It took, however, another four years before concrete proposals had been developed. These proposals were based upon a much more pragmatic approach and encompassed, among others, marine mammals, fish species with low reproduction potential and bird indicator species (Nordic Council 1999). Also ecological quality objectives (EcoQOs) for eutrophication (EcoQOs-eutro) were contained in this set. The development of proposals for EcoQOs for the North Sea had been carried out under the responsibility of the OSPAR Biodiversity Committee (BDC), which had requested the OSPAR Eutrophication Committee (EUC) to elaborate the EcoQOs for eutrophication. At INSC-5 (2002, see further 5.3.1) the EcoQO proposals were adopted for the North Sea as a pilot project. EcoQOs-eutro were formulated for five so-called "EcoQ elements." The EcoQ elements, together with their objectives are in Table 5.4.

The EcoQOs-eutro are the basis for further work within the OSPAR Strategy to Combat Eutrophication. The aim of the EcoQOs-eutro is to evaluate whether the 50% nutrient (N and P) reduction target will actually be sufficient to achieve, by the year 2010, a healthy marine environment where eutrophication does not occur (EUC 2001). The relationship of the EcoQOs-eutro with the assessment criteria of the Common Procedure is that the EcoQOs-eutro are a subset of the harmonised assessment criteria, indicated by the bold-lined boxes in the flow chart of Fig. 5.5. The non-eutrophication status of the year 2010 will have been achieved if the values of the EcoQOs-eutro are below the assessment levels, set for the assessment criteria, i.e. will be less than the 50% exceedance level.

Table 5.4. Ecological Quality Elements and Ecological Quality Objectives for eutrophication. Source: INSC (2002)

Ecological Quality Element	Ecological Quality Objective
Changes/kills in zoobenthos in relation to eutrophication	There should be no kills in benthic animal species as a result of oxygen deficiency and/ or toxic phytoplankton species
Phytoplankton chlorophyll *a*	Maximum and mean chlorophyll *a* concentrations during the growing season should remain below elevated levels, defined as concentrations > 50% above the spatial (offshore) and/or historical background concentration
Phytoplankton indicator species for eutrophication	Region/area - specific phytoplankton eutrophication indicator species should remain below respective nuisance and/or toxic elevated levels (and increased duration)
Winter nutrient (DIN and DIP) concentrations	Winter DIN and/or DIP should remain below elevated levels, defined as concentrations > 50% above salinity related and/or region specific natural background concentrations
Oxygen	Oxygen concentration, decreased as an indirect effect of nutrient enrichment, should remain above region-specific oxygen deficiency levels, ranging from 4–6 mg oxygen per litre

Scientific criticism. Because the EcoQOs-eutro are a subset of the common assessment criteria from the Comprehensive Procedure, and because their validation is directly linked to the validation of the assessment criteria, it follows that the criticism on the assessment criteria is also valid for the EcoQOs-eutro. This becomes very clear in the 2002 report of the ICES Advisory Committee on the Marine Ecosystem (ACE) (ICES 2002a). ICES had provided advice on EcoQOs in general because INSC-5 had invited OSPAR to work with ICES to make EcoQOs operational (see 5.3.1). Already in its 2001 report, ACE had given substantial criticism on methodological aspects of the EcoQO concept under development (ICES 2001). In the 2002 report ACE warned: "without substantial improvements in the rigour of the EcoQO framework, there is a risk that the framework may achieve no more than past scientific advisory and management frameworks" (ICES 2002a). According to ACE, EcoQOs should be:

- Relatively easy to understand by non-scientists and those who will decide on their use;
- Sensitive to a manageable human activity;
- Relatively tightly linked in time to that activity;
- Easily and accurately measured, with a low error rate;

- Responsive primarily to a human activity, with low responsiveness to other causes of change;
- Measurable over a large proportion of the area to which the EcoQ element is to apply;
- Based on an existing body or time-series of data to allow a realistic setting of objectives.

ACE referred, in particular, to Annex 3 of the INSC-5 Declaration, in which it was stated that the EcoQOs for eutrophication were an integrated set, and could not be considered in isolation. According to ACE, it would be necessary to develop rules for integrating the individual EcoQOs into a single clear message on necessary management action. That this was considered not to be easy was expressed with a phrasing with a high understatement character: "These rules may prove challenging to develop." An assessment of the eutrophication EcoQOs on the basis of the above requirements showed that most parameters were not considered very suitable for their intended purpose. ACE criticised that in the agreed harmonised assessment criteria from the Comprehensive procedure inadequate provision had been made for transboundary nutrient transport. According to ACE, this could dominate greatly over local anthropogenic inputs for some nutrients, particularly inorganic nutrients. ACE (ICES, loc.cit.) concluded:

"The assessment criteria appear likely to be difficult to put into practice on local scales, certainly in a consistent manner and possibly at all. No consistent rationale appears to have been developed for setting the boundaries of 'background concentrations' and 'elevated concentrations', and the values as currently tabulated may not form a basis for consistent action."

According to the ACE, chlorophyll *a* was the best proxy currently available for the amount of phytoplankton in water. At the same time, however, the chlorophyll *a* concentration showed considerable variation due to the species composition and the growth conditions of the phytoplankton. With regard to historical/background levels of chlorophyll *a*, the frequency of occurrence of elevated levels, and the presence and abundance of indicator species, ACE concluded that the information was spatially scattered, from inconsistent historical time periods, and likely to be of variable quality. The choice of a 50% criterion for elevated levels was probably set for reasons of pragmatism, and not based on scientific (risk) evaluation. ACE furthermore stated that the species, suggested as suitable indicators, were all normally occurring phytoplankton species in the North Sea, and there was no information available to show that these species did not occur under non-eutrophic conditions. ACE regarded the EcoQ element "Oxygen concentration" a sensitive metric for the production and mineralization processes in a water body, without, however, a clear link to eutrophication.

In 2003 Osparcom made a detailed request to ICES to provide advice on ecological quality objectives for eutrophication (EcoQOs-eutro). It concerned, among others, a review of the five ecological quality elements related to eutrophication (see table 5.4), as a means for their use as an integrated set, and as assessment criteria for the Common Procedure (see 5.2.3), as well as a reconsideration of the formulation of the EcoQOs and the consideration of new elements and objectives.

To this end, a meeting of an ad-hoc expert group, the Study Group to review ecological quality objectives for eutrophication (SGEUT) was convened in 2004. The outcome of this meeting was the basic material for the formulation of the ICES advice to OSPAR, by the 2004 ACE meeting. The advice, documented in the 2004 ACE report (ICES 2004a), basically contained the same criticism as formulated in the 2002 ACE report (see above). It was explicitly stated that the advice was a technical evaluation of the EcoQ elements only, and not an evaluation of the EcoQOs. The development of all five EcoQ elements into EcoQOs was supported but, according to ACE, additional work was required for each of the elements. Because the EcoQOs were to take account of area-specific aspects, such as historical background concentrations (compare table 5.4), ACE's major concern was that the widespread implementation would pose "major logistic challenges." For many areas, relevant data were lacking, and there was only partial guidance on the conditions under which data could be extrapolated from source areas to other areas (ICES, loc.cit.). ACE also noted that there was uncertainty about the appropriate spatial scale to be applied in the implementation of area-specific EcoQOs, and furthermore underlined to need to apply the EcoQ elements as an integrated set. ACE had, therefore, made specific proposals for each of the EcoQ elements, how to use existing information to set area-specific EcoQOs and, in case of lacking information, how to develop an appropriate information basis as soon as possible. The latter was motivated as follows: "With many management initiatives to address eutrophication planned for the near future, it is important to develop rapidly the scientific basis for EcoQOs to address eutrophication, and for approaches which use them formally as an integrated set."

Categorising eutrophication: Conclusions

The activities addressed in this section must be regarded as those with the highest political relevance. Whereas monitoring and assessment, described in Sects. 5.2.1 and 5.2.2, are to a high degree straightforward activities, in which the emphasis is on techniques and logistics (monitoring), or the integration and assessment of relevant scientific information (assessment), the development of criteria for judging whether or not there is a eutrophication problem is an activity with a high value content.

It had taken more than 15 years of intensive negotiations within OSPAR working groups to arrive at a common strategy to combat eutrophication, including (quantified) criteria, on the basis of which an analysis of the "status quo" of eutrophication problem areas can be carried out, and criteria which can be used to judge whether the strategy has been or will have been successful in the future. The overriding driving force in developing these criteria has been the desire to make them scientifically sound. The ICES advisory bodies (ACMP, ACME and ACE) have been a watchdog in this process, criticising developments and decisions that could not meet the criteria of scientific rigour. The end products of all this hard work do, however, hardly meet the criteria of scientific soundness. Like the 1993 map, the 2003 map of eutrophication problem areas seems to a high degree to be determined by national borders, rather than common scientific criteria. Also the ecological quality objectives, intended to determine when and where eutrophication must be regarded as problematic, both now and in the future, are the result of bureaucratic compromises, forced by political time pressure. This comes as no surprise, considering the complexity of the marine ecosystem, the increasing number of possible causes of phenomena, originally linked with eutrophication, and, most of all, the high political content of the questions, relevant for the management of marine eutrophication. In itself this must not necessarily be a negative outcome, as long as the results are useful and applicable in management. The criteria applied in the Strategy to combat eutrophication are, to a large extent, based upon pragmatic considerations, as is best illustrated by the 50% exceedance levels. On the other hand, however, the impression is given that both the choice of the criteria and their quantification have a sound scientific basis, as shown by the assumed causal links between nutrient inputs, concentrations, proliferation of toxic algae and oxygen depletion, in the flow scheme (figure 5.5). By doing so, it seems as if the Strategy is based upon a robust, scientifically sound qualification system. In reality, however, most of the parameters are not (directly) interlinked, nor do the value ranges assigned to them have a bearing on events observed in the ecosystem.

In a comprehensive study, aiming at developing proposals for Wadden Sea specific eutrophication criteria, it was concluded that, because of the complex interactions within the ecosystem, it was almost impossible to find causal relationships between so-called eutrophication phenomena and increased nutrient levels. The only plausible assumption that could be made was that increasing nutrient inputs would increase the risk of events, such as oxygen depletion or excessive macroalgal growth (Van Beusekom et al. 2001). Consequently, fixing ranges for nutrient concentrations, based upon pragmatic considerations, is probably the only practical solution to the value-laden question of how much eutrophication is acceptable. Moreover, practical experiences have shown that the only parameters that can be monitored relatively easily are nutrient inputs, nutrient concentrations and chlorophyll concentrations. As set out in the ICES advice about EcoQOs-eutro (ICES 2004a; see above), the five proposed EcoQ elements are usable in principle. There are, however, several conditions set to this usefulness. For some elements (zoobenthos, phytoplankton), the scientific basis has to be improved. Furthermore, for all elements, detailed area-specific information is necessary, among others to fix reference conditions. Another prerequisite is that all five elements are monitored in coherence, so as to be able to assess the links between them. Considering the experiences with the applicability of new scientific information and the development and application of a monitoring programme for eutrophication, comprehensively documented in this chapter, it must be questioned whether the EcoQOs-eutro will become practicable in the near or medium-term future.

5.2.4 Predicting eutrophication effects

What would be the effects of a 50% reduction of nutrient inputs? Already at the end of the 1980s this central question had been addressed by NUT (see 4.2.4), and it is also an important aspect of the Strategy to Combat Eutrophication (see 5.2.3).

The first NUT meetings had concluded that the then existing models were not sufficiently sophisticated to provide INSC-3 (1990) with reliable estimates. At the end of the 1980s also NSTF had begun to inventorise available models for their suitability for answering questions, related to the assessment of the status of the North Sea ecosystem, to be included in the 1993 QSR (5.2.2). The main fields in which models were to be applied were hydrography, and the role of nutrients in causing eutrophication effects (NSTF 1991a).

In 1991 lead country Belgium presented the results of a comprehensive inventory of available models, containing some 100 entries about local, regional and North Sea wide models. It was concluded that a quality assurance programme for models was necessary, and a proposal was made to compare five major models (NSTF 1991b). This testing was done in 1992 in a workshop under the auspices of NSTF. The main objectives of the workshop were to evaluate to which extent models could be used as tools in the preparation of the 1993 QSR, to predict the influence of various sources of nutrients in problem areas, and to model the transport of nutrients and the effects of nutrients throughout the North Sea (De Vries 1992). The results of the workshop, summarised in Chap. 2 of the 1993 QSR, very well reflected the limitations of modelling (NSTF 1993):

"The modelling exercise has resulted in rather consistent predictions of reductions in nutrient concentrations in the Dutch coastal zone, following a 50% reduction in inputs. The predicted reduction in concentration was 16–24% for winter nutrient levels with the annual primary production decreasing by a similar amount."

This text was, however, followed by a quite critical appraisal: "Careful interpretation of these results is needed, however, as some models do not simulate important processes involved in nutrient dynamics or physical features such as coastal fronts, which must have an influence on nutrient reduction scenarios." The QSR also addressed the "Achilles heel" of numerical modelling, being the validation of the models. According to NSTF (loc.cit.):

"Validation of models remains one of the most important issues that must be dealt with in the next few years. [.....] As a result of initiatives by the NSTF, the preparation of standardised data sets for the North Sea bathymetry, temperature and salinity is already under way [.....] However, there is a need for similar information on contaminants and nutrients inputs and concentrations. There is specifically a lack of data simultaneously observed in space and also in the form of time-series that are on scales comparable to those used in models."

What was addressed here was the problem of the lack of long-term data series with good geographical coverage, an issue also valid for the items discussed in the foregoing parts of this and the previous chapters. In 1996 a follow-up workshop on eutrophication modelling was held, this time in the framework of the preparation of the Convention-wide 2000 QSR, the development of the JAMP (see above), and the work on the development of the Common Procedure (5.2.3). The following specific requests, raised in the different OSPAR groups, were summarised in the draft report of the workshop (ASMO 1997):

- Where do elevated nutrient concentrations and fluxes cause an increase of phytoplankton blooms?

- Do changed nutrient concentrations cause changes in species composition and an increase of toxic/nuisance blooms?
- Do increased phytoplankton abundance or changed phytoplankton species composition result in altered zooplankton and zoobenthos communities, and ecological disturbance?
- How does zooplankton influence phytoplankton abundance and species composition?
- To what extent are temporarily low oxygen concentrations, caused by eutrophication, harmful to marine life?

This catalogue apparently exceeded the capabilities of the models, considering the actual issues addressed at the workshop, being an evaluation of model validation and an evaluation of the responsiveness of the models to natural variability and reductions in anthropogenic loads. Ten models were compared in the workshop, and it was concluded that all models had a "good to reasonable" fit for nutrient results when compared with coarse, synoptic data. There was, furthermore, a reasonable fit for chlorophyll results, although some models produced poor results for the spring bloom. With regard to the responsiveness of the models to a 50% reduction of nutrient inputs, it was concluded that the reduction in nutrient concentrations was maximally 35%, and the maximum reduction in both chlorophyll and primary production would be 30%. Most models could not make predictions about changes in the oxygen status. The calculated response for species composition varied considerably, both in extent and direction. With regard to potential contributions to the 2000 QSR, it was concluded that the workshop had shown that this was principally possible, but that specific questions would have to be asked by policy makers and translated into specific questions to be answered by the model. This is somewhat peculiar because several specific questions had already been addressed to the workshop. The specific input of modelling studies to the 2000 QSR, at least as far as eutrophication is concerned, appeared to be limited to the workshop results regarding the response of concentrations to the 50% input reduction (OSPAR 2000). This seems a rather poor result, the more so, since such a prediction had already been part of the 1993 QSR (see 5.2.2).

Parallel to the preparation of the 2000 QSR, work was done to implement the requirement from the Strategy to evaluate the situation of the maritime area after the fulfilment of the 50% reduction goal (5.2.3). In addition, such an assessment had to be carried out for the North Sea, in particular in the framework of the preparation of INSC-5, scheduled for 2002. In 1998 NEUT discussed a work programme for this particular action. A representative of the North Sea Conference secretariat pointed out that the information to be provided to the ministers would have to go beyond the QSR (NEUT 1998). It was realised that modelling work was required to provide answers to this request, but that there was insufficient time to build new models. It was, therefore, agreed to make use of the results of the 1996 workshop (see above) and the outcome of mesocosm studies. In this discussion, the political relevance of this activity was once again highlighted. According to the NEUT summary record "it was noted that the extremely large costs of nutrient/eutrophication control measures provided an incentive to develop reliable means for predicting the effects of nutrient reduction measures in order to reduce the risk of misdirecting resources" (NEUT 1998). The delegations of Denmark, Germany and The Netherlands, however, expressed concern that too much emphasis would be placed on the costs of measures, and stated that failure to produce suitable models should not be used as a justification for not preparing measures. The working programme, established by NEUT, listed several purposes of models. In the short term, numerical models should be used to support:

- Describing the situation that could be expected after reaching the 50% reduction target (see 5.2.3);
- The implementation of the Common Procedure (see 5.2.3);
- Further work on ecological quality objectives (see 5.2.3).

In the longer term, models should assist in defining additional reduction targets, in order to achieve a healthy marine environment where eutrophication does not occur, in describing and defining EcoQOs, and evaluating results in conjunction with the Common Procedure. According to the work programme, in the long term an OSPAR modelling system should be designed in order to have "a scientifically and administratively accepted tool to assist in the further implementation of the Strategy to Combat Eutrophication" (NEUT, loc.cit., Annex 10).

Following the NEUT decisions, the 50% reduction scenario was elaborated under the lead of Germany and The Netherlands and adopted by the OSPAR 2001 meeting. The results of the evaluation were also submitted to INSC-5 (see 5.3.1). The evaluation made use of the results of model studies, mesocosm experiments and observations of developments in the real world (EUC 2000). The effects on the assessment parameters of the Comprehensive Procedure were summarised as follows:

"1. for direct causative factors, a reduction of up to 25–30% in N and P concentrations in coastal waters is expected. Due to reduction measures being more effective for P than for N, the current increased N/P ratios in these waters will move towards normal ratios when reductions for N match those for P;

2. for the direct effect parameters, the expected effects are:
   - (i) up to 25–30% reduction of chlorophyll in coastal waters: and up to ca. 30% reduction in primary production in coastal waters;
   - (ii) for phytoplankton indicator species, a reduction in *Phaeocystis* bloom levels and in duration of its bloom and a decreased risks of toxic blooms;
   - (iii) for macrophytes including macroalgae in shallow waters, an improvement in occurrence and depth limits for long lived species (such as eelgrass and brown algae);

3. for the indirect effect parameters, the expected effects are:
   - (i) no pronounced oxygen depletion in normal climatic years and decreased risk of oxygen depletion in stratified coastal waters as well as in stratified offshore waters and sedimentation areas;
   - (ii) hence, decreased risk for benthic life;

4. the following effects are anticipated:
   - (i) a food supply that is still sufficient for higher trophic levels;
   - (ii) an improved quality of food supply (lower risks of nuisance and toxic algal blooms and oxygen deficiency);

(iii) an increased ecological efficiency."

As shown above, the conclusions for nutrient concentrations and chlorophyll *a* were mainly based upon model calculations, and reflected the general picture that had already emerged at the beginning of the 1990s. For the other parameters, however, the outcome of modelling exercises was not so consistent. For the evaluation of these parameters, use was made of the outcome of specific models, mesocosm experiments, regression analyses and field observations. The reduction of *Phaeocystis* levels was predicted in the Belgian MIRO model, which had been developed specifically for *Phaeocystis*. Hoever, at the 1996 modelling workshop (see above), the results of this model had not been proven to be consistent with most other models. Predictions for *Phaeocystis* in the Marsdiep were based upon regression analyses by Van Beusekom et al. (2001), which, seen in the light of the results of scientific research as presented in 5.1, must be judged with caution. The predicted improvement in depth limit for macrophytes was mainly based upon field observations in Danish waters where, due to low runoff in the years 1996 and 1997, an increase in visibility had been measured. Also for the indirect effects of nutrient reduction very little substantial support was available. The prediction of improvement in oxygen situation and the consequent improvement in benthic life, was based upon Dutch and Danish models and the above mentioned observations in Danish waters. For the anticipated effects, no scientific basis was given at all. Of the above predictions, only the reduction in nutrient concentrations and chlorophyll *a* in coastal waters were based upon comparable outcomes of several models and, at least partly, backed by field observations. The predictions for the other parameters, especially the anticipated effects, were not or very poorly substantiated. In the Progress Report to INSC-5 (Nilsen et al. 2002) the outcome of the evaluation was worded as follows: "Predictive methods suggest that the environmental conditions in the OSPAR area may improve by up to 25–30% as a result of a 50% reduction of nutrients for many coastal waters." According to the report this assumption was reinforced by the field observations in Danish waters in the years 1996 and 1997. It was, furthermore, acknowledged that considerable work would be needed to obtain more precise predictions.

As shown in this section, getting predictions about the 50% reduction targets at all, had already been very difficult. For the future work it would, however, not only be necessary to refine the predictions about the effects of a 50% reduction. As stated in the final paragraph of the OSPAR Evaluation Paper (EUC 2000), the 50% reduction target was based upon a policy decision. It was expected that a "more precise level of required reduction" would be established by implementing the Strategy to Combat Eutrophication and, in particular, the application of harmonised assessment criteria and the further elaboration of ecological quality objectives. According to the Evaluation Paper, the further development and use of predictive tools (modelling and mesocosm studies) would assist in this process. Considering the poor progress in the development of numerical models in the foregoing 15 years, especially with regard to predicting biological parameters, it must be doubted whether a substantial contribution to assessment criteria and

EcoQOs is plausible, the more so since the ecological basis for the latter is, as argued in 5.2.3, very meagre.

5.3 Political developments

In the previous sections, reference was made to political developments, in particular to the North Sea Conferences and OSPAR ministerial meetings. In order to be able to put the foregoing analyses into the proper political perspective, this section provides a consistent overview of relevant political developments in 1991–2003. It is divided into three main parts, addressing the main political actors with regard to combating and preventing marine pollution in the North Sea and the Northeast Atlantic Ocean. It concerns the North Sea conferences, described in 5.3.1, the ministerial meetings of the OSPAR Convention (5.3.2) and the European Union (5.3.3).

5.3.1 The North Sea Conferences

The 1993 North Sea Interministerial Meeting

At INSC-3 (4.4.4) it had been decided to convene a ministerial working group meeting in 1993, to address progress in the implementation of the INSC-2 and INSC-3 decisions, and problems encountered with regard to the nutrient and pesticides agreements. Therefore, also ministers of agriculture were invited. Another aim of the meeting was to discuss the 1993 QSR. At this so-called Intermediate Ministerial Meeting (IMM), which was poorly attended by both environment and agriculture ministers, it was acknowledged that the 50% phosphorus reduction goal would probably be reached by most countries, but that this was not the case for the nitrogen goal, mainly due to insufficient progress in reductions from agriculture (IMM §22). The meeting also discussed the so-called administrative map of eutrophication areas in the North Sea, which had been elaborated after a decision by INSC-3 (5.2.3; figure 5.4). Because the identification of the areas was based upon national criteria, Parcom was requested to develop a common identification procedure. IMM furthermore requested Parcom to develop a further strategy to combat eutrophication and to "consider the size and nature for further reduction targets for nutrients, in light of the above strategy, the QSR and additional scientific knowledge" (IMM §36–38).

Whereas INSC-3 had still been firm on the need for reducing nutrient inputs, the outcome of the IMM showed several signs of decreasing political interest in the issue of marine eutrophication. The IMM did not tighten the 50% reduction goal, as hoped for by non-governmental environmental organisations. Also the fact that the administrative map of eutrophication areas was not adopted as a common starting point for reduction policies, must be valued as an indication that marine eutrophication had gone down on the political agenda of the North Sea Conferences. At the IMM, the trend of referring important activities to Osparcom, which had begun at INSC-3, continued. Not only the decisions regarding the strategy, the need for further reduction goals and the development of common criteria support this observation, also the decision by the IMM to terminate the work of NSTF is consistent with this trend.

The fourth North Sea Conference

At the fourth North Sea Conference (INSC-4, Esbjerg 1995), the decrease in political interest in marine eutrophication and the transfer of actions to Osparcom became even more obvious than during the IMM. In §29 of the INSC-4 Declaration, it was made clear that the North Sea states would not meet the 50% reduction for nitrogen inputs within the period 1985–1995, as agreed at INSC-2 in 1987. Instead, reductions of 20 to 30% were expected. With regard to phosphorus, the North Sea states expected to meet the 50% reduction goal by 1995, with the exception of France with a reduction percentage of 25%. The UK had no obligation to report on nutrient reductions, since it was not committed to the INSC-2 decisions.

At the IMM, Osparcom had been requested to further develop a strategy to combat eutrophication in the North Sea and to report on the outcome of this work at INSC-4. However, within the working groups of OSPAR progress was slow, among others on the development of criteria for the identification of eutrophication problem areas (see 5.2.3). In §32 of the Declaration, the Ministers, therefore, only noted progress made by OSPAR and agreed on the main principles and elements of the strategy to comprise both a source-oriented approach and a target-oriented approach.

The increasing relevance of EU legislation, also for the marine environment, was illustrated by the initiative to designate the North Sea catchment area as a vulnerable zone under the EU Nitrates Directive and the North Sea as sensitive area under the EU Waste Water Treatment Directive (5.3.3), taken by Denmark, and supported by the European Commission, Germany, the Netherlands and Sweden. However, under pressure of the UK and France, this text was considerably weakened by the adoption of the clause that the decision does not apply to those parts of the North Sea "where comprehensive studies, to be delivered by 1997, demonstrate [....] that nutrient inputs do not cause eutrophication effects or contribute to such effects in other parts of the North Sea" (ED, §31). It is in this respect relevant to mention the decision, taken by INSC-3, that sewage treatment plants with secondary treatment would be mandatory, unless "comprehensive scientific studies demonstrate to the satisfaction of the competent international authorities, that this discharge will not adversely affect the North Sea environment on a local or regional level." At the IMM, it had been reported that only limited information from Norway, France and the United Kingdom was available, and that the assessment had not yet been completed. A report should therefore be delivered to INSC-4. However, there was no mentioning of such a report at INSC-4, nor was this the case at INSC-5 in 2002.

The decreased political interest in marine eutrophication is not only apparent in the lack of concrete results. It is also reflected in the increase in interest in other issues, most notably the protection of species and habitats and the impact of fisheries on the ecosystem, a development which had already started at INSC-3. That both fisheries and the protection of species and habitats were high on the agenda, is underlined by the fact that it was decided to hold an intermediate ministerial meeting in 1997, which should exclusively address these issues. Also for hazardous substances new political action was taken. At INSC-2 it had been agreed to reduce the inputs of hazardous substances by 50% between 1985 and 1995. At INSC-4 the so-called "One Generation Target" for hazardous substances was adopted, implying the cessation of discharges, emissions and losses of hazardous substances within one generation (25 years) (ED §17).

The fifth North Sea Conference

In 2002 the fifth North Sea Conference (INSC-5) was held in Bergen, Norway. It is the last International North Sea Conference addressed in this study and it was the last North Sea Conference to be held at all, underlining the decreasing political interest in North Sea matters and the increasing role of the OSPAR Convention and the European Union. The latter is well illustrated by the fact that almost the whole section on eutrophication in the INSC-5 declaration, covering only two pages out of 35, referred to relevant EU legislation, i.e. the Urban Waste Water Directive and the Nitrates Directive (see further 5.3.3). Seven years after INSC-4, it had to be acknowledged that the 1987 reduction goal for nitrogen had still not been achieved. This commitment was, however, reaffirmed (Bergen Declaration [BD], §60). Generally, the reduction of nitrogen inputs would have to be achieved through the implementation of the above mentioned EU Directives. There was one reference to Osparcom, in which OSPAR was called upon to complete the first application of the Common Procedure (BD, §61iii).

The main issues discussed by the ministers were "Integrated Ecosystem Management," the protection of species and habitats, sustainable fisheries and environmental impacts of shipping. In the framework of the political wish to further develop integrated ecosystem management, the importance of a coherent and integrated set of ecological quality objectives (EcoQOs) was stressed (BD, §4). It was agreed to develop EcoQOs for ten issues, among which four related to nutrients and eutrophication effects. For the North Sea, a pilot project was decided upon, in which ten parameters would be tested. Five of these were parameters relevant for marine eutrophication. OSPAR was the body responsible for the project, but it was explicitly stated that the work would have to be co-ordinated with relevant developments within the EU. It concerned the development of marine indicators by the European Environment Agency (EEA) and the development of environmental objectives in the framework of the implementation of the EU Water Framework Directive (see further 5.3.3). OSPAR was, furthermore, invited to review progress, in collaboration with ICES and other relevant bodies, "with the aim of adopting a comprehensive and consistent scheme of EcoQOs" (BD, §4vi).

For the first time in the history of the International Conferences on the Protection of the North Sea, which had started in 1984 in Bremen, no follow-up Conference was agreed upon. In Sect. XII of the INSC-5 Declaration, dealing with future co-operation, it was acknowledged: "for some issues the North Sea process can efficiently be continued in an equally fruitful way but on a much broader geographical scale in other fora such as OSPAR, and the EU thematic strategy on the protection and conservation of the marine environment of European seas" (BD, §81). Only for a few themes specific emphasis on the North Sea could be useful. Such a theme was the environmental impact of shipping, and it was therefore agreed that a ministerial meeting, dealing with this issue, would be hosted by Sweden in 2006 at the latest.

5.3.2 The OSPAR Convention

The 1992 Ministerial Meeting

In 1992 for the first time a ministerial meeting of the Oslo and Paris Commissions was held. The meeting formally agreed upon the merging of the two Conventions into one Convention for the protection of the marine environment of the Northeast Atlantic, the OSPAR Convention. The final declaration of the ministerial meeting contains a chapter, in which the priorities and objectives for the future work are set out. One of the priorities, which was a new political initiative and, interestingly, not one originating from the North Sea Conference framework, was the agreement to reduce, by the year 2000, the inputs to the marine environment of toxic, persistent and bio-accumulating substances to levels that are not harmful to man or nature, with the aim of their elimination. A second priority made was about the reduction of nutrients, for which discharges to areas "where the inputs are likely to cause eutrophication," would be reduced. The actions to be carried out were specified in an Action Plan. For nutrients, the Plan contained the well-known decision of INSC-2 of 1987 and the Parcom recommendation of 1988 (4.4.4), "to reduce by 50%, between 1985 and 1995, the inputs of nutrients from human activities to areas where these inputs are likely, directly or indirectly, to cause pollution." It is stressed here that this was a political decision only, just like the one taken at INSC-2, and a repetition of the intentions, already expressed in the Parcom recommendation of 1988. Other decisions taken at this meeting concerned the development of a new Joint Assessment and Monitoring Programme (5.2.1) and the restructuring of the working groups. An outline of the changes in the structure of the working groups is in Fig. 5.3.

The 1998 Ministerial Meeting

In 1998 again a ministerial meeting was held. This time the focus of the meeting, which was held in Sintra, Portugal, was on the protection and conservation of ecosystems and biological diversity, for which a new Annex to the Convention was adopted. For the guidance of the future work of the Commission, long-term strategies were adopted for hazardous substances, radioactive substances, eutrophication and the conservation of ecosystems and biological diversity. The Strategy to combat eutrophication has the objective to combat eutrophication in the OSPAR maritime area, in order to achieve and maintain a healthy marine environment, where eutrophication does not occur. According to the timeframe of the Strategy, such a situation should be achieved by 2010. To this end, the Commission would take

> "steps necessary to achieve by 2005, in parallel with the adoption of an integrated set of Ecological Quality Objectives for application in a pilot project for the North Sea, an agreement on any additional programmes and measures deemed necessary, including, as appropriate, further intermediate targets for specific areas and the further development of ecological quality objectives" (OSPAR 1998).

The Strategy is reproduced in full in Annex 3.

The increasing importance of European Union legislation and policies becomes clear from the statements in the Strategy that the implementation would take place within the framework of the obligations and commitments of the various contracting parties, among others the developing European Marine Strategy to Protect and Conserve the Marine Environment, the Water Framework Directive, the Urban Waste Water Directive and the Nitrates Directive.

The 2003 Ministerial Meeting

Above, the increasing influence of the European Union in matters of marine protection has been addressed several times. In 2002 a communication on the development of an EU Marine Strategy was published (see further 5.3.3), which was one of the main reasons for a joint meeting in 2003 of the ministers of the OSPAR and Helcom countries, to discuss their roles in relation to that of the EU.

In the Final Declaration of the joint meeting, this possible role and the potential for joining forces in the light of the expanding EU power, are cautiously worded as follows:

> "Changes in Europe – the enlargement of the European Union; the increasing interdependence of the marine environments of different countries; the ever-growing public interest in, and concern for the seas; the European Union initiative to develop a strategy to conserve and protect the marine environment – make it essential for us to develop and improve the ways in which we work together in HELCOM and OSPAR. In particular, we commit ourselves to work with the European Union initiative and, in collaboration with the other marine conventions, to extend and develop it, within our fields of competence, into a European Marine Strategy for the seas around Europe, which can receive the commitment of other Conventions and their Contracting Parties. Through such developments, we must exploit the possibilities for synergy between all the international bodies and national authorities involved. [....]" (OSPAR 2003b. JMM, §3).

Potential fields for co-operation between Helcom and OSPAR, identified at the joint meeting, were the ecosystem approach, conserving biological diversity, species and habitats, and the environmental impacts of fisheries and of shipping. For other major issues – eutrophication, hazardous substances, the environmental impact of the offshore oil and gas industry, radioactive substances, and monitoring and assessment of the marine environment – it was acknowledged that there were more substantial differences between the two regions.

The joint OSPAR-Helcom meeting was held in conjunction with an OSPAR ministerial meeting. One of the major issues of this meeting was the review of progress made in the implementation of the four OSPAR strategies, adopted at the 1998 Sintra Conference. With regard to the Eutrophication Strategy, the 2003 integrated report on the eutrophication status of the OSPAR maritime area was discussed. The ministers acknowledged that the identification of the eutrophication status was only partly based on shared criteria (see further 5.2.3), and that there was room for further improvement. It was, furthermore, concluded that there was a need for better delivery of the 50% reduction commitments that applied to problem areas identified, and that were necessary for achieving the 2010 goal of the Eutrophication Strategy. It was also agreed to review, how greater consistency could be achieved in identifying these areas and in quantifying the various anthropogenic contributions to inputs of nutrients to these areas. Interestingly, the development of ecological quality objectives (by 2005 for the North Sea and thereafter for other areas) was seen as one of the ways to integrate the OSPAR work on eutrophication with activities in the framework of the development of the European Marine Strategy.

The ministers also discussed monitoring and assessment matters. The 2000 QSR was welcomed and a number of serious problems, revealed in the QSR, summed up. These were the environmental impacts of fisheries, hazardous substances, the impact of climate change, the introduction of non-native species through ballast-water discharges, and the need to improve the knowledge base. Eutrophication was not mentioned, despite the fact that the 2000 QSR had classified it as one of the high impact issues (see 5.2.2). The strategy for a new Joint Assessment and Monitoring Programme (JAMP) was adopted, and the aim expressed that the implementation of the JAMP would be consistent with the monitoring of coastal waters under the EU Water Framework Directive. Finally, it was agreed that a new QSR would be published in 2010.

5.3.3 The European Union

Following the 1988 *Chrysochromulina* bloom, the European Council of Ministers had adopted a "Resolution on the protection of the North Sea and other waters of the Community" (see 4.4.4). This event can be regarded as the start of the active involvement of the European Community in marine protection matters. In 1991 two Directives were adopted on the reduction of nutrient inputs to all waters of the Community: the Urban Waste Water Directive and the Nitrates Directive. In 2000 the Water Framework Directive (WFD) was adopted, providing an integrated legal framework for the protection of the Community waters, which include all groundwater, surface water and coastal water systems, the latter extending up to 1 nautical mile from the baseline. Because of the major role of land-based inputs in marine eutrophication, these Directives are of direct relevance for the quality of the marine environment. In 2002 the European Commission published the communication to the Parliament and the Council "Towards a Strategy to Protect and

Conserve the Marine Environment," intended as the first step in the development of an EU Marine Strategy. A proposal for such a Directive was published 2005.

Below, these legal instruments are presented in more detail.

The Urban Waste Water Directive and the Nitrates Directive

The Urban Wastewater Directive. The objective of the Council Directive 91/271/EEC, concerning urban waste water treatment (Urban Waste Water Directive), is to protect the environment from the adverse effects of waste water discharges. The Directive was motivated by the above mentioned Council Resolution and the fact that there are member states with insufficient waste water treatment, which may affect waters of other member states. Generally, the Directive obliges all member states to ensure that all agglomerations are provided with secondary waste water treatment. For agglomerations with more than 15,000 p.e[13] this requirement must be implemented by the end of 2000, and for those with 2000–15,000 p.e. the deadline is the end of 2005. For so-called sensitive areas, to be designated by the member states, "more stringent" treatment than secondary treatment should be installed by the end of 1998 at the latest, for agglomerations with more than 10,000 p.e.. The reduction percentage for this type of treatment should be at least 80% for phosphorus and 70–80% for nitrogen. However, member states may also designate less sensitive areas. For waste water, discharged into such areas, primary treatment is regarded sufficient, together with proof from comprehensive studies that such discharges "will not adversely affect the environment" (article 6). This article is particularly relevant for the marine area because it concerns only discharges to coastal areas from agglomerations of 10,000–150,000 p.e. and to estuaries from agglomerations of 2000–10,000 p.e..

The Nitrates Directive. The Council Directive 91/676/EEC concerning the protection of waters against pollution caused by nitrates from agricultural sources, in short, the Nitrates Directive, has a much more specific scope than the Urban Waste Water Directive, and deals with nitrate emissions from agriculture only. The Directive reflects the growing concerns about nitrate pollution of, in particular, groundwater intended to be used as drinking water. But also the reduction and prevention of marine eutrophication is an objective of the Directive. Member states are obliged to identify waters, which are polluted or could become polluted, if no action is taken. It concerns freshwater systems used for drinking water extraction, groundwater, containing more than 50 mg/l nitrate, and freshwater lakes, estuaries and marine waters, which are, or could become eutrophic. All areas which drain into the identified waters should be designated as vulnerable zones. In these vulnerable zones agriculture must be carried out according to good agricultural practice, for which codes must be established. Basic elements of the code are contained in the Directive and are about the prevention and reduction of nitrate

[13] p.e. = population equivalent: the organic biodegradable load having a five-day biochemical oxygen demand (BOD_5) of 60g of oxygen per day.

emissions to water and air through, for example, regulations for applying manure and fertilisers and the storage of manure.

Implementation. The implementation of both Directives has, according to evaluations of the Commission, not been very successful. This is especially the case for the Nitrates Directive, which, contrary to the Urban Waste Water Directive, deals with the much more complicated problem of diffuse pollution. In an evaluation, carried out in 2002, the status of implementation of the Urban Waste Water Directive by the end of 1998 was presented (CEC 2002a). It was concluded that "considerable efforts" had been made to achieve compliance with the Directive, and that these efforts had already led to "significant improvements in the quality of a large number of European rivers and lakes." However, the verification had also revealed major shortcomings in most of the member states as regards compliance with the obligations of the directive, as follows from the quote below:

"For a large number of agglomerations, sometimes very large ones such as London and Paris, the level of treatment required for wastewater has been underestimated. Many of the Member States have not recognised the sensitive nature of the aquatic environments which receive wastewater. Apart from a failure to identify properly the sensitive nature of waters close to the point at which effluent is discharged, some of the Member States have ignored the fact that the pollutants contained in wastewater, which has not been properly treated, could migrate via the river basin into the marine environment. They have therefore not provided for the necessary treatment measures to tackle the problem of the pollution of estuaries or downstream stretches of rivers, caused by cities often situated far upstream in the river basin or to reduce the overall problems of marine eutrophication which are increased by all discharges from river basins, which flow directly or indirectly into marine waters. The North Sea, the Baltic and the Adriatic are, therefore, severely eutrophicated, but some of the Member States have not taken all necessary measures to reduce the pollution."

The implementation of the Nitrates Directive has proven to be even more problematic. In the implementation report of 1998 (CEC 1998) it was concluded that "Six years after the adoption of the Nitrates Directive most member states have failed to implement it." In the second implementation report from 2002 the picture was more positive. According to a commentary in the Journal Europe Environment it is thanks to legal pressure by the Commission that a "tangible improvement" in the implementation could be documented (EE 2002). In the report itself it was concluded that at least 30 to 40% of rivers and lakes showed eutrophication symptoms, or transported high nitrogen loads to coastal waters and seas, and that 50 to 80% of these loads were of agricultural origin (CEC 2002b). The report continued with the conclusion that "Following a delay of 5 years or more by Member States to fulfil their commitments for implementation of the Directive and an effective reduction of N losses from agriculture to water, a real improvement can be pointed out in the sensibilisation of Member States during recent years." However, also a word of caution was given: national action plans now covered 40% of the surface of the EU, but they were often inadequate, due to the absence of proper measures and the long time lapse between the introduction of measures and improvement in water quality.

The problematic implementation of both directives has probably been the main reason for the Commission to become more actively involved in the North Sea Conference and OSPAR frameworks (compare 5.3.1). The Commission representatives have tried several times to tighten the measures and time schedules for nutrient reduction policies in these bodies, so far, however, without much success. But, as will be shown in the following, the EU started to develop more direct competencies in the field of marine environmental protection.

The Water Framework Directive

The Directive "2000/60/EC of the European Parliament and of the Council, establishing a framework for Community action in the field of water policy," in short, the Water Framework Directive (WFD), covers all waters of the Community, i.e. inland surface waters, ground water and coastal waters. Since the beginning of the environmental protection activities of the European Community in 1972 (compare 2.6.2), many Directives and regulations dealing with water quality had been adopted. The overall purpose of the Directive is to place these sectoral instruments in a common framework with common objectives, monitoring and administration. The overall aim of the Directive is to achieve "good water status" for all waters by 2015. The WFD is mainly concerned with water quality, but also water quantity is addressed. The Directive will promote the sustainable use of water and contribute to the mitigation of the effects of floods and droughts. One particular objective of the WFD is the prevention and elimination of pollution of the marine environment, by phasing out discharges of hazardous substances "with the ultimate aim of achieving concentrations in the marine environment near background values for naturally occurring substances and close to zero for man-made synthetic substances" (CEC 2000, Article 1e). The activities to be undertaken to implement the WFD must be organised within so-called river basin districts, consisting of a river with its catchment area, including ground water and the associated coastal water.

An important element of the Directive is that environmental objectives for the different water types must be developed. These objectives must be achieved within 15 years after the entry into force of the Directive. To this end, the WFD contains a comprehensive Annex, in which normative definitions of high, good and moderate quality status are given, together with the so-called quality elements, i.e. the parameters to be used in the assessment and monitoring. The annex furthermore comprises detailed guidance for monitoring and the classification of the ecological status of different water types. Also a time frame is fixed, within which the quality elements must be quantified.

The implementation of the basic administrative requirements of the Directive, i.e. the selection of rivers basin districts, the setting up of co-ordinating bodies, the development of monitoring and assessment programmes, including the selection of suitable parameters and the quantification of the parameters, is now in full progress in the member states, and subject to a tight time schedule. In 2006 the monitoring networks must be operational and in this year an intercalibration of the ecological status classification systems must be carried out.

Towards an EU Marine Strategy

The increasing role of the EU in marine protection policies is best illustrated by the publication of a communication paper by the Commission in 2002, in which the main elements of a "Strategy to protect and conserve the marine environment" were outlined (CEC 2002c). Following extensive consultations about the communication paper, a proposal for an EU Marine Strategy Directive was published in October 2005 (CEC 2005). In the explanatory memorandum to the proposal the following rationale for establishing a marine strategy Directive is given:

"While measures to control and reduce pressures and impacts on the marine environment do exist, they have been developed in a sector by sector approach resulting in a patchwork of policies, legislation, programmes and actions plans at national, regional, EU and international level, which contribute to the protection of the marine environment. [...]. The general picture that emerges from this policy framework is a mixed one. On the positive side, some progress has been made in certain areas, e.g. in reducing nutrient inputs or pollution from hazardous substances in particular heavy metals. However, overall, the state of the marine environment has been deteriorating significantly over recent decades. [...] The current policy framework is not delivering a high level of protection of the marine environment. A strong, integrated, EU policy on marine protection is therefore required."

In article 1 of the proposed Directive, the Marine Strategy's aim is spelled out:

"This Directive establishes a framework for the development of Marine Strategies designed to achieve good environmental status in the marine environment [by the year 2021 at the latest], and to ensure the continued protection and preservation of that environment and the prevention of deterioration."

In order to achieve this goal, each member state must develop a marine strategy for its European waters (article 4). European waters have been divided into three marine regions, the Baltic Sea, the Northeast Atlantic Ocean and the Mediterranean Sea. For the preparation of the strategies, first an initial assessment of the current status of the waters concerned must be undertaken, together with the determination of what consists "good environmental status," as well as the establishment of a series of environmental targets and a monitoring programme for assessing these targets. Finally, a programme of measures must be developed, designed to achieve "good environmental status." It is, however, not the Directive's intention that member states start from scratch. According to article 5, member states shall coordinate their actions and use existing institutional structures. The establishment of environmental targets shall take into account the continuing application of existing environmental targets (article 9), while monitoring programmes shall build upon existing programmes. Interesting questions for the coming years are to what extent existing policies and programmes (for the Northeast Atlantic marine region those developed by OSPAR) will have to be adapted to new EU requirements, and what role science is to play in the implementation of the Directive.

5.4 Summary and conclusions

In this chapter, developments related to the implementation of the political decisions taken at INSC-2 and INSC-3 have been described and analysed. More in particular, the following questions have been addressed:

- How has the knowledge basis with regard to marine eutrophication developed after the second and third North Sea conferences (INSC-2 and INSC-3)?
- Has ecology been used as a means for justifying the decisions taken at INSC-2 and INSC-3?
- Has ecology contributed to the fine-tuning of decisions and the elaboration of management instruments, i.e. monitoring, prediction, assessment and validation?
- How has the science-policy network, in particular the science-policy interface, functioned with regard to the use of science in policy making?
- To what extent has new knowledge influenced the political status quo?

In the second half of 1980s the results of marine ecological research had played an important role in the formulation of political agreements to reduce inputs of nutrients to the North Sea (chapter 4). It was concluded that the political decision to reduce nutrient inputs to the North Sea by 50% was based upon instantly available knowledge, not tuned with the international scientific community, and with Danish researchers and administrators playing a central entrepreneurial role. A central question discussed in the present chapter is whether new scientific knowledge, that is knowledge that has become available after INSC-2 and INSC-3, was in support of these decisions. A related question is whether new knowledge has contributed to a reduction of uncertainty regarding the eutrophication process. The main conclusions with regard to these questions are presented in Sect. 5.4.1.

The second issue analysed in this chapter is to what extent new knowledge has been used in the development of management instruments. At the time the political decisions were taken, the knowledge available was limited and coarse, i.e. not yet sufficient for the fine-tuning of the political agreements. These political agreements were, therefore, accompanied by a catalogue of requests to OSPAR and ICES working groups to stimulate the generation of new knowledge, and to integrate and apply scientific information in the implementation work. The results of the analysis are summarized in 5.4.2.

In Sect. 5.4.3 a summary is given of the main conclusions with regard to the role of the science-policy interface, i.e. the relevant ICES and OSPAR working groups, in the implementation process. It concerns, in particular, the question how these groups have used their discretionary powers in dealing with value-laden issues.

In 5.4.4 the main conclusions with regard to political developments during the period 1991–2003 are presented, in particular to what extent new scientific findings have influenced political decision-making.

5.4.1 New knowledge, justification and uncertainty

The main conclusion with regard to the role of new knowledge (section 5.1) is that the new knowledge that became available in the 1990s and the beginning of the 2000s, generally put the role of anthropogenic nutrients in marine eutrophication, and the negative impacts of marine eutrophication into a perspective, quite different from that of the 1980s. Instead of supporting the general assumption that increased anthropogenic nutrient inputs had caused an increase in phytoplankton growth, the results of analyses of long-term data series underlined the importance of several other factors, among which relative changes in the composition of nutrients, the role of other growth factors, the role of light, denitrification and, most of all, climatic changes.

Politically even more relevant were new scientific insights concerning the scope and severity of eutrophication effects. The results of scientific reviews (5.1.2) showed that the impacts of marine eutrophication, or rather the phenomena linked with marine eutrophication, occurred on a small scale only, in bays and fjords. The predicted increase in the proliferation of toxic and nuisance blooms, as a result of changes in the N/P ratio, could not be substantiated. It also became clear that, generally, a fast recovery had occurred of the impacts of oxygen depletion events and toxic blooms. Moreover, evidence was presented of the relevance of factors other than increased nutrient loading, for the development of these phenomena. By far the most important is in this respect the impact of changes in climatic conditions, among others through changes in the overall wind speed and direction, the light regime and precipitation in the catchment areas of rivers. It is concluded that the implicit assumption of the rational management model, that more knowledge will lead to more support for policies, proved to be wrong for the case analysed in this study. More than that, new knowledge provided a basis for questioning the seriousness of marine eutrophication and the relevance of nutrient enrichment for eutrophication impacts.

Another assumption of the rational management model is that in the course of time new knowledge will lead to a decrease in uncertainty (compare figure 1.1). Also this assumption proved to be wrong: uncertainty about the relationship between nutrient loading and eutrophication effects increased, rather than decreased, as a result of new knowledge becoming available. The relevance of increased uncertainty for the implementation of the political decisions, i.e. the development of a nutrient monitoring programme, the development of criteria for the designation of eutrophication problem areas and the development of models to predict eutrophication effects, will be discussed in more detail in the following sections.

5.4.2 New knowledge and implementation

Following the political decisions of INSC-2 and INSC-3, several requests for clarification and specification of these decisions were forwarded to the responsible administrative working groups. Essential questions for which new scientific knowledge was considered necessary were:

- Which of the nutrients nitrogen or phosphorus was limiting primary production;
- Would the agreed 50% reduction of nitrogen and phosphorus be sufficient to prevent adverse eutrophication effects;
- Which areas were most affected by or potentially sensitive to eutrophication impacts;
- What would be the effects of agreed reduction measures on the marine ecosystem.

Four categories of implementation tasks have been distinguished in this chapter, based upon the working methodologies within OSPAR, ICES and North Sea Conference working groups. These are:

1. Monitoring nutrients and eutrophication effect parameters;
2. Assessment of the scope and causes of the problem;
3. Structuring and categorising the problem;
4. Prediction of the effects of reduction measures.

There is an important fifth category dealing with the development of specific reduction measures, amongst which sewage treatment and best environmental practices in agriculture. This category has not been covered because it has, generally, little direct bearing on marine ecology.

Monitoring

The development of an international programme for the monitoring of nutrients and other eutrophication parameters (5.2.1) started in 1993, eight years after the recognition of marine eutrophication as an international political issue. An important reason for this delay were disagreements about the technical and scientific contents of such a programme. Throughout the elaboration process, ICES had supplemented the responsible OSPAR working groups with scientific advice and criticism, the core of which was that the measurement of nutrient concentrations in seawater, which was the backbone of the draft programme, hardly provided information about the impact of nutrient loading on primary production. ICES was also critical about the monitoring frequency and the spatial resolution of sampling points. These factors continued to play a role after the 1993 decision to develop a mandatory nutrient monitoring programme, and are directly related to the scientific uncertainty about the fate and impact of nutrients in the marine ecosystem, which is a large, open and highly variable system.

Despite the ICES criticism, a nutrient monitoring programme was adopted in 1995, in which the measurement of nutrient concentrations was the central element. After the adoption of the programme, it took another two years before common guidelines had been developed, be it for nutrient concentrations only, and not for the other parameters, such as chlorophyll and phytoplankton species composition. But, as became clear in the following years, the guidance was insufficient with regard to the monitoring frequency and the spatial resolution.

It is concluded that in 2005, 10 years after its initiation, a common eutrophication monitoring programme was still not in full operation. As argued above, the problems with the development and execution of the programme were, first of all, related to scientific uncertainty. But also communication and co-ordination difficulties between several different national systems hampered the development and implementation of the programme. In addition, problems with data handling, i.e. the timely exchange and validation of monitoring data, occurred. Finally, the relevance of differences in the scientific and political "life-cycles" must be mentioned. With the decreasing political interest in marine eutrophication (see further 5.4.4), combined with cuts in most national budgets, the available resources for monitoring generally became less in the 1990s (compare De Jonge et al. 2006). This development contradicted the scientific demand for more quality of the programme, i.e. a higher monitoring frequency and resolution, necessary to account for the high variability of the marine ecosystem. Also relevant in this respect is the changing political playing field (5.4.4): As made clear above, in 2005 the OSPAR nutrient monitoring programme was not yet in full operation. Adaptations will, however, have to be undertaken because tuning with the requirements of the Water Framework Directive and Marine Strategy Directive will be necessary. In conclusion: because of the changing political constellation and, consequently, changing political needs, the programme will have to be amended before it has become fully operationable.

Assessment

Quality Status Reports (QSRs) are the official carriers for the transfer of knowledge from science to politics within the North Sea Conference and OSPAR frameworks. In 1993 the third North Sea QSR was published. This QSR had been prepared under the responsibility of the North Sea Task Force (NSTF), a liaison group of OSPAR and ICES, established in 1990 by INSC-3, with the task of enhancing the understanding of marine pollution.

The information presented in the 1993 QSR, only partly reflected the status of new knowledge concerning marine eutrophication, available at the beginning of the 1990s, including some of the more critical conclusions about the relation between nutrients and the proliferation of toxic algal blooms or oxygen depletion events, mentioned above. In the QSR's overall conclusions and recommendations to the North Sea ministers, the differences with the scientific state of the art were even bigger, and the status quo about nutrients and eutrophication effects, established in the 1980s, was largely confirmed.

In 2000 again a QSR was published, this time under the responsibility of Osparcom. For the various OSPAR regions, amongst which the North Sea, individual assessments were produced. In the North Sea QSR, drafted under the responsibility of lead country The Netherlands, there was a good reflection of the latest scientific findings, according to which the earlier assumed close relationships between nutrients and eutrophication effects and the seriousness of eutrophication impacts, were questioned. However, none of these new insights were part of the overall conclusions of the report. On the contrary: in the overall assessment chapter, eutrophication in the North Sea was placed in the category "highest impact on the ecosystem."

It is concluded that the overall conclusions of the QSRs, in particular the 2000 QSR, did not, or insufficiently, reflect the changing scientific insights about the seriousness of the marine eutrophication problem and, thus, provided responsible politicians with an assessment clearly in contradiction with the outcome of the latest scientific research. It is, furthermore, concluded that the conclusions and recommendations from both the 1993 and the 2000 QSRs were the result of negotiations between civil servants within the responsible working groups, rather than based upon an analysis and assessment of new knowledge. This finding underlines the increasing importance of "administration scientists" or "civil-servant scientists" within the science-policy network (5.4.3).

Structuring marine eutrophication

The political wish to structure the marine eutrophication problem, dates back to INSC-2 and INSC-3. It consists of two main elements, the designation of eutrophication problem areas and the elaboration of ecological quality objectives (EcoQOs). These two elements are the core of the OSPAR Strategy to combat eutrophication, adopted in 1998. It was also politically agreed that both the designation of eutrophication problem areas and the elaboration of EcoQOs would have to be based upon sound science and a common, internationally harmonised approach.

The central aim of the strategy to combat eutrophication is "to achieve, by the year 2010, a healthy marine environment, in which eutrophication does not occur." In the OSPAR Common Procedure, which is a formalised common approach to the designation of eutrophication problem areas, such areas have been defined as "areas for which there is evidence of undesirable disturbance to the marine ecosystem due to enrichment by nutrients." Both "undesirable disturbance" and a "healthy ecosystem" are value-laden concepts.

The necessity to designate the maritime area into areas affected by eutrophication or sensitive to eutrophication and areas for which this is not the case, has its roots in the decision of INSC-2 (London, 1987) to reduce by 50% inputs of nutrients to the North Sea "into areas where these inputs are likely, directly or indirectly, to cause pollution." In 1988 this decision was adopted by Parcom as a recommendation for the whole OSPAR Convention area. The practical relevance of the clause "into areas where these inputs are likely, directly or indirectly, to cause pollution" and, thus, its heavy political load, lies in the fact that nutrient discharges do not have to be reduced when they do not cause pollution.

The analysis in Sect. 5.2.3 has revealed the problems that occur with the application of science in the management of value-laden issues with a high political relevance. These problems were aggravated by the scientific uncertainty about the causes and effects of eutrophication. On the basis of the analysis, the following three conclusions can be drawn:

1. Science has been used as a means to delay negotiations. After the 1987 INSC-2 nutrient reduction decision, it took 10 years of negotiations to arrive at a Common Procedure and five more years to implement the Common Procedure. The use of scientific knowledge in resolving political disputes with high stakes is problematic (Miles, 1989; Nelkin 1987; Brickman 1987). This is even more so if

there is much uncertainty and the available knowledge not consensual, as is the case for marine eutrophication. Controversies about knowledge can in such settings be used to delay decision-making (Boehmer Christiansen 1994).

2. New knowledge has interfered with the goal of developing harmonised criteria. It had been the political wish of the IMM (1993) to base the designation upon a common approach. However, in the course of time new scientific knowledge became available, supporting the UK, which had all along the way maintained its position that many of the phenomena, connected with increased nutrient inputs, had multifactorial, and in many cases, natural causes. It was, therefore, agreed that criteria for eutrophication would have to be region-specific, which is, from a scientific perspective, the "correct" outcome. However, as has become clear from the first application of the procedure, regional criteria mean, in fact, national criteria. From the perspective of international eutrophication policies, this must be judged as a defeat because it had been the political intention to develop harmonised international criteria. This underlines the fact that new knowledge is not necessarily in support of running policies (5.4.1), as also observed by Miles (1989), who warned that, over the long run, the accumulation of knowledge may produce unanticipated consequences for management

3. The Common Procedure designation criteria and the eutrophication EcoQOs are based upon a pragmatic administrative, rather than a scientific approach. This goes for the choice of the parameters, as well as their quantification. Several of the selected parameters have little direct bearing on eutrophication impacts. With regard to the criterion for differentiating between non-eutrophication and eutrophication levels, a general exceedance level of 50% above so-called reference levels was adopted. Whereas reference levels can to some extent have a scientific backing, this is certainly not the case for the 50% exceedance level which is, as criticised by ICES, based on pragmatic, rather than scientific considerations. This is not necessarily a wrong approach, given the fact that a certain level of discretion by policy-makers is necessary when dealing with value issues (Jasanoff 1990). However, throughout the process of developing and applying the Common Procedure, the impression has been held upright that we deal here with a policy instrument with high scientific rigour. The following example illustrates the relevance of this observation. In a recent assessment of the quality status of the Dutch North Sea, carried out under the responsibility of the Dutch Ministry of Transport and Public Works, it was concluded that almost the whole Dutch part of the continental shelf was a eutrophication problem area, because of increased nutrient levels, nuisance and toxic algae, occasional oxygen deficiency, mortality of benthos and mussel infections (Zevenboom et al. 2003). This assessment is certainly not in conformity with new scientific knowledge about the eutrophication problem (compare 5.1). In fact, the only reference for these conclusions was the OSPAR document on the outcome of the first application of the Comprehensive Procedure (compare 5.2.3 and figure 5.6). The results of this first application have actually been used as an implicit source of scientific authority, although, as comprehensively described in 5.2.3, the procedure had been mainly an administrative exercise.

The EcoQOs for eutrophication have a direct relationship with the assessment criteria of the Common Procedure: the aims of the Strategy will have been achieved if levels in the marine ecosystem are below the 50% exceedance level. As agreed by INSC-5 (5.3.1), the EcoQOs will be tested in the North Sea. A comprehensive working programme for the pilot project has been set up by OSPAR. The programme foresees close co-ordination with the implementation of the EU Water Framework Directive (WFD) (5.3.3). It must be feared that the very complex process of developing EcoQOs for the OSPAR area will become even more complicated when also the WFD requirements will have to be accounted for. Another potential pitfall is the role of science: Like in the development of the assessment criteria for the Common Procedure, the emphasis of the pilot study is on science, with an important advisory role for ICES. The work programme does, however, contain a paragraph on communication with stakeholders, underlining the societal element of ecological quality objectives which are, in fact, constructs at the interface of science and society. Experiences with the use of EcoQOs in the Wadden Sea have shown that these may serve as a communication instrument between authorities and stakeholders in the discussion about conservation management (De Jong 1998; 2003). A prerequisite is that the objectives are understandable and their relevance for the ecosystem is supported by all parties. The eutrophication EcoQOs are still far removed from this ideal: on the one hand, they have mainly scientific appeal, on the other hand, the relationship with eutrophication-related phenomena is unclear. Finally, the implementation of EcoQOs in terms of monitoring and integrated assessment will very likely be too complex to be usable in practice, due to the high scientific demands.

Prediction

A political wish, expressed already by the end of the 1980s, was to know the effects of a 50% reduction of nitrogen and phosphorus. In 1998 this particular question became one of the requirements of the OSPAR Strategy to combat eutrophication. In Sect. 5.2.4 attempts by OSPAR working groups to provide answers to this question have been described and evaluated. It was concluded that, generally, models were able to give fairly comparable and consistent predictions of the expected concentrations of nutrients in the marine environment, following a 50% input reduction. The effects of these reduced concentrations on phytoplankton and zooplankton communities, the occurrence of toxic blooms or oxygen depletion events, however, could not be predicted with existing numerical models. Whereas hydrodynamic models have a good predictive capacity, the development of numerical models for the whole ecosystem, i.e. including biological parameters, has proven to be too complex. This is, of course, directly related to the complexity of the marine ecosystem, in which a host of different forcing factors influences the parameters to be investigated. As comprehensively described in 5.1, the results of new knowledge had added several new possible forcing factors for eutrophication events, in particular climatic changes, and thus increased the complexity of modelling.

A major problem in improving the reliability of models is the lack of long-term data of high spatial resolution, which are necessary for the validation of the models. With a view to the above mentioned monitoring problems, it is not to be expected that the data problem will be solved in the short to medium term. Despite the problems with ecosystem models, OSPAR is still optimistic about their possible role in the management of marine eutrophication, in particular in the implementation of the Strategy to Combat Eutrophication.

5.4.3 The science-policy interface

One of the most striking observations of this chapter is the increasing importance of the science-policy interface within the science-policy network, following the period of political decision-making. It is concluded that the civil servants working in the responsible working groups took over the ownership of the marine eutrophication problem (compare Hannigan 1995), at the same time excluding the scientific community and responsible politicians. The major consequence of this development has been a distortion of the feedback from science to politics through the blocking of the direct flow of new scientific information from the scientific community to responsible politicians. The scientific criticism on both the original political decisions and the implementation of these decisions has never reached the political realm. To this it must be added that there has never been massive criticism on the political status quo or the way the administration dealt with the marine eutrophication problem. Contrary to the global warming problem, stakes were apparently not high enough for a heated scientific debate.

What could have been the reason for withholding politicians or the general public the message that marine eutrophication was not as serious as originally assumed? The most plausible answer to this question is that already much political prestige and much money had been invested into implementing the decision to reduce nutrient inputs to the marine environment, and that, thus, these investments might seem to have been "in vain." Secondly, measures for reducing nutrient inputs to the sea are also relevant for improving the quality of freshwater and groundwater and these aims might become endangered by a possible relaxation of marine eutrophication policies.

Because new knowledge may not be in support of running policies (5.4.2), Miles (1989) pleaded for a system allowing iterative decision-making, i.e. learning by trial and error. It is obvious from the case of marine eutrophication that this was politically not feasible because high investments and political reputations might become questionable. The science-policy interface acted as a filter, rather than a feedback for messages, which were not politically opportune. On the other hand, the exaggerated picture of marine eutrophication, presented in the overall conclusions of the QSRs, which probably did reach the political realm, had not prompted additional political action in the form of a tightening of the reduction measures. In 4.4.5 it was argued that this was most probably the result of political fatigue: reducing nutrient emissions, in particular nitrogen, had turned out to be a very tough and time-consuming action without much political appeal. Moreover, because there had not been large-scale catastrophic events since 1988, there was no public pressure to introduce additional measures.

5.4.4 New knowledge and new politics

In the past 15 years, international eutrophication politics in the North Sea and Northeast Atlantic Ocean have been determined by three factors. First, there has been a decrease in political interest in marine eutrophication; second, there has been a shift of the political centre of marine activities from North Sea Conferences, via the OSPAR Commission to the European Union and, third, a broadening from sectoral to integrated policies has occurred.

At first sight the decreasing political interest in marine eutrophication seems a logical consequence of the fact that, after the political agreements taken in the 1980s, it was now time to implement these decisions, which is mainly a management and not a political activity. However, this was also the case for the issue of hazardous substances, for which in the 1990s new political action was taken. It does not seem plausible that the decreasing political interest was caused by the outcome of scientific research. The 1993 and 2000 QSRs both concluded that marine eutrophication in the North Sea was a serious problem, despite a substantial amount of new knowledge pointing into the opposite direction (5.1.3). It may be so that the latter information did reach responsible politicians, who thus did not consider additional measures necessary. But it is considered more plausible that the decreasing interest has been the result of the absence of catastrophic events, such as the 1988 *Chrysochromulina* bloom, as well as the highly problematic implementation of the 50% reduction target for nitrogen. The latter appeared to be closely connected with the massive problems of intensive agriculture in Northwest Europe, for which responsible ministers generally had no mandate. But also the introduction of nitrogen removal in treatment plants turned out to be a heavy burden for most administrations. With the exception of Denmark and Sweden, none of the North Sea and OSPAR countries have gone beyond the 50% reduction goal. The Netherlands originally had a policy goal of a 70% reduction of nutrient inputs to the North Sea, but this goal silently disappeared from policy papers in the course of the 1990s. Marine eutrophication had, apparently, developed into an issue without much public appeal, but loaded with heavy administrative burdens.

The second political development concerned the shifting of political responsibilities with regard to marine matters from the North Sea Conferences to the OSPAR Convention, which occurred in the course of the 1990s, followed by an increasing role of the European Union, which started at the beginning of the 2000s. These changes in the political constellation have negatively influenced marine eutrophication management, in particular the development of a eutrophication monitoring programme because, on several occasion, new political wishes with regard to the implementation process were put on the table before this process could be finalised. It will be particularly interesting to see whether it will be possible to develop a harmonised international nutrient monitoring programme, common ecological objectives and a common classification of the maritime area within the framework of the EU Water framework Directive and the EU Maritime Strategy Directive, which are, or will become, binding legal instruments, contrary to the North Sea Conference political decisions and the OSPAR recommendations.

The third political development concerns the increasing importance of integrated environmental policies. At the last regular International North Sea Conference (Bergen, 2002) an ecosystem approach to the management of the North Sea ecosystem was adopted. Also the European Marine Strategy will be based upon an ecosystem approach. Ecological science is to play an important role in the development and implementation of these strategies, underlining the fact that the political belief in the rational management model is still very much alive. In the next and final chapter, the possible role of science in the development and implementation of integrated ecosystem policies will be critically evaluated, on the basis of the findings from the previous chapters.

6 Summary, discussion and conclusions

"Science and technology are serious activities. They must not be diminished by accountants' pens or by politicians seeking unrealistic goals. Nor should they be distorted by scientist' lurid claims or promises." (De la Mothe and Dufour, 1995)

In this final chapter, the main findings from this study will be summarized and placed into an integrated perspective. In the foregoing chapters, I have attempted to answer several questions about the role of science in policy-making with regard to marine eutrophication. These were questions related to normative, structural and temporal aspects of the interaction between science and policy.

The normative aspect is about the model of rational policy-making, according to which scientific information is a necessary condition for decision-making and management (chapter 1). This model has been the central paradigm in environmental policies in the past 50 years (compare Brooks 1987; Nowotny 1987). A basic assumption of the rational policy-making model is that through the generation of scientific knowledge, the uncertainty about the problem under consideration is reduced, and at the same time information becomes available that can be used as a basis for political decision-making, for the justification of decisions taken and for the fine-tuning and implementation of decisions. The central question, addressed in the foregoing chapters, was whether marine ecology has indeed contributed to decision-making, and to the fine-tuning of political decisions in the implementation phase.

The analysis of the structural aspect of the science-policy interactions has focused on the development and functioning of a network for the exchange of information between the scientific and the political communities. The science-policy network consists of scientific, political and management bodies and actors, and is the structure for the transfer of knowledge from science to politics, and for political requests for scientific information into the other direction.

The policy life-cycle (chapter 1) has been the temporal framework for the analysis of the role of scientific information in the policy process. The policy life-cycle consists of three phases, i.e. the discovery phase, the decision-making phase and the management phase. In each of these phases science may serve different purposes.

In the following Sect. 6.1, the main conclusions with regard to the three aspects of the interaction between marine ecology and marine eutrophication policy are presented. The section starts with a description of the development of the different phases of the policy life-cycle for marine eutrophication, and developments within the science-policy network. What follows, are the main conclusions with regard to the actual role of marine ecology in marine eutrophication policies. In Sect. 6.2 the main conclusions are discussed in more detail, for the three phases of the policy life-cycle. In this section, also general conclusions about the rational policy-making model are drawn, and compared with other cases of international environmental problems. In the final Sect. 6.3 alternative models for the use of science in policy-making are presented, and discussed from the perspective of the main conclusions from the analysis of the marine eutrophication case.

6.1 Marine eutrophication in perspective

6.1.1 The temporal aspect

When was marine eutrophication discovered, and when did the issue enter the political agenda? What was expected of science in the different phases of the development of the marine eutrophication issue? In order to be able to answer these questions, it is first of all relevant to know that the history of marine eutrophication is closely linked to that of marine pollution. In the history of marine pollution two phases of increased international political interest can be distinguished. It concerns, roughly, the period 1970–1975 and the period 1984–1990 (figure 6.1). The first phase was preceded by a long period, during which the interest in marine pollution was confined to "discharge engineers" and marine scientists (chapter 2). After a series of serious pollution incidents, amongst which mercury poisoning in the Japanese Minamata, political and societal interest in pollution issues increased. This not only concerned marine pollution, but environmental pollution in general. The culmination of this interest, and a reflection of the changing attitude of society, was the 1972 Stockholm Conference on the Human Environment. Stimulated by the Stockholm Conference, a series of international regulations to control dumping and discharges of hazardous substances to the marine environment was agreed upon in the first half of the 1970s (figure 6.1). According to these regulations, there were substances for which dumping or discharges should be forbidden or eliminated, and substances for which discharges should be regulated. This clearly reflected the general feeling within the scientific community that the sea could be used as a medium to receive wastes, under certain scientific premises. For the Northeast Atlantic Ocean, the Oslo and Paris Conventions became responsible for regulating pollution from sea-based, respectively land-based sources.

A second "wave" of international political action on marine pollution occurred in Northwest Europe during the period 1984–1990, with a series of international political conferences of the North Sea littoral states (figure 6.1). These political conferences, of which the first three were held 1984, 1987 and 1990, had a strong impetus on the further development of the science-policy network in the Northeast Atlantic Ocean and the Baltic Sea.

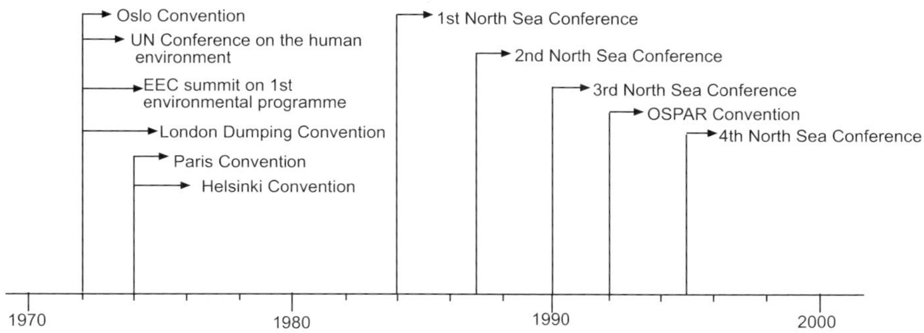

Fig. 6.1. Main political events with regard to marine pollution

Whereas during the first period of political interest in marine pollution, eutrophication was not considered an issue for which action was to be undertaken, such was clearly the case for the second phase. In 1987, the North Sea states decided to substantially reduce the inputs of both hazardous substances and nutrients to the North Sea. The nutrient reduction, which was to be in the order of 50% for the period 1985–1995, concerned the inputs of phosphorus and nitrogen compounds. In 1988, such a measure was adopted by the Paris Commission, as a recommendation for the Northeast Atlantic Ocean. In that same year, a similar decision for the Baltic Sea was taken by the Helsinki Convention. Also the European Commission became active in the field of eutrophication. Important legal instruments, adopted in the beginning of the 1990s, were the Nitrates Directive and the Urban Waste Water Directive. The changing political interest in marine eutrophication is schematically illustrated in Fig. 6.2, showing a steep increase in the mid 1980s, followed by a rapid decrease in the first half of the 1990s. This development is, first of all, relevant for North Sea eutrophication policies, for which in 1987 and 1990 concrete political decisions about nutrient reductions were taken. At the following North Sea political meetings in 1993, 1995, 1997 and 2002, no new political initiatives were taken. But also within the framework of the OSPAR Convention, the eutrophication policy of the end of the 1980s was merely consolidated in the years thereafter. Figure 6.2 also outlines the impacts of the increasing political activities on science and management. As analysed in Chap. 4, marine eutrophication research intensified as a result of political demands in the 1980s, following a long period of relatively low activity in the 1960s and 1970s (chapter 2). The political activities also initiated the policy field, as illustrated by the increasing management activities in the 1990s (chapter 5).

It is concluded that the history of marine eutrophication can be divided into three distinct periods, the discovery or pre-political phase, the decision-making phase and the management phase (figure 6.2), which is in good agreement with the policy life-cycle.

The second question asked with regard to the temporal aspect, was which type of contribution ecology has delivered – or was expected to deliver – in the three phases of the policy life-cycle. The various uses of ecology throughout the policy process, as identified and analysed in this study, are shown in Fig. 6.2. To what extent and in which way ecology has played a role in the different phases of the policy life-cycle is addressed in Sects. 6.1.3 and 6.2.

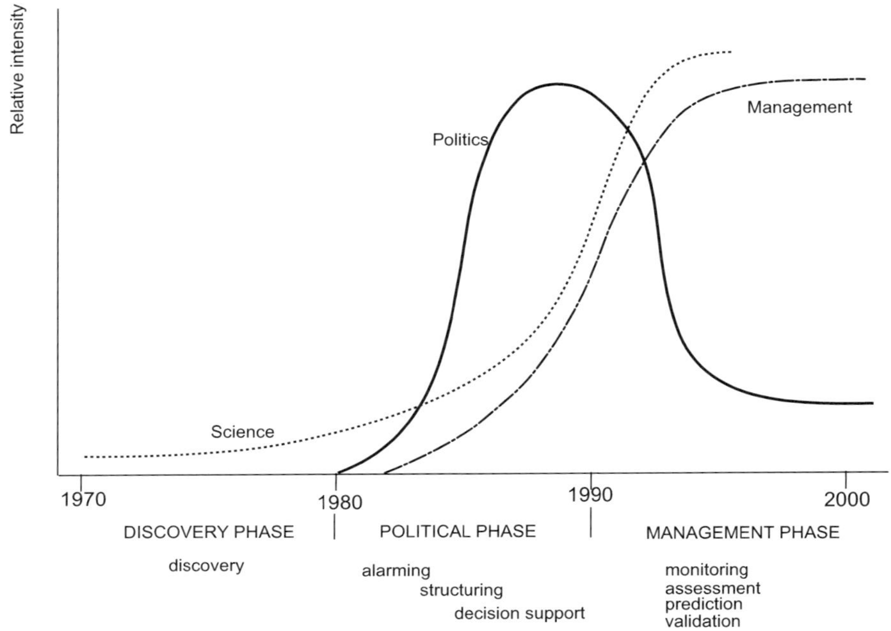

Fig. 6.2. The marine eutrophication policy life-cycle. The curves indicate the relative intensity of activities of science, politics and management. The politics and management curves are qualitative and based upon the analyses in Chaps. 3, 4 and 5. The science curve is semi-quantitative and based upon information presented in Chaps. 3, 4 and 5, supplemented with information from Nixon (1995) and Vidal et al. (1999). The main functions of science in the three phases of the science-policy cycle are specified below the time-axis

6.1.2 The structural aspect

The first wave of political action (see above) resulted in the establishment of a marine pollution science-policy network for the Northeast Atlantic Ocean, which centred around the Oslo and Paris Commissions (Osparcom), and in which ICES was the main body responsible for the delivery of scientific information (chapter 2). The science-policy network basically consists of three components. These are the scientific community, the political community and, in between, the science-policy interface. All three components, as well as their interactions, have undergone substantial changes during the period covered by this study. The structure of the network basically remained the same until the mid of the 1980s, but the second wave of political action initiated several changes (chapter 4). Most relevant for this study was the establishment of a marine nutrients and eutrophication working group of the Paris Commission (Parcom), the NUT group (NUT), in 1986. A second important change was the establishment of the North Sea Task Force (NSTF), a body under the joint responsibility of ICES and Parcom, in 1988. The NSTF was the result of a political decision of the second North Sea Conference in 1987 (INSC-2), to strengthen the science-policy interface with respect to North Sea pollution matters. In addition, in the 1980s some ICES technical working groups, dealing with issues relevant for marine eutrophication, were established, as a result of the increasing political interest in the issue. The creation of these new bodies at the interface of science and politics illustrates how, as a result of political action, the "normal" distance between science and politics was reduced and the science-policy connectivity enhanced. An important observation with regard to the bodies at the science-policy interface is that its membership consists mainly of civil servants with a scientific or technical background, i.e. ecologists and engineers. The science-policy network dealing with marine eutrophication, as it existed at the end of the 1980s, is in Fig. 4.4. It should be noted that several bodies within this network, among which NSTF, not only dealt with marine eutrophication, but also with other fields of marine pollution. But not only the structure, also the interactions within the network changed. In Chaps. 3, 4 and 5 this was illustrated for the interactions with regard to marine eutrophication during the period 1970–2003. As a result of strong political pulling by the North Sea Conferences, the demand for scientific knowledge increased. Consequently, there was an increasing need to integrate, elaborate and transfer this knowledge, as a result of which the activities within the network intensified. The changes in the network and the type of interactions between the three components in the network are schematically shown in Fig. 6.3.

In the discovery phase (chapter 3), awareness arose within part of the scientific community that excess nutrients might cause unwanted effects (figure 6.3a). The warning signal was, however, too weak to provoke political action. After a series of oxygen depletion incidents in Danish, Swedish and German coastal waters during the period 1981–1983, strong alarm signals were directed at politics, and the issue was placed on the international political agenda (figure 6.3b). One of the results of political action was the strengthening of the science-policy interface, by the creation of new working groups (chapter 4). The increasing importance of the science-policy interface, initiated by the political decisions of the second and third North Sea Conferences of 1987 and 1990 (INSC-2 and INSC-3), is illustrated in Fig. 6.3c. An important observation is that in the final phase of the policy life-cycle, the management phase, the main flow of activities is from the science-policy interface to politics. A second relevant conclusion is that there is no longer a direct link between science and politics (chapter 5).

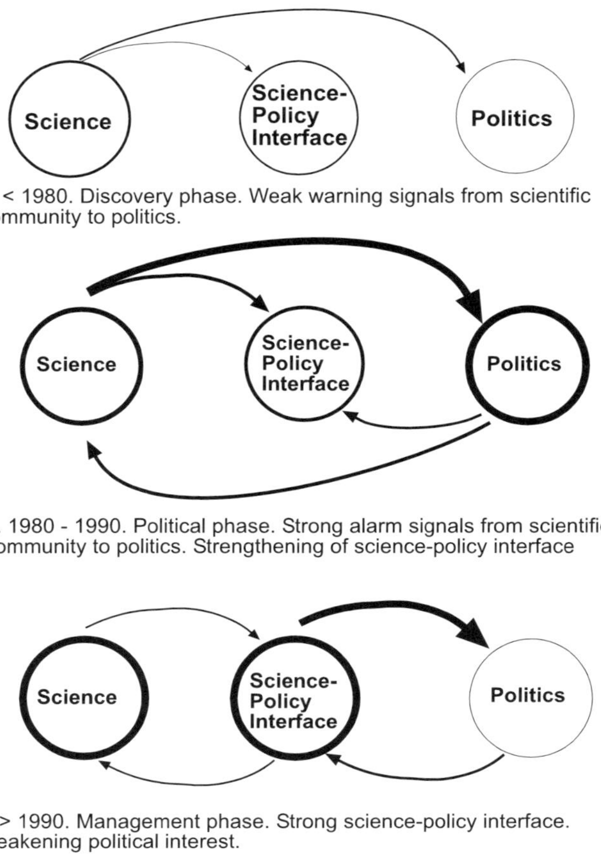

Fig. 6.3. Interactions within the science policy network. The line width of the arrows and circles is indicative of the relative intensity of the interaction within the science-policy network, respectively the relative activity of the separate elements of this network (compare figure 6.2)

The main function of the science-policy interface is to improve communication between science and politics (chapter 1). During the period 1986–2003, several inconsistencies and frictions between science and policy were successfully managed by the science-policy interface. My general conclusion is that the bodies, working at the science-policy interface, functioned well in making results of scientific research transparent for policy makers and a broader public. The most important carriers for scientific information to politics were the Quality Status Reports (QSRs), of which four were published in 1986–2000. Contrary to the first two reports, published in 1986 and 1987, the last two QSRs, published 1993 and 2000 and elaborated in the framework of, respectively, NSTF and the OSPAR Assessment and Monitoring Committee (ASMO), had an attractive lay-out and were relatively easy-to-read for laypersons. However, as shown in this study, the science-policy interface was much less effective in handling inconsistency and incompatibility problems between science and politics, related to uncertainty and complexity, different time scales and value-laden issues (see below).

6.1.3 The normative aspect

The analysis of the normative aspects of the interaction between science and policy has been the central aim of this study. For the normative aspect, the following questions have been investigated:

1. How has the notion of rational decision-making with regard to marine pollution and marine eutrophication developed in the course of time?
2. What has been the impact of science on the policy process?
3. Which contextual factors have influenced the impact of science?

Below, the main conclusions with regard to these three questions are presented. In Sect. 6.2 these conclusions will be discussed in more detail, and within the perspective of the temporal and structural aspects of marine eutrophication.

1. Changes in the rational decision-making model

During the period investigated in this study, some important changes have occurred in the notion of the use of science in the policy process. Above, it was concluded that there were two main phases of political interest in marine pollution, the first from1970 to 1975, the second during the period 1984–1990. Also with regard to the use of scientific knowledge, two main phases can be distinguished, which are closely related to the political developments. Whereas in the 1960s, there was a strong believe by scientists and engineers that marine pollution was mainly a scientific and technological matter, increasing pollution problems made clear that also a legal regime was necessary to control pollution (chapter 2). However, scientific knowledge was still considered a prerequisite for the implementation of these legal regimes, for example in the drafting of discharge licenses. By the end of the 1970s, the limits to the use of science in protecting the environment started to become questioned in a much more principal way. This discussion centred around two principles: the assimilative capacity and the precautionary approach (chapter 4). The assimilative capacity principle reflects the classical scientific conviction that the oceans have a certain capacity to assimilate wastes, and that the limits to this capacity can be determined on the basis of scientific research. The advocates of the precautionary approach questioned the ability of science to determine these limits, among others because of the complexity of (marine) ecosystems. They, therefore, favoured a precautionary attitude in dealing with waste discharges and dumping. After years of scientific and political battles, the precautionary approach was formally adopted at the second North Sea Conference (1987). The principle of precautionary action reads as follows:

"the principle of safeguarding the marine ecosystem of the North Sea by reducing polluting emissions of substances that are persistent, toxic and liable to bioaccumulate [....] especially when there is reason to assume that certain damage or harmful effects on the living resources of the sea are likely to be caused by such substances, even when there is no scientific evidence to prove a causal link between emissions and effects."

But this event can not yet be regarded as the end of the model of rational decision-making, or the start of a new era of precautionarity. As will be made plausible below, it must rather be seen as a transition phase. The transition aspect is clearly reflected in the 50% nutrient reduction decision of INSC-2. Although formally not taken under the aegis of the precautionary principle, the decision was, at least partly, the result of the spirit of the precautionary approach[14]. There was, however, a condition connected with the 50% reduction agreement that is still in line with the classical notion of rational decision-making and the importance of scientific proof. The 50% reduction of nutrient discharges only applied to those areas where such inputs might cause pollution. The how, where and what of nutrient pollution would have to be determined on the basis of scientific investigations. The 50% nutrient reduction agreement is, thus, a hybrid construction, consisting of two principally different views on the use of science in decision-making. As shown in the following section, the dualistic nature of the decision had a substantial impact on what happened in the management phase.

2. The impact of ecology.

Science has not shaped policies, but policies have shaped science so as to fit the policy process. This is the central conclusion with regard to the question whether ecology has had an impact on the marine eutrophication process. More in particular, the following questions have been addressed:

1. Has ecology reduced uncertainty with regard to the causes and impacts of marine eutrophication?
2. Has ecology been used as a basis for decision-making and for the justification of these decisions?
3. Has ecology contributed to the fine-tuning of decisions and the elaboration of management instruments?

A central assumption of the rational decision-making model is that, in the course of the policy life-cycle, scientific uncertainty about the problem under consideration is reduced. For the marine eutrophication case it was found that, generally, new knowledge, that is knowledge that has become available after decision-making, did not reduce the uncertainty about causes and effects. On the contrary,

[14] At INSC-2 the precautionary approach formally only applied to the decision to reduce the reduction of hazardous substances. There was much controversy about the seriousness of the marine eutrophication problem (chapter 4), and the 50% reduction decision as such can, thus, be valued as a reflection of the new precautionary approach. In later years, also marine eutrophication policies were officially placed within a precautionary framework.

the number of possible causes of the observed phenomena, in particular increased algal growth, increased.

A second assumption of the rational decision-making model is that decisions be based upon the best available scientific information. The analysis has shown that marine ecology has been very relevant for the discovery of the marine eutrophication, and that it has contributed to the agenda-setting of the problem and to the formulation of the decision to reduce nutrient inputs to the North Sea by 50%. This, despite the fact that there was scientific and political controversy about the seriousness of marine eutrophication and about the role of nutrients in causing eutrophication effects. There are three reasons for the fact that a decision on nutrient reduction was reached, despite these controversies. In the first place, it was the international "political mood" of the 1980s, which was very much in favour of a precautionary approach to environmental decision-making. The second reason was the above described dualistic nature of the decision, which provided opponents, in particular the United Kingdom, with an escape clause. Thirdly, the uncertainty surrounding the causes of marine eutrophication was artificially reduced by narrowing the scientific horizon, through focusing on nutrients only, and excluding other causes (see further below). This mechanism also contributed to the justification of the nutrient reduction decision, by giving it scientific authority.

The contribution of ecology to the fine-tuning of the political decisions and to the management of marine eutrophication has been very limited. Politicians had called upon marine ecology to provide a sound scientific basis for the causes and effects of the problem, and for criteria for the selection of areas sensitive to nutrient loading, as well as for ecological quality objectives for marine eutrophication. Also management instruments had to be put in place for the monitoring of nutrients and eutrophication impacts, and for the prediction of effects of nutrient reduction measures. Fulfilling these expectations proved to be very problematic, time consuming and, in some cases, almost impossible.

3. Contextual factors

The main reason for science not being able to contribute substantially to the policy process, lies in the incompatibility of the scientific and political processes. In Chap. 1, several contextual factors were discussed, which may potentially hamper the use of science in the policy process. It concerns complexity, consensus within the scientific community, differences in time scales between the scientific and political processes and dealing with values. Of these four factors, only the consensus within the scientific community proved not to be very relevant for the marine eutrophication case. This was, because in the course of time, the scientific community became increasingly excluded from the decision-making and management processes, while the influence of the science-policy interface increased (figure 6.2).

Complexity, differences in time-windows and dealing with values proved to be genuine incompatibility factors, although their relevance differed throughout the policy life-cycle. Whereas in the discovery phase only complexity was a relevant incompatibility factor, the decision-making phase had to deal with both complexity and time pressure. In the management phase, all three incompatibility factors hampered the expected use of science. Several processes and instruments to dealing with these incompatibility problems were identified. In the decision-making phase, problems with complexity and time constraints were solved by focusing on nutrients as the major cause of eutrophication problems, at the same time excluding other potentially relevant factors, most notably variations in climate. This process is called the narrowing of the scientific horizon. It was, to an important degree, facilitated by Danish scientists and civil servants, who played a central entrepreneurial role in the beginning of the decision-making phase.

As concluded above, all three identified incompatibility factors were relevant in the management phase, in which the science-policy interface was the dominant element in the science-policy network. It had the complicated task of combining demands for scientific rigour with demands for pragmatic management solutions, and it used much discretion in attempting to fulfil this task. The science-policy interface dealt with the increasing uncertainty about causes and effects of marine eutrophication by maintaining the limited "nutrient view," and excluding other possible causes. This was done by largely excluding the scientific community, and by disregarding new scientific knowledge, not in support of official policies (see also 6.1.2). Value-laden issues, mainly concerning the development of science-based objective yardsticks for judging the seriousness of marine eutrophication, were solved by negotiation, rather than applying new scientific facts. The results of this negotiation process, amongst which criteria for the designation of eutrophication problem areas and ecological quality objectives, were presented as based upon sound science, thus using science as a source of authority for justifying nutrient reduction policies.

The discrepancy in time frames between the scientific and political processes was most tangible in the management phase, during which political attention for marine eutrophication dwindled (figure 6.2). This caused pressure on resources to be used for management purposes, which affected, in particular, the development of an international eutrophication monitoring programme. A second relevant development was the changing political constellation, causing new demands to management instruments, while these had not even been finalised.

6.2 The discrepancy between science and politics

Above, three incompatibility factors between the scientific and political processes were briefly presented. In the next section, these three incompatibility factors are discussed in more detail for the three phases of the policy life-cycle.

6.2.1 The discovery phase

The discovery of marine eutrophication was, as with many environmental issues, mainly the result of scientific research. At first sight, the discovery phase may not seem to be part of the policy life-cycle. It is the science-dominated phase, preceding the policy process and, therefore, not impacted by frictions due to incompatibility problems. Only complexity is relevant in the discovery phase, and dealing with complexity is a typical feature of ecological research. But marine eutrophication science in the discovery phase was not exclusively academic science (chapter 3). There were several cases of problems with coastal eutrophication, among others in the Oslo Fjord and the Baltic Sea, and the science of marine eutrophication also dealt with practical questions, related to solving the excess nutrient problem. From this perspective, dealing with complexity is no longer an issue free of obligations. Complexity has to be matched with practical solutions and questions of feasibility. The experiences from the discovery phase, i.e. the period before 1980, appeared to be very relevant for the formulation of political decisions in the 1980s. From a policy and management perspective, the most important scientific questions dealt with in the discovery phase, were the seriousness of marine eutrophication as a pollution issue and the limiting factors for algal growth.

The seriousness of marine eutrophication

In the first half of the 1960s, organic pollution was an emerging marine pollution issue (chapter 2). It is closely related to marine eutrophication because through the breakdown of organic material, nutrients are released, which may stimulate phytoplankton growth. In the second half of the 1960s, there was a rapid increase in scientific interest in heavy metals and organic micropollutants (DDT, PCBs), and these pollution themes continued to be the dominant ones in the 1970s and the beginning of the 1980s (chapter 3). In the 1970s, marine eutrophication research comprised only 2% of the publications in the journal Marine Pollution Bulletin and 3% in the journal Ambio. The relatively low scientific interest in marine eutrophication was due to the fact that, contrary to heavy metals and organic micropollutants, so far no serious pollution cases resulting from anthropogenic nutrient inputs had been found. Negative effects of excess nutrient inputs had only been documented for small coastal areas, amongst which the Oslo Fjord. For large marine water bodies, such as the Baltic Sea and the North Sea, it proved to be very hard to separate the effects of anthropogenic nutrient inputs from natural causes. In the Baltic Sea, the oxygen deficits in the deepest parts seemed to be related to long periods of lacking water exchange with the North Sea, rather than increased phytoplankton growth due to nutrient loading. In the North Sea, substantial increases in nutrient concentrations could be documented in the 1970s, but these could hardly be related to changes in phytoplankton development. Data, collected by the Continuous Plankton Recorder (CPR), showed similar developments in phytoplankton and zooplankton stocks for the whole Northeast Atlantic Ocean. Finally, several researchers underlined the potential benefits of increased nutrient inputs in terms of increased primary and secondary production, rather than the negative aspects. For these reasons, marine eutrophication was not regarded a large-scale serious international marine pollution issue in the 1970s and the beginning of the 1980s.

The limiting factor

Contrary to the assessment by the scientific community about the (potential) impact of marine eutrophication, there was much controversy about whether phosphorus or nitrogen was limiting primary production. The scientific controversy was related to the complexity of the interactions between nutrients and phytoplankton growth. This was aggravated by problems with measuring primary production and the analysis of nitrogen compounds. Moreover, the nitrogen cycle is much more complicated than the phosphorus cycle. There was also a substantial influence on marine eutrophication theory from the much further advanced body of knowledge about freshwater eutrophication. Consequently, there was a scientific emphasis on dealing with P in the search for the effects of marine eutrophication. At the same time, part of the scientific community stressed that it was nitrogen that was limiting primary production. The political consequences of the emphasis on P were not unwelcome. Without doubt, P-removal was technically the easiest thing to do and, thus, the cheapest solution. There was, furthermore, already much experience with P-removal because it was the agreed policy for combating freshwater eutrophication. The political desire for P removal put much pressure on scientists and the scientific debate about the limiting factor (chapter 3). My conclusion is that, at the end of the 1970s, there existed a strong controversy about the limiting factor, inspired by both scientific and political points of view.

6.2.2 The decision-making phase

The general perception that marine eutrophication was a pollution issue of local interest and limited magnitude, rapidly changed in the first half of the 1980s, after the occurrence of large-scale oxygen depletion events in Danish, Swedish and German waters. It was especially in Denmark that the events created much public uproar and, consequently, the need for political action (chapter 4), but also in Germany and Sweden they reached the political agendas. In the middle of the 1980s, political pressure to start combating marine eutrophication was building up, also because The Netherlands and Belgium joined the Danish-Swedish-German stance on the need for nutrient reductions. Already in 1987, this resulted in the decision at the second North Sea Conference to reduce the inputs of nutrients to the North Sea by 50%. Seen from the general perspective of international decision-making, this was a remarkably rapid development (compare Döös 1994), even more so because at the start of the 1980s, marine eutrophication was not regarded a serious large-scale pollution issue, and because there was much scientific uncertainty about the role of anthropogenic nutrient inputs in changes in phytoplankton stocks. How, then, had it been possible to arrive at an international agreement to reduce nutrient inputs to the North Sea by 50% within such a brief period of time? As will be argued below, it was the result of a combination of the political mood of the 1980s, the narrowing of the scientific horizon and the postponement of unresolved conflicts about the seriousness of marine eutrophication.

The political mood

The oxygen depletion events would most probably not have had such international political impact, if they had not coincided with the second wave of environmental policies in Northwest Europe (6.1.1). The Scandinavian countries and the Federal Republic of Germany (FRG) played the central role in this new environmentalism, whereas the United Kingdom can be regarded as its main opponent. Stimulated by a comprehensive assessment of the quality status of the North Sea by the German Expert Council on Environmental Affairs of 1980, the FRG became the host of the first international Conference on the protection of the North Sea in 1984 (chapter 4). But marine pollution was not the only international environmental issue at stake. One of the major transboundary environmental themes was the acid rain problem, which gained international momentum at the beginning of the 1980s after the FRG joined Sweden and Norway in the desire for international action to reduce SO_2 emissions (Wetstone 1987). The second wave of environmentalism had three main characteristics. First, changes observed in the environment were associated with anthropogenic influences and were judged as negative human impacts. Second, precautionary action was to be taken. Third, there was a strong tendency towards coming to international agreements. Thus, the political climate for international environmental action was favourable, and it was within this climate that an agreement on nutrient reduction could develop, despite scientific uncertainties and disagreements, and political opposition by the UK. In the following, it is explained how the scientific and political hurdles were overcome.

The narrowing of the scientific horizon

The political process asks for clear-cut answers, delivered within rigid time frames. The complexity of the ecosystem, together with technical and coordination problems, generate uncertainty, controversy and time delays. The political "answer" to solving this principal discrepancy is through the narrowing of the scientific horizon and the use of "unripe" or "premature" scientific information. That was what happened in 1987, with the political decision to reduce nutrient inputs to the North Sea by 50%. The 50% nutrient reduction decision implied a narrowing of the scientific horizon by focusing on nutrients as the main cause of certain changes observed in the marine environment, most notably oxygen deficits and increase in phytoplankton blooms. This narrowing does not follow from the body of knowledge available by the end of the 1970s and the beginning of the 1980s (chapter 3). But although much scientific knowledge was available about phytoplankton dynamics, the role of anthropogenic nutrient inputs on phytoplankton development was not well understood. It was, however, this type of information that was most relevant for the policy process.

After the oxygen depletion events, research programmes were initiated by Denmark, Sweden and Germany, with the aim of finding causes and developing remedies. First results, which became available during the period 1984–1987, pointed to land-based anthropogenic nutrients inputs as the main cause of, or at least an important factor in the development of oxygen depletion. The Danish and

Swedish results underlined the importance of nitrogen for primary production, whereas the results of the German research programme emphasised the relevance of phosphorus (chapter 4). The suggested relation between excess nutrient inputs and increased algal growth was supported by scientific findings in Dutch and Belgian coastal waters of increasing blooms of the nuisance algae *Phaeocystis*, and increased secondary production in the Dutch Wadden Sea. The political decisions, taken at INSC-2, were based mainly upon the knowledge, generated in the above research programmes because these provided suitable starting points for political decisions, by focusing predominantly on nutrients. This type of knowledge may be termed "instant" knowledge, i.e. knowledge delivered within a very brief period of time and, contrary to academic science, not subject to extensive discussions within the scientific community or, in some cases, even to international scientific review.

An important implication of complexity is that multidisciplinary scientific information has to be integrated, aggregated, condensed and abstracted, before it can be used in the political process. It is in this process, which generally is subject to strong time pressure, that the spectrum of possible causes and solutions to a problem, proposed by science, is narrowed down to one cause and one cure. Most important for the process of aggregating the available scientific information was not the international scientific community, but the evolving marine eutrophication science-policy interface, in particular the nutrient working group (NUT). Danish civil servants played an essential role in integrating scientific knowledge into international political decisions. It concerned, in particular, the importance of international nutrient transport, the relevance of nitrogen in causing algal blooms and the fixing of a 50% reduction target for nutrients. This finding supports the conclusions of Hannigan (1995) about the importance of an entrepreneur in the construction of an environmental problem and the relevance of national scientific action (section 4.3).

Although focusing on nutrients only was indeed a narrowing of the spectrum of possible causes of increased phytoplankton growth, the 50% nutrient reduction must in one respect be valued as a broadening of the scope of policy options because it involved both phosphorus and nitrogen compounds. From a political point of view a reduction programme for phosphorus only would have been the easiest solution; first, because phosphorus removal is relatively easy and, second, because such programmes were already in operation in most North Sea countries for combating freshwater eutrophication. But the scientific evidence, especially from Danish research, for the relevance of nitrogen as the limiting factor for primary production, was too strong to be ignored, and nitrogen was included in the political decision.

An important conclusion of this study is that the narrowing of the perception of marine eutrophication was not based upon broad scientific consensus. In fact, the contrary was the case. In particular the ICES Advisory Committee on Marine Pollution (ACMP) had underlined the existing scientific uncertainty about the role of nutrients in phytoplankton dynamics. Also in the 1986 and 1987 North Sea QSRs, uncertainties about the role of nutrients and the influence of natural factors were underlined. In many scientific publications, especially those written jointly by British and continental scientists, there was consensus about the important role of other factors, most notably climatic changes. There was, however, a clear difference in opinion as to the need for precautionary action. Most continental scientists were in favour of reducing nutrient inputs, if only for precautionary reasons. They acknowledged the potential dangers of marine eutrophication, in particular oxygen depletion as a result of increased phytoplankton blooming, fundamental changes in species composition and the proliferation of toxic blooms. Most UK scientists did not share these fears and pointed to natural changes in the marine ecosystem and the strong tides along the UK coast, causing rapid dilution of nutrients. The dualistic stance of many scientists from the European mainland with regard to their scientific and political points of view, is reflected in publications in scientific journals and opiniating statements at conferences and in newspaper articles. In scientific journals, they agreed with their British colleagues that eutrophication effects must be regarded as multi-causal phenomena; in newspaper articles, they pleaded for the application of nutrient reduction. The 50% nutrient reduction decision is, thus, first of all, a political decision, though supported by a large part of the scientific community, inspired by the mood of precautionary action.

The narrowing of the scientific horizon has two political functions. It creates the scientific authority for decision-making and, by focusing on one cause and one cure, makes the process manageable or creates the impression that the process can be managed. An important implication of the narrowing process is that it must be sustained in the following management phase. The reason is that if new knowledge proves not to be in support of current policies, these may loose their rational justification. Especially in the 1990s, the phase of implementation of the nutrient reduction decisions of INSC-2 and INSC-3, there was a clear need for consolidation of the limited (nutrient-related) view of marine eutrophication. It was, in particular, the science-policy interface that played a central role in maintaining the limited view of marine eutrophication (chapter 5). How this was done, is further explained in 6.2.4

Postponing conflicts

The 50% nutrient reduction decision not only implied a focusing on nutrients only. It also acknowledged that marine eutrophication was an issue, serious enough for international political action. In other words: marine eutrophication had evolved from a minor problem of local relevance to a serious transboundary pollution issue. This was not in line with the general scientific perception of the issue (6.2.2), and can only be explained by the national impact, created by the oxygen depletion events. An important driver in the internationalisation of the issue was the outcome of Danish research, stressing the importance of transboundary nutrient transport (4.1.5), after which Denmark initiated international political action by calling for a consultation meeting under the Paris Convention in 1985.

Combating marine eutrophication has important societal implications. The costs of reducing nutrient inputs to the sea could only be justified if there existed a real threat and, as pointed out above, if this threat could be reduced and possibly taken away by the reduction measures imposed. The magnitude and scope of the problem, and the question whether the imposed reduction percentage of 50% during the period 1985–1995, would be sufficient for solving it, were issues for which there was severe political disagreement, and for which the "instant" knowledge had not delivered sufficient proof. They were, therefore, postponed by the second North Sea Conference, by the adoption of an additional paragraph, stating that the 50% reduction commitment only applied to those areas, where the nutrients would cause pollution (6.1.3). New scientific research would have to provide answers to the outstanding questions, where in the marine environment nutrients would cause pollution and, most importantly, what actually was meant by "pollution." In this case, the scientific dispute had not preceded the political decision, but was incorporated in the political decision to be solved in the future. In the next section, it will be shown that especially the last question proved to be hard to answer by scientific research because it involves value judgements about the seriousness of phenomena ascribed to increased nutrient inputs, such as nuisance algal blooms and oxygen depletion.

6.2.3 The management phase

The phase of political decision-making of the second half of the 1980s was followed by a phase of implementing these decisions. This so-called management phase was characterised by an intensification of the activities of the working groups of the science-policy interface, the generation of new knowledge about marine eutrophication and a decreasing political interest (figures 6.2, 6.3). The management process was concerned with monitoring nutrient inputs and concentrations, assessing the quality status of the marine environment at regular intervals, predicting the effects of nutrient reduction programmes and, finally, developing a process for validating the eutrophication problem. For all these tasks, new knowledge was considered necessary. After all, the 50% reduction decision was based upon limited existing knowledge. Targeted research programmes had only started in the beginning of the 1980s, and an internationally co-ordinated and scientifically sound nutrient monitoring programme was not yet available. The desire for new knowledge was clearly expressed in the political decisions of the North Sea Conferences (4.2.5 and 4.4.4). How the factors complexity, different time frames and dealing with values interfered with the use of knowledge in the management phase, will be further explained below for the management tasks monitoring and prediction, assessment, and validation.

Monitoring and Prediction

Monitoring is supposed to be the basis for decision-making and management. The regular collection of marine environmental data, and the assessment of these data, must provide policy makers with an overview of spatial and temporal developments in the marine environment, in this case about nutrients and eutrophication-related phenomena. Monitoring is, thus, a central element in rational policy-making. The setting up and running of an international monitoring programme for eutrophication appeared to be extremely difficult. This is due to two facts. First, the marine environment is a large and dynamic ecosystem, the monitoring of which requires sustained financial support and a high co-ordination and evaluation effort. Second, monitoring is an activity without political appeal.

The scientific requirements to an international eutrophication monitoring programme, provided by ICES, were hardly compatible with the national reluctance to continue to invest sufficient time and money in the sampling, analysis, quality control and evaluation of large amounts of data. The results of these conflicting interests were delays in the development of programmes and insufficient quality of programmes, the latter in terms of insufficient spatial and temporal coverage and, partly, inappropriate monitoring parameters. The incompatibility problem between scientific requirements and political interests increased with time: During the political decision-making phase, the support for developing a monitoring programme was high, but already after some five years political interest decreased, and with it the national willingness to continue to further develop and sustain such a programme. Especially in large and dynamic ecosystems, it is essential that monitoring is sustained over a longer period of time, typically more than ten years, in order to be able to detect changes in the measured parameters. A second factor interfering with sustained monitoring are changes in political and administrative structures. The most recent development is the increasing role of the European Commission in marine environment matters, imposing new and different requirements upon current OSPAR programmes. Due to the above described factors, a functioning international marine eutrophication monitoring programme in the North Sea and the Northeast Atlantic could not yet been established. The most important sources of information about developments in nutrients and phytoplankton were, and still are, the Helgoland Reede, the Marsdiep and the Continuous Plankton Recorder (CPR) programmes. None of these were originally intended to serve marine eutrophication policies, but had been set up for scientific research purposes.

Prediction is closely coupled with monitoring because the validation of numerical models, the main instrument used for predicting developments in the marine ecosystem, requires accurate field data, the latter to be supplied by monitoring programmes. Because of the problems with the development and execution of these programmes, the data, necessary to validate models, have proven to be insufficient in quality and quantity. The greatest problem for numerical modelling, however, is the complexity of the marine ecosystem. Despite the enormously increased computing and storage capacities of computers, it still is not possible to predict phenomena, such as algal blooms, with sufficient accuracy and reliability, to be applicable in management. MacGarvin (1995) has critically discussed the development of ecosystem theories for the marine environment, and concluded that the optimistic attitude of the 1950s and 1960s (compare chapter 2) has changed into a much more sceptical mood in the last two decades. According to MacGarvin (loc.cit.)

> "The fact that theoretical ecologists, working in far easier fields than marine ecology, are now asking such searching questions of their methods highlights how unreasonable it is to expect that we can predict the effect of human actions upon marine ecosystems with any accuracy."

Assessment

As argued above (6.2.2), there was a need to justify the relevance of nutrient reduction and the seriousness of the marine eutrophication problem. New knowledge was to deliver the sustained scientific basis for the political decisions. The analysis of the results of scientific research into marine eutrophication (section 5.2), clearly revealed that the generation of new knowledge does not automatically lead to suitable contributions to the policy process. In the 1990s, scientific information became available about many other possible causes of the observed changes. These were changes in nutrient ratios, changes in the light regime, polluting substances interfering with the grazing ability of zooplankton and, in particular, climatic changes. The scientific uncertainty about the relevance of increased nutrient inputs to the marine environment, thus, increased, instead of decreased, as a result of more research. Moreover, new research and new observations in the 1990s did not support the earlier assumed increase of negative impacts of marine eutrophication, and some researchers even underlined the positive aspects of increased nutrient levels. Finally, in the 1990s, increasing evidence became available that eutrophication problems were mainly caused by local inputs, thus questioning the relevance of international nutrient transport and, consequently, the need for international nutrient reduction policies. While these new scientific facts were becoming available, an increasing number of scientists expressed their doubts about the assumed high relevance of nutrients for the observed phenomena. Also the attitude to refer to the precautionary principle, slowly decreased. In fact, the scientific community again widened the horizon of possible causes and consequences of marine eutrophication. The widening of the scientific perspective was, however, not communicated to politics, and it was especially the science-policy interface that maintained the restricted perspective. The instruments used were the Quality Status Reports (QSRs), the main official vehicles for transferring scientific research results to the responsible politicians. Although the North Sea QSRs that were published in 1993 and 2000, contained elaborate scientific information, documenting both the limited role of nutrients and putting the seriousness of eutrophication effects into perspective, such was not or poorly reflected in the overall assessment chapters of both reports (5.2.2). On the contrary, the need for further nutrient reduction was stressed, and the 2000 QSR even qualified marine eutrophication as one of the major threats to the marine ecosystem. This finding underlines the fact that new knowledge is expected to contribute to the justification of political agreements. It also shows the central role of the science-policy interface in the delivery of this justification. Boehmer-Christiansen (1994) has stressed the importance of the science-policy interface to critically test scientific information before it enters the policy process. She stated to have more trust in the bargaining model of policy-making than in expert advice, provided the decision-making process is open, eclectic and pragmatic. Boehmer-Christiansen (loc.cit.) and Barisich (1989) warned of giving more mandates for creating regulations to experts. These could hide behind a "veil of rationality" (Boehmer-Christiansen 1994), while in the policy-making process many other factors are at stake, which should be made clear in the public debate. By increasing the number of political actors and types of expertise, more options can be debated (Nelkin 1987). However, an important finding of this study is that for the case of marine eutrophication, academic scientists, contrary to civil-servant scientists working in the science-policy interface, were very well able to view the problem in a broader perspective, and to change their judgement on the basis of new information. It was the science-policy interface that had used science as a means of justification and rational support for its activities. Young (1989) has underlined the relevance of the international scientific community for monitoring and compliance purposes, while Jasanoff (1990) and Lambright (1995) have stressed the importance of participation of the external scientific community for the certification of new knowledge. However, in the marine eutrophication case, the consensus within the scientific community about the limited relevance of nutrients in creating eutrophication problems, did not turn out to be a very relevant factor. The battle was fought within the science-policy interface, from which both academic science and other social actors were largely excluded.

What can be learned from the marine eutrophication case, is that it is important that the scientific and policy assessments are clearly separated. Ideally, the scientific assessment should be done by an independent scientific body, in this case ICES. This scientific assessment may be supplemented with a policy assessment, to be carried out by the science-policy interface. Both products must be submitted to the political and public arenas. By doing so, it will be clear for all stakeholders where and to what extent the science-policy interface has used its discretion in formulating policy recommendations.

Values and conflicts

The central conflict in the management of international marine eutrophication is not about the role of nutrients (see above), but about the scope of eutrophication effects. In other words: where does increased nutrient input cause problems and how can problems be defined? These questions were the legacy of the second North Sea Conference, at which the political disagreement about the seriousness of marine eutrophication and the need for nutrient reduction had not been solved, but postponed by transferring it to the science-policy interface. It was, in particular, the UK that considered eutrophication not to be a problem in its marine waters. The questions about the where and what of marine eutrophication effects were to be answered in the first place on the basis of scientific information (5.2). What followed after INSC-2, was a 15 year period of negotiations within OSPAR working groups, during which a Common Procedure for the selection of eutrophication areas was elaborated and applied. The result of all this hard work was a map with eutrophication problem areas, eutrophication non-problem areas and potential eutrophication problem areas, adopted by Osparcom in 2003. The political wish to base the designation of these areas on an internationally harmonised approach had, however, not been fulfilled. The map, produced in 2003, clearly shows (figure 5.6) that national considerations must have played the dominant role in the designation process, as was already the case 10 years earlier when a first map of eutrophication areas in the North Sea was rejected by the Interministerial North Sea Meeting because it was not based upon a harmonised approach (5.2.4).

The results of scientific research of the past 15 years (5.2,) and the critical stance of relevant ICES advisory committees (5.2.4), underline that the criteria and methodology of the Common Procedure are the outcome of administrative negotiations, rather than scientific review. This is not surprising, given the value-laden content of the problem. However, both the methodology applied and the results of the negotiation process, have been presented as being based upon sound science. Science is, thus, used as an authoritative basis for the continuation of the nutrient reduction policies. Recent examples of the use of this authority are the assessment of the eutrophication status of Dutch coastal waters (Zevenboom et al. 2003) and the German Wadden Sea (SRU 2004). In both reports, eutrophication is valued as a major pollution problem, using the OSPAR classification as an important basis for this assessment.

6.2.4 Comparison with other cases

The results of this study have made clear that the central assumptions of the rational policy-making model, i.e. that scientific information can be used to formulate decisions, based upon objective scientific information (rational decision-making), and, secondly, to help implementing these decisions (rational management), were clearly not valid for the case of marine eutrophication. In general terms, the following can be concluded about the use of ecology in international marine eutrophication policy:

- The generation of knowledge has increased rather than reduced uncertainty;
- In order to handle the problem of dealing with complexity and uncertainty at the political level, a simplification of facts has occurred, in this case focusing on nutrients as the main cause of the problem, at the same time excluding other possible causes;
- Both the limited scientific view (i.e. the nutrient view) and the exaggeration of the seriousness of the problem (impacts, scope) have been used as an authoritative basis for the justification of political decisions. Both were not supported by the majority of the scientific community;
- New scientific knowledge, not in support of existing policies, has been excluded from the policy process;
- The science-policy interface has been the central element in the simplification and exclusion processes.

Also for other large-scale environmental issues, the application of science in decision-making and management has proven to be problematic. In the first half of the 1980s, the so-called forest dying (Waldsterben) in central Europe was attributed to sulphur dioxide, although only limited data supporting this view were available (Boehmer-Christiansen and Skea 1991). This is a clear example of the narrowing of the scientific horizon, as became clear in the course of the 1980s, when several other causes of the phenomenon were found. Boehmer-Christiansen and Skea (loc.cit.) described the narrowing process as follows: "The phenomenon of sifting and selectivity is reinforced by the multi-layered process of review and interpretation. At each stage the information retained is likely to appear more conclusive, with countervailing indications and results being suppressed." The seriousness of the problem was furthermore exaggerated by statements that forests would die within five to fifteen years. Zierhofer (1999) has documented an ambiguous behaviour of researchers, on the one hand supporting such statements in the popular media, and, on the other hand, putting them into perspective in peer-reviewed scientific publications, a situation similar to what was found for the marine eutrophication case.

A second example of the narrowing of the scientific horizon, also from the 1980s, is the case of fish diseases and marine pollution. Especially German researchers emphasised that polluting substances, in particular heavy metals, were responsible for certain types of fish diseases, whereas UK scientists underlined the relevance of natural factors. In the run-up to the second North Sea Conference, a mood developed, in which supporting the latter view was "politically incorrect" (compare 4.2.3). But, as demonstrated several years later by Vethaak (1992), factors other than pollution, such as stress due to changes in salinity, were very relevant for the development of certain types of fish diseases.

Overfishing is another case for which a narrowing of the scientific horizon can be documented. According to Corten (2001), the increase in fishing intensity after World War II led to a situation, in which very little attention was given to the subject of natural variability. This is reflected in political decision-making, which is almost exclusively focused on the role of fisheries in the development of fish stocks. For the North Sea herring, Corten (loc.cit.) has shown that natural factors, in this case changes in climate, are very relevant as well.

Climate change is a topical case, showing many similarities with marine eutrophication, in terms of the use of scientific knowledge. Also here, a narrowing of the spectrum of possible causes occurred, focusing almost exclusively on "greenhouse gases," most notably carbon dioxide, and largely excluding natural factors, such as solar forcing. In the technical summary of the IPCC 2001 scientific report (IPCC 2001), only a few sentences are spent on the relevance of solar forcing. In recent years, the indications of the importance of solar forcing for climate change have been building up (Haigh 2001; Karlén 2001; Shindell et al. 2001, 2003; Zorita et al. 2003). Despite such clear indications of natural forcing factors, there seems to be a reluctance to seriously discuss this aspect of climate change. In 2001 Karlén published a debate article in the journal Ambio, in which he concluded that changes in solar irradiation had been the dominant causes of changes in climate, and that CO_2 had become an additional forcing factor in the past two decades (Karlén 2001). Comments to his contribution have, to my current knowledge, not been submitted. In an article in the German weekly journal "Die Woche," a stratosphere researcher of the Freie Universität Berlin complained that the solar forcing theme was "taboo" in Germany (Verseck 2000). The heavy critique on Lomborg's critical stance on official climate change policies (compare Lomborg 2001), is another indication of the narrowing process, attempting to exclude alternative and deviating positions, forced by the political need to support the Kyoto process.

There is no doubt about the fact that, in the cases described, the human factor is relevant. Nutrients are essential for primary production, acid deposition does influence soil biology, and CO_2 is a forcing factor in climatic processes. However, as abundantly made clear in this study, complex processes in large ecosystems are not determined by one parameter only. The ideal of H.P. Odum that such processes can be managed by operating the proper valves (chapter 2), has proven to be an illusion. As shown above, this illusion is still being held upright by many academic and regulatory scientists, and used by administrators and politicians as an authoritative basis for decisions.

6.3 The future of rational policy-making

The weaknesses of the model of rational policy-making have since long been made clear by science analysts (Collingride and Reeve 1986; Brooks 1987; Nowotny 1987; Jasanoff 1990; Funtowicz and Ravetz 1993; Collins and Evans 2002). Some researchers even speak of a crisis in the use of scientific expertise (Horlick-Jones and De Marchi 1995), and an eroding trust in science by policy makers and the general public (Nowotny 2005). However, politicians and policy makers continue to ask for and apply scientific information, as has become abundantly clear from the political initiatives within the EU with regard to the development of a marine strategy and the implementation of the Water Framework Directive (5.3.3). In the Communication of the European Commission to the Council and the Parliament about the establishment of a European research area (CEC 2000a), a paper basically concerned with strengthening the global position of EU scientific research, the following is stated with regard to the public research effort:

> "Research plays a central role in the implementation of public policy and it is also at the heart of the policy-making process. In areas such as health, sustainable development or industrial, food and nuclear safety, policy options and decisions must be based on more solid scientific knowledge and a full and proper understanding of the social and economic aspects surrounding the problems in question."

Interestingly, also "full and proper understanding" of social and economic aspects is required. In recent years, politics have called for an ecosystem-based approach to management (see for example the fourth North Sea Conference [5.3.1] and the EU [5.3.3]). As documented in this study, research of large and open ecosystems is connected with high uncertainty and high monitoring and assessment efforts. For the management of marine eutrophication these efforts have, so far, turned out to be too high to result in a functioning monitoring and evaluation approach. When also social and economic aspects must be integrated into management, this will lead to requirements to knowledge and the application of knowledge, which go far beyond those of today, and which must be judged as unrealistic. Alternatives to the rational management model, which may possibly accommodate the requirements of future public environmental policies, will be critically discussed in the following, using the outcome of the marine eutrophication analysis as a reference.

6.3.1 From normal to post-normal science

The problems with using "normal" science for public policies have proven to be so fundamental that substantial changes to the rational decision-making model have been proposed. Several authors distinguish between normal science (science carried out within the regular academic settings, or applied science) and science for policy, that is scientific research with the purpose of generating information for the policy process (Brooks 1987; Nowotny 1987; Jasanoff 1990, Funtowicz and Ravetz 1993; Funtowicz et al. 2000). For the latter type of science, several proposals have been developed, which are presented below.

Regulatory Science

Jasanoff (1990) has identified three main contextual differences between regulatory science and research science. First, the institutional and cultural setting, which is the major factor from the standpoint of political legitimisation. This is especially the case with the production and certification of knowledge in which government and industry are heavily involved: "Science carried out in non-academic settings may be subordinated to institutional pressures that critically influence researchers' attitudes to issues of proof and evidence." Second, time is a critical factor in regulatory science because decisions must often be taken before a consensus has been formed about the acceptability of evidence. Third, academic science works within established paradigms, whereas regulatory science often works "at the margins of existing knowledge where science and policy are difficult to distinguish and claims are backed by few, if any allies." Therefore, Jasanoff (loc.cit.) concluded that the guidelines for judging regulatory science are fluid, controversial and more politically motivated than in academic science.

Adaptive management

In the course of the 1990s, several fundamental changes to the role of science in policy-making have been proposed, based upon the increasing awareness of the necessity to deal with uncertainty and values, and of involving stakeholders in decision-making and management. The social sciences were to play an important role in new approaches to environmental policy-making. Mangel et al. (1996) published a comprehensive catalogue of principles for conservation management, based upon these changing views. The new principles were to replace principles from 1978, a time in which "resource managers behaved as if it were possible to manage the use of living resources in a relatively sustainable and predictable way" (Mangel et al., loc.cit.). One basic element, underlying the new principles, is the recognition that ecosystems are open systems in constant flux and without long-term stability, rather than stable, closed and deterministic, as assumed in the 1970s. A second basic element is the notion that "science, by itself, is not capable of making judgements about esthetics or ethics." Scientist should, therefore, take care to avoid mixing values and knowledge. Mangel et al. (loc. cit.) furthermore proposed to include the full range of knowledge and skills from the natural and social sciences at the earliest possible stage, and to take account of the motives, interests and values of all users and stakeholders. In order to accommodate the above elements and principles, Mangel et al. (loc.cit.) proposed to promote the model of adaptive management. Adaptive management implies that it must be possible to amend policies and practices as quickly as possible on the basis of new insights. Management should be a process of "learning by doing," instead of based upon prescriptive paradigms. Also the involvement of stakeholders is an important element of rational management: "The management process must always be accountable to the full range of stakeholders, and should be continually appraised according to biological, social and economical targets" (Mangel et al., loc.cit.). Adaptive management is not only relevant for the conservation and management of wildlife. On the basis of a comprehensive evaluation of the impact of hazardous substances and nutrients on the Baltic Sea and the policy answers to these impacts, Elmgren (2001) concluded the following:

> "Our ability to predict the effect of environmental management decisions will thus remain limited, in spite of progress in modelling and monitoring techniques. [....] Thus, future management decisions will continue to be made without scientific certainty that they will have the intended effect. They should therefore be viewed as experiments, with their effects carefully monitored, evaluated and learnt from, and the decisions then modified as needed."

Post-normal science

In 1993, Funtowicz and Ravetz coined the term "post-normal science," a new type of science for problem solving in which "uncertainty is not banished but managed and values are not presupposed but are made explicit" (Funtowicz and Ravetz 1993). These basic ingredients do not differ from the concept of adaptive management but, as will be shown below, post-normal science is more explicit in how to deal with uncertainty and the involvement of stakeholders. According to Funtowicz and Ravetz (loc.cit.), post-normal science is applied in situations, where both decision stakes and system uncertainties are high, as is often the case in environmental problems. Earlier, Collingridge and Reeve (1986) had analysed the role of science in decision-making in situations with high stakes and uncertainties, and concluded that scientific information is, in such cases, not suitable as conflict solver (chapter 1). For the case of marine eutrophication, the findings of this study largely confirm their conclusions. Funtowicz and Ravetz (1993) proposed the use of so-called extended peer communities in far-reaching societal policies, which they motivated by the following statement:

> "When problems lack neat solutions, when environmental and ethical aspects of the issues are prominent, when the phenomena themselves are ambiguous, and when all research techniques are open to methodological criticism, then the debates on quality are not enhanced by the exclusion of all but the specialist researchers and official experts. The extension of the peer community is then not merely an ethical or political act; it can positively enrich the processes of scientific investigation."

Funtowicz and Ravetz (loc.cit.) underlined that post-normal science should be complementary to applied science and professional consultancy, and that it was not intended to replace traditional forms of science. Only in cases where uncertainties and ethical aspects are more relevant than in applied science or professional consultancy, should post-normal science be applied by extending the peer community, implying that stakeholders become involved in the process of quality assurance of scientific input to the process. In this respect they used the term "extended peer review." Funtowicz and Ravetz (loc.cit.) furthermore proposed that in certain cases the work of extended peer communities could go even further than quality assessment, by actively involving stakeholders in the production of knowledge. Also Nowotny (1999) has underlined the need for involving non-experts in the production of knowledge. She has called this type of knowledge "socially-robust knowledge."

Post-normal science is one of the many approaches to the analysis of the use of science in decision-making, the so-called social studies of science (Brooks 1987; Nowotny 1987; Jasanoff 1990, Funtowicz and Ravetz 1993; De Marchi and Ravetz 1999; Funtowicz et al. 2000; Collin and Evans 2002; Nowotny 2005). Such critical analyses also exist for nature conservation (Mangel et al. 1996; Robertson and Hull 2001) or fisheries management (Jentoft 2005). These analyses, several of which cited in this study, share the common conclusion that science is not the objective deliverer of truth it claimed to be and, thus, can and should not be the sole basis for decision-making. For reasons of convenience, I will apply the term policy-related science as a common denominator for these various approaches to both the production and certification, as well as the application of science in the policy process. The main ingredients of policy-related science have been formulated by Funtowicz et al. (2000), in a contribution to the debate on the European research area (CEC 2000a) and the European Commission's guidelines for applying the precautionary principle (CEC 2000b). Funtowicz et al. (2000) identified several elements in which policy-related science, which they also called precautionary science, differs from normal science. These differences concern:

1. Purpose. Contrary to basic research, the purpose of which is the advancement of knowledge, or applied research, which is intended to develop techniques or devices, the output of policy-related science is, according to Funtowicz et al. (2000), "but one input among many in a policy process run by people, and the scientific contribution is rarely conclusive." The purposes include quality of the outcome, legitimacy of the process, acquiescence of the public and extension of democracy.
2. People. Policy-related science is not exclusively done by scientists. As stated above, also stakeholders must be involved in peer review and, in certain cases, in the production of knowledge.
3. Problems: How can problems be framed, so as to reflect the purposes, and what reliable answers can be hoped for? Problem formulation not only concerns technical questions, but has also a policy dimension. Therefore, also in this part of the process, stakeholders must be involved.
4. Procedures. The procedures for developing and applying policy-related science must take account of working with uncertain, inadequate and contradictory in-

formation. Therefore, procedures must accommodate participation of non-scientists in the production, certification and application of knowledge.
5. Product. The outcome of policy-related research should, ideally, be socially-robust knowledge. It should reflect all relevant aspects of the process, amongst which uncertainty, policy considerations, and ethical aspects, and should have been subject to extended peer review.

According to Funtowicz et al. (2000), the rebuilding of trust should be the prime objective of policy-related science, which they worded this as follows:

"It must be accepted that all stakeholders including scientists, have interests. Respect and trust should be developed on this honest basis. Scientists who dogmatically reject alternative views in their work do not inspire trust. Neither do governments who treat all sensitive information as confidential."

In the next and final section, the question will be discussed what policy-related science may mean for the three elements of the science-policy network, investigated in this study, i.e. scientific advice, the science-policy interface and politics.

6.3.2 Marine eutrophication policies and policy-related science

Preparing decisions

The work done within the marine eutrophication science-policy network (6.1.2) has been, and still is, heavily dominated by the natural sciences. This has not fundamentally changed, after non-governmental organisations were admitted observer status to OSPAR working groups in 1992 and the preparatory groups of INSC-4, which started their work in 1991. Also the occasional workshops with stakeholders about the development of quality objectives have not fundamentally altered the science-dominated character of the process. This is not surprising, given the historical roots of the marine eutrophication case. The main political decisions were taken in 1987 and 1990, after which the science-policy interface was mainly concerned with implementing and consolidating these decisions. In other words: there was little room for fundamental changes. Also the heavy emphasis on the scientific rigour in the implementation process, the fundaments of which also originate from the 1987 and 1990 North Sea Conferences, left little room for alternative approaches. It must, in this respect, be realised that many of the problems the working groups were faced with in implementing the political decisions, had been created by the working groups themselves in the phase of preparing the political decisions.

What can be learned is that the pre-decision-making phase is essential for a policy-related science approach. It is in this phase that problems are formulated and decisions prepared. It is very relevant that problems and decisions are formulated realistically, i.e. in such a way that uncertainties are made explicit, as well as the role science can play or not in taking away these uncertainties. In the case, investigated in this study, the decision taken at INSC-2 in 1987, to determine, on scientific grounds, which areas were vulnerable to pollution by nutrients, followed by the decision of INSC-3 in 1990, to establish a scientifically based map of eutrophication areas, proved not to be realistic from the perspective of the possibilities of marine ecology, certainly not within the set time frames (chapter 4). Also the feasibility of long-term monitoring should be addressed in the pre-decision-making phase. In particular in the light of the topical fashion of ecosystem management, it is essential that practical aspects of implementation are accounted for. Moreover, decisions should leave room for adaptation to new scientific and socio-economic developments, which is the basis for adaptive management.

Scientific advice and realistic problem formulation

An important element of realistic problem formulation is the delivery of useful scientific advice. This task may be carried out by the scientific advisory part of the science-policy network, in this case ICES. What has become very clear from the analysis in Chap. 5, is that ICES advice has had little impact on the policy process. The heavy criticism by ICES on the eutrophication monitoring programme, the Common Procedure and the EcoQOs (see 5.2), have hardly had an effect. In order to improve this situation, ICES should not only be responsible for providing sound scientific advice, but should also estimate the feasibility of implementing the advice in practice, for example with regard to monitoring, by taking account of national differences, financing, maintenance, analyses and other practical matters. In other words: not only the scientific content is relevant, it is also the context in which it is to be used that is important.

The question arises whether such tasks would undermine the scientific "objectivity" of ICES. Jasanoff (1990) has stressed that a strict separation of science and policy is an artificiality, and that some sort of negotiation and construction regarding facts, values and their context may enhance the usability of scientific knowledge and reduce conflicts. Arentsen et al. (1999) underlined the importance of so-called epistemic communities for the policy process because these scientific fora can achieve a certain amount of stability and integrity. However, according to these researchers, the contribution of science to the policy process will remain limited without exchange with other social actors. This study has shown that the ICES advisory bodies have a very good record as to the delivery of independent scientific advice, i.e. independent of the policy interests at stake. ICES has also been able to maintain its respected position as supplier of sound science, despite the fact that the composition of ICES advisory committees is now according to national representation. The importance of "good" science, i.e. science with a high credibility within the scientific community, for use in the policy process, is underlined in several analyses (compare Jasanoff 1990; Boehmer-Christiansen 1989). Equally important, however, is the transfer from relevant "good" science to useful "good" science, i.e. science that is applicable in decision-making and management (Boehmer-Christiansen 1994). In this respect, scientific advice delivered by ICES, has in many cases proven not to be very policy-practice oriented. An improvement of the current situation could be for relevant ICES advisory committees, in particular ACME and ACE, to extend their scientific analyses with an assessment of the practicability of the advice. To this end, independent and respected scientists from other disciplines than marine ecology, in particular the social sciences (e.g. economists, experts in political science and science analysts), should be included in the relevant advisory bodies. This conclusion is supported by the findings of Rice (2005), who has compared the advisory process of the Canadian Science Advisory Secretariat (CSAS) with that of ICES. He has underlined the relevance of inclusiveness and transparency in the preparation of scientific advice, two principles which are a fixed part of the CSAS procedure. According to Rice (loc.cit.), inclusiveness and transparency imply the presence of the full range of disciplinary experts, as well as academic scientist and those "whose lives may be affected by the advice."

Realistic problem formulation and the ecosystem approach

Recent political developments (compare 5.3) do not seem to go into the direction of pragmatic, implementable policies. On the contrary: With the introduction of the ecosystem approach, the burden on the science-policy interface, as well as the dependence on scientific advice, will increase rather than decrease. The ecosystem approach has been defined as follows[15]:

> "The comprehensive integrated management of human activities based upon the best available scientific knowledge about the ecosystem and its dynamics, in order to identify and take action on influences which are critical to the health of marine ecosystems, thereby achieving sustainable use of ecosystem goods and services and maintenance of ecosystem integrity."

Important supplementary elements of the ecosystem approach are the integration with social and economic goals and the participation of stakeholders. In 2004 a so-called ICES dialogue meeting was held, in which scientists, administrators and stakeholders discussed the future of scientific advice, in the framework of the ecosystem approach (ICES 2004b). The gap between the political desire for an ecosystem approach and the practical implementation consequences was clearly worded by FAO representative Garcia (ICES, loc.cit.):

> "Managers, scientists, and stakeholders together will need to determine how to turn a set of ethically and politically correct, but fuzzy, principles into operational plans [....]. The scientific, administrative, and institutional capacity to implement conventional management is already insufficient in many places. This capacity is a fortiori very insufficient to implement the more complex ecosystem approach."

What was required, according to Garcia, were the modernisation and creation of new institutions, processes and interactions, the further development of human resources, among which scientists, information specialists, managers and advisers. With regard to the science to support the ecosystem approach, Garcia made the following statement:

[15] Definition adopted by European stakeholders at the Conference on the Development of a European Strategy for the Protection and Conservation of the Marine Environment, Køge, Denmark, 4–6 December 2002.

"It will need to focus on better understanding of ecosystem functioning, variability and change; on uncertainty and risk assessment and management; on improvement of forecasting capacity; on identification and elaboration of key indicators; on provision of ex-ante and ex-post assessments of policy and management options."

When compared with earlier expectations to the progress in our knowledge of ecosystems (chapter 2), several of these requirements seem like a "déjà vu." After almost half a century of failing attempts to develop yardsticks to judge human impacts on ecosystems, or instruments for predicting ecosystem developments, to be used by politicians and administrators, the above requirements seem utterly naive. This is even more so in the light of new and more pragmatic insights in the nature of ecosystems and the capacities of ecological science (Schrader-Frechette and McCoy 1993; Mangel et al. 1996; (6.3.1); MacGarvin 1995 (6.2.4); De Jong 1998; Sagoff 2003). Moreover, realising the additional demands on the co-ordination and integration of increasing amounts of scientific information from an increasing number of disciplines, seems, in the light of the experiences with the management of marine eutrophication, highly unrealistic, especially in times of decreasing capacities of public administrations. Finally, integrating public and stakeholder participation into the existing administrative system, will already be a tremendous task in itself, demanding much effort and financial investment. What is needed most, is the setting of clear and pragmatic priorities for improving "conventional" management, with an emphasis on the development of a functional approach to stakeholder participation.

The science-policy interface and policy-related science

The science-policy interface is responsible for the preparation and the implementation of political decisions, and is the central element of the science-policy network. As was shown for the case of marine eutrophication, the science-policy interface uses much discretion in its work and has, thus, considerable powers. This is not only true for marine eutrophication. De Marchi and Ravetz (1999), for example, have investigated the Bovine Spongiform Encephalopathy (BSE) case, and concluded: "The control of knowledge and of ignorance enables the bureaucracy to exercise quite considerable power, complementary to and sometimes exceeding that of the legislature." As documented in Chap. 5, the marine eutrophication science-policy interface has blocked the flow of new scientific information to politics, by maintaining the view that marine eutrophication was a serious large-scale problem, causing undesired effects, while actually these effects seemed to be limited in extent, and only partly connected with anthropogenic nutrient inputs. With the introduction of additional requirements to management (see above), uncertainties in scientific assessments will probably increase in future, and it can, thus, be expected that also the discretionary powers of the science-policy interface will increase. It is exactly against this background that the science-policy interface should become more transparent and more democratic, and that the working groups of the science-policy interface should apply the main elements of policy-related science. As argued above, it is essential that a policy-related science approach is already applied in the pre-decision-making phase, in order to frame decisions in such a way that the technical, social and ethical aspects are taken into consideration. Such is a prerequisite for the practical and pragmatic implementation of the decisions, which proved to be hardly possible in the case analysed in this study because of too high expectations to science in dealing with uncertainties and values.

Collins and Evans (2002) raised the question which additional groups to admit to the process of providing and evaluating knowledge, and on which grounds. They put much emphasis on the presence of additional knowledge in new actors, which may supplement the official knowledge. As shown in the analysis of marine eutrophication, it was not the lack of appropriate knowledge that appeared to be a problem, but the selection and interpretation of available knowledge, both in the decision-making and the implementation processes. Most relevant, therefore, is to make these processes more democratic and transparent. For decision-making, it is necessary to involve relevant stakeholders; for implementation, to engage independent (i.e. non civil-servant) experts with knowledge of methodological aspects of the natural sciences.

Political decision-making, the Precautionary Principle and policy-related science

Although decision-making is the prime responsibility of governments and parliaments of democratic societies, it is the administration that exerts considerable powers in the preparation and implementation of political decisions. This is especially the case for complex environmental issues where, in the model of rational decision-making, scientific backing is considered a prerequisite and, thus, much technical preparation is required. Hinssen (1995) has investigated the role of the Dutch parliament in North Sea policy-making, and concluded that this role is limited and the process dominated by the lower echelon of civil servants, in which most technical expertise is available. The implementation of policy-related science, especially in bodies at the science-policy interface, is, therefore, in the interest of politics, since it extends democratic control of political decision-making. This is, of course, unless politics wishes to maintain a situation in which, on the grounds of scientific uncertainty, delay of action is considered more appropriate (compare Boehmer-Christiansen 1994).

The Precautionary Principle is a political instrument, originally regarded as a way of dealing with scientific uncertainty. However, the application of the precautionary principle does not imply that scientific information is no longer relevant in the policy process. In the EC Communication on the Precautionary Principle it is stated that monitoring and research should continue, in order to improve basic understanding, and that it must be possible to review measures on the basis of new scientific insights (CEC 2000b). When there is political and scientific controversy, as for the case of marine eutrophication, a decision, made on the basis of the Precautionary Principle, may even increase the demand for scientific information for the posterior scientific justification of the decision. Decisions on the basis of the Precautionary Principle, in fact, postpone the scientific debate, as is the case with the narrowing of the scientific horizon. Therefore, also for those cases in which the precautionary principle is applied, the extension of the science-policy network with additional stakeholders and independent experts, is relevant and desirable. When looking at some of the other conditions in the EC paper, set to the application of the Precautionary Principle, the need for extending the peer review community becomes even stronger. These conditions are:

- Proportionality, i.e. tailoring measures to the chosen level of protection;
- Non-discrimination, meaning that comparable situations should not be treated differently;
- Based upon a cost-benefit analysis.

The case of marine eutrophication has shown how complicated it is to deal with proportionality and non-discrimination in large marine ecosystems, and that such questions can certainly not be solved on the basis of scientific expertise alone. This is even more so, when cost-benefit considerations are added to the decision-making and implementation process.

The consequences of a policy-related science approach

The application of a policy-related approach to science will most probably lead to a situation with more incremental decision-making. For marine eutrophication in the North Sea, the 50% nutrient reduction decision might never have been taken, had the elements of policy-related science been applied. Admitting additional stakeholders – i.e. both independent experts from other fields of science and interest groups – to the policy-making process, will have a levelling and balancing effect on decision-making, and will make far-reaching decisions less probable. Incremental decision-making was already proposed by Collingridge and Reeve (1986) as the most pragmatic way of introducing new rules and regulations, in the light of actual or potential environmental problems, accompanied by a sustained and open scientific debate.

Arentsen et al. (1999) have pointed to some disadvantages of incremental decision-making resulting from policy-making on the basis of consensus building with stakeholders, which they illustrated for the case of the Dutch manure problem. Problems with excess manure, in particular acidification of soil and water, were prolonged for a long period of time because of denial by the agriculture sector, contra-expertise and insufficient solutions. However, the disadvantages of far-reaching decisions, based upon uncertain scientific grounds, probably outweigh those of policies based upon small steps. Big steps, to be taken within relatively short periods of time, usually favour short-term, technocratic solutions. For the marine eutrophication case, for example, nitrogen reduction was, first of all, to be achieved by introducing technical measures with regard to sewage treatment, although the introduction of structural measures in agriculture would have had much more effect. Also the political decision of INSC-3 to equip all coastal settlements of more than 5000 p.e. with extended sewage treatment, must be judged as a technocratic decision, hardly contributing to the eutrophication situation of the North Sea. In some European countries, current CO_2 reduction policies have led to the large-scale introduction of wind energy without sufficient spatial planning and without sufficiently taking into consideration the consequences for the regular electricity supply. Also the nuclear industry again sees chances for a revival of a nuclear-power society, on the grounds of the urgent need to reduce CO_2 emissions. More drastic technocratic solutions to solving the global climate change problem were given in an overview by Pearce (2004), amongst which a mirror to shield the earth from solar radiation. Decisions on the basis of premature science, furthermore carry the risk of invoking conflicts, after new information has become available, which is contradicting earlier knowledge. According to Lambright (1995), the speeding up of the communication of available science worked well for the CFC case, but he also showed how delicate the process of using "unready" knowledge is, and that in this respect the CFC case was an exception.

The application of a policy-related science approach in policy-making may overcome the inherent disadvantages of the use of science in the classical model of rational policy-making. Within the boundaries of a policy-related science approach, there may be a positive future role for ecological information to play in public environmental policies, although it will be a more realistic one than today, and to be shared with inputs from other scientific disciplines and from non-scientific sources.

References

Aebischer NJ, Coulson JC, Colebrook JM (1990) Parallel long-term trends across four marine trophic levels and weather. Nature 347: 753–755

Ærtebjerg Nielsen G, Carstensen J, Axe P, Druon J-N, Stips A (2003) The 2002 oxygen depletion event in the Kattegat, Belt Sea and western Baltic. Baltic Sea Environment Proceedings No.90, Helsinki Commission, Helsinki

Ærtebjerg Nielsen G, Schelde Jacobsen T, Gargas E, Buch E (1981) Evaluation of the physical, chemical and biological measurement. The Belt Project. NAEP, Copenhagen

Akkerman I (1997) National evaluation report of the joint assessment and monitoring programme 1995 of the Netherlands. Report RIKZ-97.011, RIKZ, The Hague

AKN (1987) The North Sea Memorandum, 1984 and Results and Demands, 1987. Aktionskonferenz Nordsee eV, Bremen

Anderson DM, Morel FMM (1979) Toxic dinoflagellate blooms in the Cape Cod region of Massachusetts. In: Taylor DL, Seliger, HH (eds) Proceedings of the 2nd conference on toxic dinoflagellate blooms. Elsevier, New York, Amsterdam, Oxford, pp 145–150

Anderson DM, Glibert PM, Burkholder JM (2002) Harmful algal blooms and eutrophication: Nutrient sources, composition and consequences. Estuaries 25(4b): 704–726

Andresen S (1989) Increased public attention: communication and polarisation. In: Andresen S, Østreng W (eds) International resource management: the role of science and politics. Belhaven Press, London, New York, pp 25–45

Arentsen M, Bressers H, O'Toole L (1999) Omgaan met onzekerheid in het milieubeleid: Een analyse met illustraties uit de Nederlandse en Amerikaanse beleidspraktijk. Beleidswetenschap 13(4): 375–400

ASMO (1995) Summary record of the 2nd meeting of the Assessment and Monitoring Committee. Annex 13. Secretariat OSPAR Commission, London

ASMO (1997) Draft report of the ASMO (OSPAR) modelling workshop on eutrophication issues within the JAMP. Document ASMO 97/7/3, Secretariat OSPAR Commission, London

ASMO (2002) Summary Record of the Assessment and Monitoring Committee, 12 April 2002. Document ASMO 02/13/18, Secretariat OSPAR Commission, London

Baalsrud K (1967a) Part III Discussion. In: Olson TA, Burgess FJ (eds) Pollution and Marine Ecology. Interscience, New York, London, Sydney, pp 89–95

Baalsrud K (1967b) Influence of Nutrient Concentrations on Primary Production. In: Olson TA, Burgess, FJ (eds) Pollution and Marine Ecology. Interscience, New York, London, Sydney, pp 159–169

Bach SD, Josselyn MN (1978) Mass blooms of the alga *Cladophora* in Bermuda. Mar Poll Bull 9(2): 34–37

Backhaus JO, Soetje KC (1988) On the three-dimensional approach to environmental modelling. In: Newman PJ, Agg AR (eds) Environmental Protection of the North Sea. Heinemann, Oxford, pp 613–627

Barisich A (1989) International cooperation for the protection of the North Sea. In: Ten Hallers CC, Bijlsma A (eds) Distress Signals, Proceedings of the 3rd North Sea Seminar. Werkgroep Noordzee, Amsterdam, pp 31–34

Bartsch AF (1967) Discussion. In: Olson TA, Burgess FJ (eds) Pollution and Marine Ecology. Interscience, New York, London, Sydney, pp 293–296

Bartsch AF (1972) Panel discussion: Nutrients eutrophication-prospects options for the future. In: Likens GE (ed) Nutrients and Eutrophication: The limiting Nutrient Controversy. American Society of Limnology and Oceanography, Allan Press, Lawrence, Kansas, pp 297–300

Bätje M, Michaelis H (1986) *Phaeocystis pouchettii* blooms in the East Friesian coastal waters (German Bight, North Sea). Marine Biology 93: 21–27

Beeton AM (1969) Changes in the Environment and Biota of the Great Lakes. In: Eutrophication, Causes, Consequences, Correctives. National Academy of Sciences, Washington, pp 150–187

Berge JA, Green, N, Rygg B (1988) The invasion of the planktonic algae *Chrysochromulina polylepis* along the coast of southern Norway in May–June 1988. Acute effects on coastal biota. Summary Report. SFT Report 339/88, SFT, Oslo

Beukema AA, Hekstra GP, Venema C (1986) The Netherlands environmental policy for the North Sea and Wadden Sea. Environmental Monitoring and Assessment 7: 117–155

Beukema JJ (1986) Eutrophication: reasons for satisfaction or concern? In: Peet G (ed) Proceedings of the 2nd North Sea Seminar, Vol 2. Werkgroep Noordzee, Amsterdam, pp 25–40

Beukema JJ, Cadée GC (1986) Zoobenthos responses to eutrophication of the Dutch Wadden Sea. Ophelia 26: 55–64

Beukema JJ, Cadée GC (1997) Local differences in macrozoobenthic response to enhanced food supply caused by mild eutrophication in a Wadden Sea area: Food is only locally a limiting factor. Limnol Oceanogr 42(6): 1424–1435

Bicknell WJ, Chapman Walsh D (1975) The first "red tide" in recorded Massachusetts history, Managing an acute and unexpected public health emergency. In: LoCicero VR (ed) Proceedings of the first international conference on toxic dinoflagellate blooms. Mass Sci Techn Found, Massachusetts, pp 447–458

Billen G, Somville M, De Becker E, Servais P (1985) A nitrogen budget for the Scheldt hydrographic basin. Neth J Sea Res 19(3/4): 223–230

BLMP (2000) Meeresumwelt 1994–1996. Bundesamt für Seeschiffahrt und Hydrographie, Hamburg, Rostock

BMEPC (1988) Guidelines for the Baltic Monitoring Programme for the 3rd stage. Baltic Marine Environment Protection Committee. Baltic Sea environment proceedings No.27, Helsinki Commission, Helsinki

BMI (1985) Internationale Nordseeschutz-Konferenz Bremen. Deklaration. Bundesminister des Innern, Bonn

Boddeke R (1978) Changes in the stock of the brown shrimp (*Crangon crangon*) in the coastal area of The Netherlands. P-v Reun Cons int Explor Mer, 172: 239–249

Boddeke R, Hagel P (1991) Eutrophication, a blessing in disguise. ICES CM/1991/E:7

Boehmer-Christiansen S (1989)The role of science in the international regulation of pollution. In: Andresen S, Østreng W (eds) International resource management: the role of science and politics. Belhaven Press, London, New York, pp 143–167

Boehmer-Christiansen S (1994) Reflections on scientific advice and EC transboundary pollution policy. Paper prepared for the Conference "Scientific Expertise in the European Public Policy Debate". London School of Economics and Political Science. 14–15 September 1994

Boehmer-Christiansen S, Skea J (1991) Acid politics: environmental and energy policies in Britain and Germany. Belhaven Press, London, New York

Bolomey JGW (1960) Effect of the river Rhine on Netherlands beaches. In: Pearson EA (ed) Proceedings of the first international conference on waste disposal in the marine environment. Pergamon, Oxford, London, New York, Paris, pp 164–174

Brickman R (1987) Commentary to Ravetz, J "Uncertainty, Ignorance and Policy". In: Brooks H, Cooper CL (eds) Science for Public Policy. Pergamon Press, Oxford, New York, pp 89–93

Brockmann U, Billen G, Gieskes WWC (1988) North Sea nutrients and eutrophication. In: Salomons W, Bayne WN, Duursma EK, Förstner U (eds) Pollution of the North Sea; An assessment. Springer, Berlin, Heidelberg, pp 348–389

Brohult S (1972) Introductory speech. In: Åkerblom A (ed) 1st Soviet-Swedish Symposium on the pollution of the Baltic. Ambio Special Report 1, p 11

Brongersma-Sanders M (1957) Mass mortality in the sea. In: Hedgpeth J (ed) Treatise on Marine Ecology and Paleoecology, pp 941–1010

Brooks H (1987) Introduction and overview. In: Brooks H, Cooper CL (eds) Science for Public Policy. Pergamon Press, Oxford, New York, pp 1–11

Brundtland GH (1997) The scientific underpinning of policy. Science 277: 457

Buttermore RE (1977) Eutrophication of an impounded estuarine lagoon. Mar Poll Bull 8(1): 13–15

Cadée GC (1984) Has input of organic matter into the western part of the Wadden Sea increased during the last decades? Neth Inst Sea Res Publ Ser 10: 71–82

Cadée GC, Hegeman J (1979) Phytoplankton primary production, chlorophyll and composition in an inlet of the western Wadden Sea (Marsdiep) Neth J Sea Res 13(2): 224–241

Cadée GC, Hegeman J (1986) Seasonal and annual variation in *Phaeocystis pouchetii* (Haptophyceae) in the westernmost inlet of the Wadden Sea during the 1973 to 1985 period. Neth J Sea Res 20(1): 29–36

Cadée GC, Hegeman J (1991) Historical phytoplankton data of the Marsdiep. Hydrobiol Bull 24(2): 111–118

Cadée GC, Hegeman J (1993) Persisting high levels of primary production at declining phosphate concentrations in the Dutch coastal area (Marsdiep) Neth J Sea Res 31: 147–152.

Cadée GC, Hegeman J (2002) Phytoplankton in the Marsdiep at the end of the 20th century; 30 years monitoring biomass, primary production, and *Phaeocystis* blooms. J Sea Res 48(2): 97–110

Carlson H (ed) (1986) Quality Status of the North Sea. Deutsche Hydrographische Zeitschrift. Erg H Nr. 16, DHI, Hamburg

CEC (1998) Measures taken pursuant to Council Directive 91/676/EEC concerning the protection of waters against pollution caused by nitrates from agricultural sources. Office for official publications of the European Communities, Luxembourg

CEC (2000a) Towards a European reseach area. Communication from the Commission to the Council, the European Parliament, the Economic and Social Committee and the Committee of the Regions. Document COM (2000)6. Commission of the European Communities, Brussels

CEC (2000b) Communication from the Commission on the precautionary principle. Document COM (2000)1. Commission of the European Communities, Brussels

CEC (2002a) Summary report on the identification of sensitive areas by the Member States; the measures implemented by the Member States with the view to the deadline of 31 December 1998; wastewater treatment in major cities; verification of the identification of sensitive areas by the Commission. Office for Official Publications of the European Communities, Luxembourg

CEC (2002b) Report on the implementation of the Coucil Directive 91/676/EEC concerning the protection of waters against pollution caused by nitrates from agricultural sources. Europe Environment Supplement 618

CEC (2002c) Towards a strategy to protect and conserve the marine environment. Communication from the Commission to the Council and the European Parliament. COM(2002) 539 final

CEC (2005) Proposal for a Directive of the European Parliament and the Council establishing a framework for community action in the field of marine environmental policy (Marine Strategy Directive) [SEC(2005)1290]. Commission of the European Communities, Brussels

Christensen PB (ed) (1996) Havmiljøet under forandring? Havforskning fra Miljøstyrelsen Nr. 61. Miljøstyrelsen, Copenhagen

Christensen PB (ed) (1998) The Danish marine environment: has action improved its state? Havforskning fra Miljøstyrelsen Nr. 62. Danish Environmental Protection Agency, Copenhagen

Clark RB (1970) Spreading news of pollution. Marine pollution Bulletin 1(1):1

Clark RB (1971) Futility of Phosphate Detergent Ban. Mar Poll Bull 2(4): 50

Clark RB (1987) Editorial. North Sea Science. Mar Poll Bull 18(5): 257

Clendenning KE, North WJ (1960) Effects of waste on the giant kelp *Macrocystis pyrifera*. In: Pearson EA (ed) Proceedings of the first international conference on waste disposal in the marine environment. Pergamon, Oxford, London, New York, Paris, pp 82–91

Cloern JE (2001) Our evolving conceptual modal of the coastal eutrophication problem. Mar Ecol Prog Ser 210: 223–253

Colijn F (1992) Changes in plankton communities: when, where and why. ICES mar Sci Symp 195: 193–212

Colijn F, Cadée GC (2003) Is phytoplankton growth in the Wadden Sea light or nitrogen limited? J Sea Res 49: 83–93

Colijn F, Dooley H, Owens NJP, Sjoldal HR (1992) Report of the expert group on nutrients to the Paris Commission Working Group on Nutrients. North Sea Task Force document HC-OSPAR/3-4-5/5

Collingridge D, Reeve C (1986) Science speaks to power. The role of experts in policy making. Pinter, London

Collins HM, Evans R (2002) The third wave of science studies: Studies of expertise and experience. Social Studies of Science 32/2: 235–296

Cooper LHN (1968) Scientific consequences of the wreck of the "Torrey Canyon". Helgoländer wiss Meeresunters 17: 340–355

Copeland BJ (1967) Biological and Physiological Basis of Indicator Organisms and Communities. Section III, Biological and Physiological Basis of Indicator Communities. In: Olson TA, Burgess FJ (eds) Pollution and Marine Ecology. Interscience, New York, London, Sydney, pp 285–288

Corten A (2001) Herring and climate. Changes in the distribution of the North Sea herring due to climate fluctuation. Ph.D thesis, University of Groningen

CPR Survey Team (1992) Continuous plankton records: the North Sea in the 1980s. ICES mar Sci Symp 195: 243–248

Cramer JM (1987) Mission-orientation in ecology: the case of Dutch freshwater ecology. Ph.D thesis. University of Amsterdam

D'Elia CF, Sanders JG (1987) Scientists don't make management decisions (and why we wish that sometimes we did). Mar Poll Bull 18(8): 429–434

De Jong F (1986) The quality of the North Sea: national points of view. In: Peet G (ed) Proceedings of the 2nd North Sea Seminar, Vol I. Werkgroep Noordzee, Amsterdam, pp 39–54
De Jong F (1987) The second ministerial conference on the protection of the North Sea; an historical event? North Sea Monitor 87.4: 2–4
De Jong F (1990) The second international conference on the protection of the North Sea: National implementation. In: Freestone D, IJlstra T (eds) The North Sea: Perspectives on regional environmental co-operation. Special issue International Journal of Estuarine and Coastal Law, Graham & Trotman, London, pp 31–46
De Jong F (1994) International environmental protection of the Wadden Sea and the North Sea and the integration of pollution and conservation policies. Ophelia Suppl 6: 37–45
De Jong F (1998) Marine ecological quality objectives: Science and management aspects. In: Müller F, Leupelt M (eds.) Eco Targets, Goal Functions and Orientors. Springer, Berlin, Heidelberg, pp 526–544
De Jong F (2003) Wadden Sea Targets: Lessons from the first six years. In: Wolff WJ, Essink K, Kellermann A, Van Leeuwe MA (eds) Proceedings of the 10th International Scientific Wadden Sea Symposium. Ministry of Agriculture Nature Management and Fisheries, Department of Marine Biology, University of Groningen, pp 207–220
De Jonge VN (1990) Response of the Dutch Wadden Sea ecosystem to phosphorus discharges from the river Rhine. Hydrobiologia 195: 49–62
De Jonge VN (1997) High Remaining Productivity in the Dutch western Wadden Sea Despite Decreasing Nutrient Inputs from Riverine Sources. Mar Poll Bull 34: 427–436
De Jonge VN, Postma H (1974) Phosphorus compounds in the Dutch Wadden Sea. Neth J Sea Res 8 (2–3): 139–153
De Jonge VN, Elliot M, Brauer VS (2006) Marine monitoring: Its shortcomings and mismatch with the EU Water Framework Directive's Objectives. Mar Poll Bull, In press
De la Mothe J, Dufour P (1995) Is science policy in the doldrums? Nature 374: 209–210
De Marchi B, Ravetz JR (1999) Risk management and governance: a post-normal science approach. Futures 31: 743–757
De Vries I (1992) Report of the NSTF modelling workshop. Report DGW-92.045
De Wit JAW, Schotel FM, Bekkers LEJ (1982) De waterkwaliteit van de Waddenzee. Rijksinstituut voor Zuivering van Afvalwater nota nr. 82.065. RIZA, Lelystad
De Wolf P, Zijlstra JJ (1988) The ecosystem. In: Salomons W, Bayne WN, Duursma EK, Förstner U (eds) Pollution of the North Sea; An assessment. Springer, Berlin, Heidelberg, pp 118–151
Degobbis D, Smodlaka N, Pojed I, Skrivanic A, Precali R (1979) Increased eutrophication of the northern Adriatic Sea. Mar Poll Bull 10(10): 298–301
Derr PF (1972) Panel discussion: Nutrients and eutrophication-prospects and options for the future. In: Likens GE (ed) Nutrients and Eutrophication: The limiting Nutrient Controversy. American Society of Limnology and Oceanography, Allan Press, Lawrence, Kansas, pp 300–303
Dethlefsen V (1986) Marine pollution mismanagement: towards the precautionary concept. Mar Poll Bull 17(2): 54–57
Dethlefsen V (1988) German Bight. In: Salomons W, Bayne WN, Duursma EK, Förstner U (eds) Pollution of the North Sea; An assessment. Springer, Berlin, Heidelberg, pp 425–440
Döös BR (1994) Why is environmental protection so slow? Global Environmental Change 4(3): 179–184
Duursma EK, Beukema JJ, Cadée GC, Lindeboom HJ, De Wilde PAWJ (1988) Assessment of environmental impacts of nutrients. In: Newman PJ, Agg AR (eds) Environmental Protection of the North Sea. Heinemann, Oxford, pp 176–188

Dybern BI (1972) Pollution in the Baltic. In: Ruivo M (ed) Marine Pollution and Sea Life. Fishing News Ltd, Surrey, London, pp 15–23

Edwards PJ, May RM, Webb NR (1994) Preface. In: Edwards PJ, May RM, Webb NR (eds) Large-scale ecology and conservation biology. Blackwell Science, pp viii–xi

EE (2002) Nitrates pollution. Room for improvement in implementing EU Directive. Europe Environment 616: I29-I31

Eisma D (1986) The North Sea. An overview. In: Peet G (ed) Proceedings of the 2nd North Sea Seminar, Vol I. Werkgroep Noordzee, Amsterdam, pp 9–28

Elmgren R (2001) Understanding human impact on the Baltic ecosystem: Changing views in recent decades. Ambio 30 (4–5): 222–228

Engelhardt HT, Caplan AL (1987) Patterns of controversy and closure: the interplay of knowledge, values and political forces. In: Engelhardt HT, Caplan AL (eds) Scientific controversies. Case studies in the resolution and closure of disputes in science and technology. Cambridge University Press, Cambridge, pp 1–23

Essink K, Beukema JJ, Brinch Madsen P, Michaelis H, Vedel G (1998) Long-term development of intertidal macrozoobenthos in different parts of the Wadden Sea. Governed by nutrient loads? Senckenbergiana maritima 29(1/6): 25–35

EUC (2000) Draft evaluation of the expected situation of the eutrophication status in the maritime area following the 50% reduction target for nutrient inputs. Annex 7 of the Summary Record of the 2000 meeting of the OSPAR Eutrophication Committee (EUC), OSPAR Secretariat, London

EUC (2001) Further Elaboration of Ecological Quality Objectives for the Greater North Sea with regard to Nutrients and Eutrophication Effects (EcoQOs-eutro) Document EUC/01/5/3. OSPAR Secretariat, London

Fahy E, Goodwillie R, Rochford J, Kelly D (1975) Eutrophication of a partially enclosed estuarine mudflat. Mar Poll Bull 6(2): 29–32

Folkard AR, Jones PGW (1975) Distribution of Nutrient Salts in the Southern North Sea during early 1974. Mar Poll Bull 5(12): 181–185

Fonselius SH (1972a) On eutrophication and pollution in the Baltic Sea. In: Ruivo M (ed) Marine Pollution and Sea Life. Fishing News Ltd, Surrey, London, pp 23–28

Fonselius SH (1972b) On biogenic elements and organic matter in the Baltic. In: Åkerblom A (ed) 1st Soviet-Swedish Symposium on the pollution of the Baltic. Ambio Special Report No. 1, pp 29–36

Fonselius SH (1976) On the nutrient variability in the Baltic. In: Åkerblom A (ed) 2nd Soviet-Swedish Symposium on the pollution of the Baltic. Ambio Special Report No. 4, pp 17–27

Fonselius SH (1977) On the distribution of nutrients in the Baltic water. In: Åkerblom A (ed) 3rd Soviet-Swedish Symposium on the pollution of the Baltic. Ambio Special Report No. 5, pp 95–102

Fonselius SH (1981) Oxygen and hydrogen sulphide conditions in the Baltic Sea. Mar Poll Bull 12(6): 187–194

Føyn E (1960) Chemical and biological aspects of sewage disposal in inner Oslofjord. In: Pearson EA (ed) Proceedings of the first international conference on waste disposal in the marine environment. Pergamon, Oxford, London, New York, Paris, pp 279–284

Føyn E (1965) Disposal of waste in the marine environment and the pollution of the sea. Ocean Mar Biol Ann Rev 3: 95–114

Foyn E (1968) Biochemical and dynamic circulation of nutrients in the Oslo Fjord. Helgoländer wiss Meeresunters 17: 489–495

Fraser JH (1973) Zooplankton in the North Sea. In: Goldberg ED (ed) North Sea Science. MIT press, Cambridge MA, London, pp 267–289

Fuhs GW, Demmerle SD, Canelli E, Chen M (1972) Characterization of phosphorus-limited plankton algae. In: Likens GE (ed) Nutrients and Eutrophication: The limiting Nutrient Controversy. American Society of Limnology and Oceanography. Allan Press, Lawrence, Kansas, pp 113–133

Funtowicz SO, Ravetz JR (1993) Science for the post-normal age. Futures September 1993: 739–755

Funtowicz S, Shepard I, Wilkinson D, Ravetz J (2000) Science and governance in the European Union: a contribution to the debate. Science and public policy 27(5): 327–336

Gerlach SA (1976) Meeresverschmutzung; Diagnose und Therapie. Springer, Berlin, Heidelberg, New York

Gerlach SA (1984) Oxygen depletion 1980–1983 in coastal waters of the Federal Republic of Germany. Berichte aus dem Institut für Meereskunde Nr. 130, Institut für Meereskunde, Kiel

Gerlach SA (1988) Nutrients, an overview. In: Newman PJ, Agg AR (eds) Environmental Protection of the North Sea. Heinemann, Oxford, pp 147–175

Gerlach SA (1990) Nitrogen, phosphorus, plankton and oxygen deficiency in the German Bight and in Kiel Bay. Kieler Meeresforschungen Sonderheft Nr.7, Institut für Meereskunde, Kiel

GESAMP (1990) Review of potentially harmful substances: Nutrients. GESAMP Reports and Studies No. 34

Gieskes WWC (1973) De massale planktonsterfte op de kust bij Callantsoog juli 1972. Waddenbulletin 8(1): 36–37

Gieskes WWC, Kraay GW (1975) The phytoplankton spring bloom in Dutch coastal waters of the North Sea. Neth J Sea Res 9(2): 166–196

Gieskes WWC, Kraay GW (1977) Continuous planton records: Changes in the plankton of the North Sea and its eutrophic Southern Bight from 1948 to 1975. Neth J Sea Res 11(3/4): 334–364

Gjøsaeter J, Lekve K, Stenseth NC, Leinaas HP, Christie H, Dahl E, Danielssen DS, Edvardsen B, Olsgard F, Oug E, Paasche E (2000) A long-term perspective on the *Chrysochromulina* bloom in the Norwegian Skagerrak coast 1988: a catastrophe or an innocent incident? Mar Ecol Progr Ser 207: 201–218

Glover RS, Robinson GA, Colebrook JM (1972) Plankton in the North Atlantic–an example of the problems of analysing variability in the environment. In: Ruivo M (ed) Marine Pollution and Sea Life. Fishing News Ltd., Surrey, London, pp 439–445

Goldberg ED (1973a) Preface. In: Goldberg ED (ed) North Sea Science. MIT press, Cambridge MA, London

Goldberg ED (1973b) Introduction and recommendations. In: Goldberg ED (ed) North Sea Science. MIT press, Cambridge MA, London, pp 1–14

Goldberg ED (1976) The health of the oceans. The Unesco Press, Paris

Goldberg ED (1981) The oceans as a waste space: the argument. Oceanus 24(1): 5–9

Goldman CR (1972) The role of minor nutrients in limiting the productivity of aquatic ecosystems. In: Likens GE (ed) Nutrients and Eutrophication: The limiting Nutrient Controversy. American Society of Limnology and Oceanography. Allan Press. Lawrence, Kansas, pp 21–33

Gray JS (1990) Statistics and the precautionary principle. Mar Poll Bull 21(4): 174–176

Gray JS, Paasche E (1984) On Marine Eutrophication. Mar Poll Bull 15(10): 349–350

Groenewegen P, Hischemöller M, Hoppe R, Midden CJH (1998) De rol van deskundigen bij politieke keuzes in milieukwesties. WTS 6(3): 70–76

Gundersen K (1981) The distribution and biological transformations of nitrogen in the Baltic Sea. Mar Poll Bull 12(6): 199–205

Hagmeier E (1978) Variations in phytoplankton near Helgoland. Rapp P-v Reun Cons int Explor Mer 172: 361–363

Haigh J (2001) Climate variability and the influence of the sun. Science 294: 2109–2111

Hall WC (1967) Welcoming remarks. In: Olson TA, Burgess FJ (eds) Pollution and Marine Ecology. Interscience, New York, London, Sydney, pp vii–viii

Hannigan JA (1995) Environmental sociology. A social constructionist perspective. Routledge, London

Hartman O (1960) The benthonic fauna of southern California in shallow depths and possible effects of wastes on the marine biota. In: Pearson EA (ed) Proceedings of the first international conference on waste disposal in the marine environment. Pergamon, Oxford, London, New York, Paris, pp 57–81

Hartwell AD (1975) Hydrographic factors affecting the distribution and movement of toxic dinoflagellates in the western Gulf of Maine. In: LoCicero VR (ed) Proceedings of the first international conference on toxic dinoflagellate blooms. Mass Sci Techn Found, Massachusetts, pp 47–68

Hartwig J (1977) Swedish legislation regarding purification and some of the techniques applied at Swedish municipal sewage treatment and industrial plants. In: Åkerblom A (ed) 3rd Soviet-Swedish Symposium on the pollution of the Baltic. Ambio Special Report No. 5, pp 231–238

Head PC (1970) Discharges of nutrients from estuaries. Mar Poll Bull 1(9): 138–140

Hempel G (1978a) Symposium on North Sea fish stocks – recent changes and their causes – Introduction and historical background. Rapp P-v Reun Cons Explor Mer 172: 5–9

Hempel G (1978b) Synopsis of the symposium North Sea fish stocks – recent changes and their causes. P-v Reun Cons int Explor Mer, 172: 445–449

Henke S (1997) (ed) Black spots in the Wadden Sea. Texte 3/97, Federal Environmental Agency, Berlin

Hickel W, Eickhoff M, Spindler H, Berg J, Raabe T, Müller R (1977) Auswertung von Langzeit-Untersuchungen von Nährstoffen und Phytoplankton in der Deutschen Bucht. Forschungsbericht 102 04 239, UBA-FB 97-057. UBA, Berlin

Hickel W, Mangelsdorf P, Berg J (1993) The human impact in the German Bight: Eutrophication during three decades (1962–1991) Helgoländer Meeresuntersuchungen 47: 243–263

Hinssen J (1995) The role of parliament in North Sea policy making. North Sea Monitor, Dec 1995: 17–19

Hischemöller M, Groenewegen P, Hoppe R, Midden CJH (1998) Kennisbenutting en politieke keuze: een dilemma voor het milieubeleid? Werkdocument 65, Rathenau Instituut, Den Haag

Hollander JM (1972) Scientists and the environment: new responsibilities. Ambio 1 (3): 116–119

Hoogerwerf A, Herweijer M (2003) Overheidsbeleid. Kluwer, Alphen aan de Rijn

Höpner T, Michaelis H (1994) Sogenannte "Schwarze Flecken" – ein Eutrophierungssymptom des Wattenmeeres. In: Lozán JL, Rachor E, Reise K, Von Westernhagen H, Lenz W (eds) Warnsignale aus dem Wattenmeer. Blackwell, Berlin, pp 153–159

Horlick-Jones T, De Marchi B (1995) The crisis of scientific expertise in *fin de siècle* Europe. Science and public policy 22(3): 139–145

Howarth MJ, Dyer KR, Joint IR, Hydes DJ, Purdie DA, Edmunds H, Jones JE, Lowry RK, Moffat TJ, Pomroy AJ, Proctor R (1994) In: Charnock H, Dyer KR, Huthnance JM, Liss PS, Simpson JH, Tett PB (eds) Understanding the North Sea system. Chapman & Hall, London, pp 5–25

Hydes DJ, Kelly-Gerreyn BA, Le Gall AC, Proctor R (1999) The balance of supply of nutrients and demands of biological production and denitrification in a temperate latitude shelf sea – a treatment of the southern North Sea as an extended estuary. Marine Chemistry 68: 117–131

ICES (1969) Report of the ICES Working Group on pollution of the North Sea. Cooperative Research Report Series A, No. 13, ICES, Copenhagen

ICES (1970) Report of the ICES working group on the pollution of the Baltic Sea. Cooperative Research Report Ser A, No.15, ICES, Copenhagen

ICES (1974a) Report of the working group for the international study of the pollution of the North Sea and its effects on living resources and their exploitation. Cooperative Research Report No. 39, ICES, Copenhagen

ICES (1974b) Report of the Advisory Committee on Marine Pollution 1974. Cooperative Research Report No. 43, ICES, Copenhagen

ICES (1975) Report of the Advisory Committee on Marine Pollution 1975. Cooperative Research Report No. 50, ICES, Copenhagen

ICES (1977a) Report of the Advisory Committee on Marine Pollution to the Oslo and the interim Helsinki Commissions 1976. Cooperative Research Report No. 57, ICES, Copenhagen

ICES (1977b) The ICES coordinated monitoring programme in the North Sea. Cooperative Research Report No. 58, ICES, Copenhagen

ICES (1977c) Studies of the pollution of the Baltic Sea. Cooperative Research Report No.63, ICES, Copenhagen

ICES (1978) Report of the International Council for the Exploration of the Sea to the Oslo Commission, the interim Helsinki Commission and the interim Paris Commission, 1977. Cooperative Research Report No. 76, ICES, Copenhagen

ICES (1979) Report of the Advisory Committee on Marine Pollution 1978. Cooperative Research Report No. 84. ICES, Copenhagen

ICES (1980) Report of the Advisory Committee on Marine Pollution 1979. Cooperative Research Report No. 92, ICES, Copenhagen

ICES (1981) Report of the Advisory Committee on Marine Pollution 1980. Cooperative Research Report No. 103, ICES, Copenhagen

ICES (1982) Report of the Advisory Committee on Marine Pollution 1981. Cooperative Research Report 112, ICES, Copenhagen

ICES (1983a) Report of the Advisory Committee on Marine Pollution 1982. Cooperative Research Report 120, ICES, Copenhagen

ICES (1983b) Report of the Advisory Committee on Marine Pollution 1983. Cooperative Research Report 124. ICES, Copenhagen

ICES (1985) Report of the Advisory Committee on Marine Pollution 1984. Cooperative Research Report 132, ICES, Copenhagen

ICES (1986) Report of the Advisory Committee on Marine Pollution 1985. Cooperative Research Report 135, ICES, Copenhagen

ICES (1987a) Report of the Advisory Committee on Marine Pollution 1986. Cooperative Research Report 150, ICES, Copenhagen

ICES (1987b) Assessment of Environmental Conditions in the Skagerrak and Kattegat. Cooperative Research Report 149, ICES, Copenhagen

ICES (1988) Report of the Advisory Committee on Marine Pollution 1987. Cooperative Research Report 142, ICES, Copenhagen

ICES (1989) Report of the Advisory Committee on Marine Pollution 1989. Cooperative Research Report 167, ICES, Copenhagen

ICES (1990) Report of the Advisory Committee on Marine Pollution 1990. Cooperative Research Report 172, ICES, Copenhagen

ICES (1992) Report of the Advisory Committee on Marine Pollution, 1992. Cooperative Research Report 190. ICES, Copenhagen

ICES (1994) Report of the Advisory Committee on the Marine Environment, 1993. Cooperative Research Report 198. ICES, Copenhagen
ICES (1997) Report of the Advisory Committee on the Marine Environment, 1997. Cooperative Research Report 222. ICES, Copenhagen
ICES (1999) Report of the Advisory Committee on the Marine Environment, 1998. Cooperative Research Report 233. ICES, Copenhagen
ICES (2001) Report of the Advisory Committee on Ecosystems, 2001. Cooperative Research Report 249. ICES, Copenhagen
ICES (2002a) Report of the Advisory Committee on Ecosystems, 2002. Cooperative Research Report 254. ICES, Copenhagen
ICES (2002b) Report of the Advisory Committee on the Marine Environment, 2002. Cooperative Research Report 256. ICES, Copenhagen
ICES (2004a) ICES Advice. Vol. 1, No.1, Book 1, Part 1. ICES, Copenhagen
ICES (2004b) Report of the 13th ICES Dialogue Meeting: Advancing scientific advice for an ecosystem approach to management: collaboration amongst managers,scientists, and other stakeholders. Cooperative Research Report 267. ICES, Copenhagen
INSC (1987a) Background to Issues of Policy. Second International Conference on the Protection of the North Sea. Background Papers for Delegates
INSC (1987b) Ministerial Declaration. Second International Conference on the Protection of the North Sea
INSC (1990) Interim Report on the quality status of the North Sea. Secretariat 3rd International Conference on the Protection of the North Sea, The Hague
INSC (2002) Ministerial Declaration Fifth International Conference on the protection of the North Sea
IPCC (2001) Climate change 2001: The scientific Basis. Intergovernmental Panel on Climate Change, Geneva
Jak RG, Michelsen BF (1996) Algenbegrazing: een nadere analyse van de invloed van toxicanten op het onstaan van eutrofiëringsproblemen. BEON Rapport Nr. 96-4, BEON, Den Haag
Jak RG, Scholten MCT (1994) Mesocosm experimenten TNO Den Helder. Indicaties voor het ontstaan van eutrofiëringsverschijnselen. In: Philippart CJM et al. (eds) Effecten van verschuivingen van nutrientenconcentraties op biota in de Nederlandse kustwateren. BEON Rapport Nr. 94-1, BEON, Den Haag
James A, Head PC (1972) The discharge of nutrients from estuaries and their effect on primary productivity. In: Ruivo M (ed) Marine Pollution and Sea Life. Fishing News Ltd, Surrey, London, pp 163–166
Jasanoff S (1990) The fifth branch. Science advisors as policy makers. Harvard University Press, Cambridge MA, London
Jaworski NA (1981) Sources of nutrients and the scale of eutrophication problems in estuaries. In: Neilson BJ, Cronin L E (eds) Estuaries and Nutrients. Humana Press, Clifton, New Jersey, pp 83–110
Jaworski NA, Lear DW, Villa O (1972) Nutrient management in the Potomac estuary. In: Likens GE (ed) Nutrients and eutrophication: The limiting nutrient controversy. Allan Press, Lawrence, Kansas, pp 246–273
Jensen A (1989) Cooperation of policy and science. Viewpoint of a scientist. In: Ten Hallers CC, Bijlsma A (eds) Distress Signals. Werkgroep Noordzee, Amsterdam, pp 197–203
Jentoft S (2005) Beyond fisheries management: The phronetic dimension. Paper presented at the III MARE Conference "People and the sea", Amsterdam, 7–9 July 2005. www.marecentre.nl

JMG (1989) Summary Record 14th meeting of the Joint Monitoring Group. Vigo, 24–27 January 1989. Secretariat Oslo and Paris Commissions, London
Johannes RE (1971) How to kill a coral reef II. Mar Poll Bull 2(1): 9–10
Johnson SP, Corcelle G (1989) The environmental policy of the European Communities. Graham & Trotman, London, Dordrecht, Boston
Johnston R (1973) Nutrients and metals in the North Sea. In: Goldberg ED (ed) North Sea Science. MIT press, Cambridge Mass, London, pp 293–307
Johnston R (1976) Marine Pollution. Academic Press, London, New York, San Francisco
Johnston R, Jones PGW (1965) Serial atlas of the marine environment. Folio 11. Amer Geogr Soc, New York
Jørgenson B, Richardson K (eds) (1996a) Eutrophication in coastal marine ecosystems. American Geophysical Union, Washington DC
Jørgenson B, Richardson K (1996b) Conclusions, research and eutrophication control. In: Jørgenson B, Richardson K (eds) Eutrophication in coastal marine ecosystems. American Geophysical Union, Washington DC, pp 243–267
Josefson AB (1990) Increase of benthic biomass in the Skagerrak-Kattegat during the 1970s and 1980s – effects of organic enrichment? Mar Ecol Prog Ser 66: 117–130
Joyce EA, Roberts BS (1975) Florida Department of Natural Resources red tide research program. In: LoCicero VR (ed) Proceedings of the first international conference on toxic dinoflagellate blooms. Mass Sci Techn Found, Massachusetts, pp 95–104
Kamlet KS (1981) The oceans as a waste space: the rebuttal. Oceanus 24(1): 10–17
Kampmann J (1981) Preface. In: Ærtebjerg Nielsen G, Jacobsen TS, Gargas E, Buch E (eds) Evaluation of the physical, chemical and biological measurement. The Belt Project. NAEP, Copenhagen
Karlén W (2001) Global temperature forced by solar irradiation and greenhouse gases? Ambio 30(6): 349–350
Ketchum BH (1967) Man's Resources in the Marine Environment. In: Olson TA, Burgess FJ (eds) Pollution and Marine Ecology. Interscience, New York, London, Sydney, pp 1–11
Ketchum BH (1969) Eutrophication of Estuaries. In: Eutrophication, Causes, Consequences, Correctives. National Academy of Sciences, Washington, pp 197–209
Kinne O (1968a) Internationales Symposion "Biologische und hydrographische Probleme der Wasserverunreinigung in der Nordsee und angrenzenden Gewässern: Eröffnungsansprache. Helgoländer wiss Meeresunters 17: 1–5
Kinne O (1968b) International Symposium "Biological and hydrographical problems of water pollution in the North Sea and adjacent waters: Closing address. Helgoländer wiss Meeresunters 17: 518–522
Kinne O (1980) (ed) 14th European Marine Biology Symposium "Protection of Life in the Sea". Helgoländer Meeresuntersuchungen 33
Kiyouria R (1964) Water Pollution and Minamata disease. In: Pearson EA (ed) Advances in Water Pollution Research. Volume 3. Pergamon, Oxford, London, New York, Paris, pp 291–301
Klein AWO, Van Buuren JT (1992) Eutrophication of the North Sea in the Dutch coastal zone 1976–1990. Report WS-92.003. Ministry of Transport and Public Works, Tidal Waters Division, The Hague
Koch P (1960) Discharges of wastes into the sea in European coastal areas. In: Pearson EA (ed) Proceedings of the first international conference on waste disposal in the marine environment. Pergamon, Oxford, London, New York, Paris, pp 122–163
Koeman JH, Veen J, Brouwer E, Huisman-De Brouwer L, Koolen JL (1968) Residues of chlorinated hydrocarbon insecticides in the North Sea Environment. Helgoländer wiss Meeresunters 17: 375–380

Korringa P (1968) Biological consequences of marine pollution with special reference to the North Sea fisheries. Helgoländer wiss Meeresunters 17: 126–140
Krebs CJ (1972) Ecology. The experimental analysis of distribution and abundance. Harper & Row, New York, London
Krom M (1986) An evaluation of the concept of assimilative capacity as applied to marine waters. Ambio 15(4): 208–214
Kröncke I, Wieking G (2003) Klimatisch bedingte Veränderungen im Benthos der Nordsee. In: Lozán JL, Rachor E, Reise K, Sündermann J, Von Westernhagen H (eds) Warnsignale aus Nordsee und Wattenmeer. Wissenschaftliche Auswertungen, Hamburg, pp 172–175
Kröncke I, Dippner JW, Heyen H, Zeiss B (1998) Long-term changes in macrofaunal communities off Norderney (East Frisia, Germany) in relation to climate variability. Mar Ecol Prog Ser 167: 25–36
Kullenberg G (1981) Editorial: The Baltic – a regional sea marine pollution case study. Mar Poll Bull 12(6): 179–182
Kullenberg G, Jacobsen TS (1981) The Baltic Sea: an outline of its physical oceanography. Mar Poll Bull 12(6): 183–186
Küppers G, Lundgreen P, Weingart P (1978) Umweltforschung, die gesteuerte Wissenschaft? Suhrkamp, Frankfurt am Main
Laane RWPM, Wolff WJ (eds) 1984. Conclusions and recommendations of the 4th International Wadden Sea Symposium "The role of organic matter in the Wadden Sea". Neth Inst Sea Res Publ Ser 10: 5–7
Lambright WH (1995) NASA, ozone and policy relevant science. Research Policy 24: 747–760
Lancelot C (1990) *Phaeocystis* blooms in the continental coastal area of the Channel and the North Sea. In: Lancelot C, Billen G, Barth H (eds) Eutrophication and algal blooms in North Sea coastal zones, the Baltic and adjacent areas: prediction and assessment of preventive actions. Water Pollution Research Report 12, CEC, Brussels, pp 27–54
Lancelot C (1995) The mucilage phenomenon in the continental coastal waters of the North Sea. Sci Total Environment 165: 83–102
Lancelot C, Billen G, Sournia A, Weisse T, Colijn F, Veldhuis MJW, Davies A, Wassman P (1987) *Phaeocystis* blooms and nutrient enrichment in the continental coastal zones of the North Sea. Ambio 16(1): 38–46
Lancelot C, Billen G, Barth H (1990) Foreword. In: Lancelot C, Billen G, Barth H (eds) Eutrophication and algal blooms in North Sea coastal zones, the Baltic and adjacent areas: prediction and assessment of preventive actions. Water Pollution Research Report 12, CEC, Brussels, pp 5–8
Leppäkoski E (1980) Man's impact on the Baltic ecosystem. Ambio 9 (3–4): 174–181
Leroy P (1989) Milieubeweging en milieubeleid. In: Glasbergen P (ed) Milieubeleid, theorie en praktijk. VUGA, 's Gravenhage, pp 57–79
Lewis JR (1972) Problems on approaches to baseline studies in coastal communities. In: Ruivo M (ed) Marine Pollution and Sea Life. Fishing News Ltd., Surrey, London, pp 401–404
Likens GE (1972) Preface. In: Likens GE (ed) Nutrients and Eutrophication: The limiting Nutrient Controversy. American Society of Limnology and Oceanography, Allan Pres, Lawrence, Kansas, pp v–vi
Likens GE (1992) The ecosystem approach: its use and abuse. Part 2: Environmental issues, scientific communication and ethics. Ecology Institute, Oldendorf/Lühe
Lindahl O, Belgrano A, Davidsson L, Hernroth B (1998) Primary production, climatic oscillations, and physico-chemical processes: the Gullmar Fjord time-series data set (1985–1996). ICES J Mar Sci 55: 723–729

Lindeboom HJ, Zijlstra JJ, De Groot SJ, Wolff WJ (1988) Eutrofiëring in de zee. Effecten en mogelijkheden voor een beter ecologisch beheer. In: Bijlsma A (ed) Integraal waterbeheer. Een nieuwe aanpak. NIBI, Utrecht, pp 83–102

Lindeboom HJ, Van Raaphorst W, Beukema JJ, Cadée GC, Swennen C (1996) (Sudden) changes in the North Sea and Wadden Sea: Oceanic influences underestimated? In: Andersen J, Karup H, Nielsen UB (eds) Scientific Symposium on the North Sea Quality Status Report 1993. Ministry of Environment and Energy, Copenhagen, pp 79–85

Livingstone DR, Moore MN, Widdows J (1988) Ecotoxicology: Biological effects measurements on molluscs and their use in impact assessment. In: Salomons W, Bayne WN, Duursma EK, Förstner U (eds) Pollution of the North Sea; An assessment. Springer, Berlin, Heidelberg, pp 624–637

LoCicero VR (1975) Oceanographic conditions associated with red tide blooms; sessions summary. In: LoCicero VR (ed) Proceedings of the first international conference on toxic dinoflagellate blooms. Mass Sci Techn Found, Massachusetts, pp 7–9

Lomborg B (2001) The skeptical environmentalist. Cambridge University Press, Cambridge

Lucht F, Gillbricht M (1978) Long-term observations on nutrient contents near Helgoland in relation to nutrient input of the river Elbe. Rapp P-v Reun Cons int Explor Mer 172: 358–360

Ludwig HF , Onodera B (1964) Scientific parameters of marine waste discharge. In: Pearson EA (ed) Advances in Water Pollution Research. Volume 3. Pergamon, Oxford, London, New York, Paris, pp 37–49

MacGarvin M (1995) The implications of the precautionary principle for biological monitoring. Helgoländer Meeresunters 49: 647–662

Machlis GE (1992) The contribution of sociology to biodiversity research and management. Biological Conservation 62: 161–170

MacKenzie D (1996) Norway's fish plan "a recipe for disaster". New Scientist 149(2012): 4

Mangel M et al. (1996) Principles for the conservation of wild living resources. Ecological applications 6(2): 338–362

Margalef R, Estrada M, Blasco D (1979) Functional morphology of organisms involved in red tides, as adapted to decaying turbulence. In: Taylor DL, Seliger HH (eds) Proceedings of the 2nd conference on toxic dinoflagellate blooms. Elsevier, New York, Amsterdam, Oxford, pp 89–94

Masood E (1996) Scientific caution "blunts efforts" to conserve fish stocks. Nature 379: 481

McErlean AJ, Reed G (1981) Indicators of estuarine enrichment. In: Neilson BJ, Cronin LE (eds) Estuaries and Nutrients. Humana Press, Clifton. New Jersey, pp165–182

McIntyre AD (1978) The benthos of the western North Sea. Rapp P-v Reun Cons int Explor Mer 172: 405–417

McKee JE (1967) Parameters of Marine Pollution – An Overall Evaluation. In: Olson TA, Burgess FJ (eds) Pollution and Marine Ecology. Interscience, New York, London, Sydney, pp 259–266

Melvasalo T, Pawlak J, Grasshoff K, Thorell L, Tsiban A (eds) (1981) Assessment of the effects of pollution on the natural resources of the Baltic Sea, 1980. Baltic Sea Environment Proceedings No. 5B, Baltic Marine Environment Protection Commission, Helsinki

Michaelis H (1977) Untersuchungen über das Sterben wirbellosen Wattentiere seit dem Sommer 1976. Unpublished report Forschungsstelle f. Insel- und Küstenschutz Norderney, Germany

Michaelis H (1997) History and distribution of black spots in the east Friesian Wadden Sea. In: Henke S (ed) Black spots in the Wadden Sea. Texte 3/97, Federal Environmental Agency, Berlin

Mihnea PE (1979) Some specific features of dinoflagellate *Exuviaella cordata* Ostf. blooming in the Black Sea. In: Taylor DL, Seliger HH (eds) Proceedings of the 2nd conference on toxic dinoflagellate blooms. Elsevier, New York, Amsterdam, Oxford, pp 77–82
Miles EL (1989) Scientific and technological knowledge and international cooperation in resource management. In: Andresen S, Østreng W (eds) International resource management: the role of science and politics. Belhaven Press, London, New York, pp 46–87
Miljøstyrelsen (1984a) Iltsvind og fiskedød-omfang og årsager. Miljøstyrelsen, Copenhagen
Miljøstyrelsen (1984b) Opgørelse af belastning fra land af de indre danske farvande med organisk stof, total-N og total-P. Miljøstyrelsen, Copenhagen
Milway CP (1968) Eutrophication in large lakes and impoundments. OECD, Paris
Moore B (1964) Water pollution and Minamata disease; Discussion. In: Pearson EA (ed) Advances in Water Pollution Research. Volume 3. Pergamon, Oxford, London, New York, Paris, pp 302–308
Nanninga HJ (1997) Invloed van stikstof en fosfor op de visstand in de Noordzee. RUG, Vakgroep Mariene Biologie, Groningen
Nehring D (1981) Phosphorus in the Baltic Sea. Mar Poll Bull 12(6): 194–198
Neilson BJ, Cronin LE (1981) Preface. In: Neilson BJ, Cronin LE (eds) Estuaries and Nutrients. Humana Press, Clifton, New Jersey, pp ix–x
Nelissen PHM, Stefels J (1988) Eutrophication in the North Sea. NIOZ-rapport 1988-4, NIOZ, Texel
Nelkin D (1975) Scientists and professional responsibility: the experience of American ecologists. Social studies of science 7: 75–95
Nelkin D (1987) Controversies and the authority of science. In: Engelhardt HT, Caplan AL (eds) Scientific controversies. Case studies in the resolution and closure of disputes in science and technology. Cambridge University Press, Cambridge, pp 283–293
NERI (2003) Evaluation of NERI's research on marine eutrophication. www2.dmu.dk/1_Nyt/2_Akt_Debat/panelets%20report-final-05-21.pdf
NEUT (1998) Working Group on Nutrients and Eutrophication. Summary Record. OSPAR Secretariat, London
Newman PJ, Agg AR (1988) Editor's preface. In: Newman PJ, Agg AR (eds) Environmental Protection of the North Sea. Heinemann, Oxford, pp xiii–xv
Nielsen A, Richardson K (1990) *Chrysochromulina polylepis* bloom in Danish, Swedish and Norwegian waters, May–June 1988 – an analysis of extent, effects and causes. In: Lancelot C, Billen G, Barth H (eds) Eutrophication and algal blooms in North Sea coastal zones, the Baltic and adjacent areas: prediction and assessment of preventive actions. Water Pollution Research Report 12, CEC, Brussels, pp 11–25
Niermann U, Bauernfeind E, Hickel W, Von Westernhagen H (1990) The recovery of benthos following the impact of low oxygen content in the German Bight. Neth J Sea Res 25(1/2): 215–226
Nilsen HG, Aarefjord H, Øverland S, Rukke J (2002) Progress Report fifth international conference on the protection of the North Sea. Ministry of the Environment, Oslo
Nitta T (1972) Marine pollution in Japan. In: Ruivo M (ed) Marine Pollution and Sea Life. Fishing News Ltd, Surrey, London, pp 77–82
Nixon SW (1981) Remineralization and nutrient cycling in coastal marine ecosystems. In: Neilson BJ, Cronin LE (eds) Estuaries and Nutrients. Humana Press, Clifton, New Jersey, pp 111–138
Nixon SW (1990) Marine Eutrophication: a growing international problem. Ambio 19(3): 101

Nixon SW (1995) Coastal marine eutrophication: A definition, social causes, and future consequences. Ophelia 41: 199–219
Nixon SW, Buckley BA (2002) "A strikingly rich zone" – Nutrient enrichment an secondary production in coastal marine ecosystems. Estuaries 25(4b): 782–796
Nordic Council. (1999) Workshop on the ecological quality objectives (EcoQOs) for the North Sea. Tema Nord 199:591, Nordic Council of Ministers, Copenhagen
North WJ (1964) Ecology of the rocky nearshore environment in southern California and possible influences of discharged wastes. In: Pearson EA (ed) Advances in Water Pollution Research. Volume 3. Pergamon, Oxford, London, New York, Paris, pp 247–261
Nowotny H (1987) A new branch of science Inc. In: Brooks H, Cooper CL (eds) Science for Public Policy. Pergamon Press, Oxford, New York, pp 61–77
Nowotny H (1999) The need for socially robust knowledge. TA-Datenbank-Nachrichten 3/4: 12–16
Nowotny H (2005) High- and low-cost realities for science and society. Science 308: 1117–1118
NSTF (1988) Summary Record 1st Meeting North Sea Task Force. The Hague, 7–9 December 1988. Secretariat Oslo and Paris Commissions, London
NSTF (1989a) Summary Record 2nd Meeting North Sea Task Force. Plymouth, 18–21 April 1989. Secretariat Oslo and Paris Commissions, London
NSTF (1989b) Summary Record 3rd North Sea Task Force. Copenhagen, 6–8 September 1989. Secretariat Oslo and Paris Commissions, London
NSTF (1991a) Summary Record 6th meeting North Sea Task Force. Secretariat NSTF, London
NSTF (1991b Summary Record) 7th meeting North Sea Task Force. NSTF Secretariat, London
NSTF (1992) North Sea Task Force document HC/OSPAR/3-4-5/5-Add.1
NSTF (1993) North Sea Quality Status Report 1993. Oslo and Paris Commissions, London
NUT (1986) Summary Record 1st Meeting Working Group on Nutrients, Berlin 11–13 November 1986. Secretariat Oslo and Paris Commissions, London
NUT (1987a) Summary Record 2nd Meeting Working Group on Nutrients, Stockholm, 27–29 October 1987. Secretariat Oslo and Paris Commissions, London
NUT (1987b) Nutrients in the southeastern North Sea, particularly in the German Bight. Document NUT 2/Info 6, Second Meeting of the Working Group on Nutrients, Stockholm, 27–29 October 1987. Secretariat Oslo and Paris Commissions, London
NUT (1988) Summary Record 3rd Meeting Working Group on Nutrients, The Hague, 24–27 October 1988. Secretariat Oslo and Paris Commissions, London
NUT (1989) Summary Record 4th Meeting Working Group on Nutrients, Oslo, 26–29 September 1989. Secretariat Oslo and Paris Commissions, London
NUT (1992) Extract from the 1992 report of the ICES working group on shelf seas oceanography. Document NUT 7/7/3. Oslo and Paris Commissions, London
Odum HT (1967) Biological circuits and the marine systems of Texas. In: Olsen TA, Burgess FJ (eds) Pollution and marine ecology. Interscience, New York, London, Sydney, pp 99–158
Oglesby RT (1967) Biological basis. In: Olsen TA, Burgess FJ (eds) Pollution and marine ecology. Interscience, New York, London, Sydney, pp 267–271
Olsen TA, Burgess FJ (1967) Preface. In: Olsen TA, Burgess FJ (eds) Pollution and marine ecology. Interscience, New York, London, Sydney, pp v–viii
OSPAR (1984) The Oslo and Paris Commissions: History and Progress (1974–1984). In: Van der Burgt C (ed) The Oslo and Paris Commissions. The first decade. Oslo and Paris Commissions, London, pp 5–39

OSPAR (1993) Eutrophication symptoms and problem areas. OSPAR Commission, London
OSPAR (1995) Document OSPAR 95/5/9, Annex 2. OSPAR Commission, London
OSPAR (1997) Summary Record of the Meeting of OSPAR 1997 (OSPAR 97/15/1), Annex 24. OSPAR Commission, London
OSPAR (1998) The OSPAR Strategy to combat eutrophication. Summary Record OSPAR 98/14/1-E, Annex 36. OSPAR Commission, London
OSPAR (2000) Quality Status Report 2000. Region II, Greater North Sea. OSPAR Commission, London
OSPAR (2002) Common assessment criteria, their (region specific) assessment levels and guidance on their use in the area classification within the Comprehensive Procedure of the Common Procedure. Document OSPAR 02/8/2. OSPAR Commission, London
OSPAR (2003a) Report on the eutrophication status of the OSPAR maritime area based upon the first application of the Comprehensive Procedure. Document OSPAR MMC 2003/2/4. OSPAR Commission, London
OSPAR (2003b) Declaration of the first joint interministerial meeting of the Helsinki and OSPAR Commissions. OSPAR Commission, London
OSPAR (2005) Summary Record OSPAR 2005. OSPAR 05/21/1-E, OSPAR Commission, London
Parker M (1983) Exceptional marine blooms and their implications for fisheries. Annex 1. In: ICES Report of the Advisory Committee on Marine Pollution 1983. Cooperative Research Report 124, ICES, Copenhagen
Paulik GJ (1967) Digital simulation of natural animal communities. In: Olsen TA, Burgess FJ (eds) Pollution and marine ecology. Interscience, New York, London, Sydney, pp 67–85
Paulsson V (1972) Preface. In: Åkerblom A (ed) 1st Soviet-Swedish Symposium on the pollution of the Baltic. Ambio Special Report No. 1, p 5
Pawlak J (1980) Land-based inputs of some major pollutants to the Baltic Sea. Ambio 9(3–4): 163–167
Pearce F (2004) A mirror to cool the world. New scientist 181(2440): 26–29
Pearson EA (1960) Introduction. In: Pearson EA (ed) Proceedings of the first international conference on waste disposal in the marine environment. Pergamon, Oxford, London, New York, Paris, pp 1–3
Pearson TH, Josefson AB, Rosenburg R (1985) Petersen's benthic stations revisited. I. Is the Kattegatt becoming eutrophic? J Exp Mar Biol Ecol 92: 157–206
Peet JG (1984) The 1984 North Sea Conference, a preview. Marine Policy 8(3): 259–270
Peet JG (1988) The anticipation principle as a basis for policy? In: Newman PJ, Agg AR (eds) Environmental Protection of the North Sea. Heinemann, Oxford, pp 503–528
Peperzak L, Van Nieuwerburg L (1998) Schuim op het strand: wat is daar aan de hand? Zoutkrant 12(4): 4
Perkins EJ, Abbott OJ (1972) Nutrient enrichment and sand flat fauna. Mar Poll Bull 3(5): 70–72
Peters RH (1991) A critique for ecology. Cambridge University Press, Cambridge
Philippart CJM, Cadée GC, van Raaphorst W, Riegman W (2000) Long-term phytoplankton-nutrient interactions in a shallow coastal sea: algal community structure, nutrient budgets, and denitrification potential. Limnology and Oceanography 45: 131–144
Planque B, Taylor AH (1998) Long-term changes in zooplankton and the climate of the North Atlantic. ICES J Mar Sci 55: 644–654
Pomeroy LR (1967) Discussion: Energy Transfer. In: Olson TA, Burgess FJ (eds) Pollution and Marine Ecology. Interscience, New York, London, Sydney, pp 184–185

Popper FE (1972) Preface. In: Ruivo M (ed) Marine Pollution and Sea Life. Fishing News Ltd., Surrey, London, p ix

Porritt J (1993) Translating ecoligical science into practical policy. In: Edwards PJ, May RM, Webb NR (eds) Large-scale ecology and conservation biology. Blackwell Science, Oxford, pp 345–353

Postma H (1967) Marine Pollution and Sedimentology. In: Olson TA, Burgess FJ (eds) Pollution and Marine Ecology. Interscience, New York, London, Sydney, pp 225–233

Postma H (1973) Transport and budget of organic matter in the North Sea. In: Goldberg ED (ed) North Sea Science. MIT press, Cambridge Mass, London, pp 326–334

Postma H (1978) The nutrient contents of North Sea water: Changes in recent years, particularly in the Southern Bight. Rapp P-v Reun Cons int Explor Mer 172: 350–357

Prakash A (1975) Dinoflagellate blooms. An overview. In: LoCicero VR (ed) Proceedings of the first international conference on toxic dinoflagellate blooms. Mass Sci Techn Found, Massachusetts, pp1–6

Pravdic V (1985) Environmental capacity. Is a new scientific concept acceptable as a strategy to combat marine pollution? Mar Poll Bull 16(7): 296–296

Provasoli L (1979) Recent progress, an overview. In: Taylor DL, Seliger HH (eds) Proceedings of the 2nd conference on toxic dinoflagellate blooms. Elsevier, New York, Amsterdam, Oxford, pp 1–14

Rachor E, Albrecht H (1983) Sauerstoff-Mangel im Bodenwasser der Deutschen Bucht. Veröff Inst Meeresforsch Bremerh 19: 209–227

Radach G (1998) Quantification of long-term changes in the German Bight using an ecological development index. ICES J Mar Sci 55: 587–599

Rasmussen MB, Andersen J (eds) (2003) Marine områder 2002 – Miljøtilstand og udvikling. NOVA 2003. Faglig rapport fra DMU nr. 467

Rat von Sachverständigen (1980) Umwelprobleme der Nordsee. Sondergutachten. Der Rat von Sachverständigen für Umweltfragen, Kohlhammer, Stuttgart, Mainz

Rawn AM (1960) Fixed and changing values in ocean disposal of sewage and wastes. In: Pearson EA (ed) Proceedings of the first international conference on waste disposal in the marine environment. Pergamon, Oxford, London, New York, Paris, pp 6–11

Raymond JEG , Shields J (1964) Toxicity of copper and chromium in the marine environment. In: Pearson EA (ed) Advances in Water Pollution Research. Volume 3. Pergamon, Oxford, London, New York, Paris, pp 275–282

Redfield AC, Ketchum BH, Richards FA (1963) The influence of organisms on the composition of sea water. In: Hill MN (ed) The sea: Ideas and observations on progress in the study of the seas. Vol. 2. Interscience Publishers, New York, London, pp 26–77

Reid PC (1978) Continuous plankton records: Large-scale changes in the abundance of phytoplankton in the North Sea from 1958 to 1973. Rapp P-v Reun Cons int Explor Mer 172: 384–389

Reid PC (1990) The dynamics of algal blooms in the North Sea. In: Lancelot C, Billen G, Barth H (eds) Eutrophication and algal blooms in North Sea coastal zones, the Baltic and adjacent areas: prediction and assessment of preventive actions. Water Pollution Research Report 12, CEC, Brussels, pp 57–73

Reid PC (1997) Discharges from hydroelectric power schemes as a trigger for marine algal blooms. Mar Poll Bull 34(9): 730–733

Reid PC, Edwards M (2001) Long-term changes in the pelagos, benthos and fisheries of the North Sea. Senckenbergiana maritima 31(2): 107–115

Reid PC, Edwards M, Hunt HG, Warner AJ (1998) Phytoplankton changes in the North Atlantic. Nature 391: 546

Reid PC, Lancelot C, Gieskes WWC, Hagmeyer E, Weichart G (1990) Phytoplankton of the North Sea and its dynamics: a review. Neth J Sea Res 26(2–4): 295–331

Reijnders PJH (1992) Retrospective population analysis and related future management perspectives for the harbour seal (*Phoca vitulina*) in the Wadden Sea. In: Dankers N, Smit CJ, Scholl M (eds) Proceedings of the 7th international Wadden Sea symposium. Neth Inst Sea Res Publ Series 20, pp 193–197

Rice J (2005) Marine science advice: learning from the Canadians. ICES Newsletter 42: 34–37

Richardson K, Cedhagen T (2001) Quantifying pelagic-benthic coupling in the North Sea: Are we asking the right questions? Senckenbergiana maritima 31(2): 215–224

Riegman R, Noordeloos AAM, Cadée GC (1992) *Phaeocystis* blooms and eutrophication of the continental coastal zones of the North Sea. Mar Biol 112: 479–484

RIZA (1983) De waterkwaliteit van de Noordzee, 1975–1982. Rijksinstituut voor Zuivering van Afvalwater nota nr. 83.084, RIZA, Lelystad

Robertson DP, Hull RB (2001) Beyond biology: towards a more public ecology for conservation. Conservation Biology 15(4): 970–979

Rohlich GA (1969) Preface. In: Eutrophication, Causes, Consequences, Correctives. National Academy of Sciences. Washington, pp iii–iv

Rosenberg R (1985) Eutrophication – the future marine coastal nuisance? Mar Poll Bull 16(6): 227–231

Rosenberg R, Larsson U, Edler, L (eds) (1984) Eutrophication in marine waters surrounding Sweden – A review. Statens naturvårdsverk PM 1808, Solna

Rosenberg R, Lindahl O, Blanck H (1988) Silent spring in the sea. Ambio 17(4): 289–290

Rosenberg R, Elmgren R, Fleischer S, Jonsson P, Persson G, Dahlin H (1990) Marine eutrophication case studies in Sweden. Ambio 19 (3): 102–108

Roskam PTh (1970) De verontreiniging van de zee. Chemisch weekblad 4 september 1970: 56–61

Rousseau V, Breton E, De Wachter B, Beji A, Deconcinck M, Huigh J, Bolsens T, Leroy D, Jans S, Lancelot C (2004) Identification of Belgian coastal zones affected by europhication (EZEUT) Federal Science Policy, SP 1349, Brussels

Royce WF (1967) Population Dynamics and Marine Ecology: Research Needs and Strategy. In: Olson TA, Burgess FJ (eds) Pollution and Marine Ecology. Interscience, New York, London, Sydney, pp 53–65

Ruivo M (1972) Section 1; summary of discussion. In: Ruivo M (ed) Marine Pollution and Sea Life. Fishing News Ltd, Surrey, London, pp 2–3

Ruud JT (1968) Changes since the turn of the century in the fish fauna and the fisheries in the Oslo fjord. Helgoländer wiss Meeresunters 17: 510–517

RWS (1985) Harmonisatie Noordzeebeleid. Waterkwaliteitsplan Noordzee. Achtergronddocument 2B. Rijkswaterstaat, Waterloopkundig Laboratorium

Ryther JH (1954) The ecology of phytoplankton blooms in Moriches Bay and Great South Bay, Long Island, New York. Biol Bull 106(2): 198–209

Ryther JH, Dunstan WM (1971) Nitrogen, phosphorus and eutrophication in the coastal marine environment. Science 171: 1008–1013

Ryther JH, Officer CB (1981) Impact of nutrient enrichment on water uses. In: Neilson BJ, Cronin LE (eds) Estuaries and Nutrients. Humana Press, Clifton, New Jersey, pp 247–261

Sagoff M (2003) The plaza and the pendulum: Two concepts of ecological science. Biology and philosophy 18: 529–552

Salomons W, Bayne WN, Duursma EK, Förstner U (eds) (1988) Pollution of the North Sea; An assessment. Springer, Berlin, Heidelberg

Sanger C (1972) The Great Lakes clean-up. Mar Poll Bull 3(9): 138–139

Schaub BEM, Gieskes WWC (1991) Eutrophication of the North Sea: the relation between Rhine river discharge and chlorophyll-a concentration in Dutch coastal waters. In: Elliot M, Ducrotoy J-P (eds) Estuaries and Coasts: Spatial and Temporal Intercomparisons. ECSA 19 Symposium. Olsen & Olsen, Denmark, pp 85–90

Schindler DW (1981) Studies of eutrophication in lakes and their relevance to the estuarine environment. In: Neilson BJ, Cronin LE (eds) Estuaries and Nutrients. Humana Press, Clifton, New Jersey, pp 71–82

Scholten MCTh, Jak RG, Foekema EM (1994) Ecological control of algal densities. TNO-MW-R94/280a. TNO Environmental Sciences, Den Helder

Schott F, Ehrhardt M (1969) On fluctuations and mean relations of chemical parameters in the northwestern North Sea. Kieler meeresf 25(2): 272–278

Schrader-Frechette KS, McCoy ED (1993) Method in Ecology. Cambridge University Press, Cambridge

Shindell DT, Schmidt GA, Mann ME, Rind D, Waple A (2001) Solar forcing of regional climate change during the Maunder minimum. Science 294: 2149–2152

Shindell DT, Schmidt GA, Miller RL, Mann ME (2003) Volcanic and solar forcing of climate change during the preindustrial area. Journal of Climate 16: 4094–4107

Simpson JH (1994) Introduction to the North Sea Project. In: Charnock H, Dyer KR, Huthnance JM, Liss PS, Simpson JH, Tett PB (eds) Understanding the North Sea system. Chapman & Hall, London, pp 1–4

Skjoldal HR, Dundas I (1991) The *Chrysochromulina polylepis* bloom in the Skagerrak and the kattegat in May–June 1988: Environmental conditions, possible causes, and effects. Cooperative Research Report 175, ICES, Copenhagen

Smayda TJ (1990) Novel and nuisance phytoplankton blooms in the sea: evidence for a global epidemic. In: Graneli E, Sündstrom B, Edler l, Anderson DM (eds) Toxic marine phytoplankton. Elsevier, Amsterdam, pp 29–40

Somer E (1988) Control options for nutrients. In: Newman PJ, Agg AR (eds) Environmental Protection of the North Sea. Heinemann, Oxford, pp 219–237

SRU (2004) Meeresumweltschutz für Nord- und Ostsee. Sondergutachten 2004. Der Rat von Sachverständigen für Umweltfragen. Nomos Verlag, Baden-Baden

Stebbing ARD (1981) Assimilative capacity. Mar Poll Bull 12(11):362–363

Stebbing ARD, Harris JRW (1988) The role of biological monitoring. In: Salomons W, Bayne WN, Duursma EK, Förstner U (eds) Pollution of the North Sea; An assessment. Springer, Berlin, Heidelberg, pp 655–665

Stein JE, Denison JG (1967) Limitations of Indicator Organisms. In: Olson TA, Burgess FJ (eds) Pollution and Marine Ecology. Interscience, New York, London, Sydney, pp 323–335

Strong MF (1972) The Stockholm Conference – where science and politics meet. Ambio 1(3): 73–78

STWG (1987) Quality Status of the North Sea. Scientific and Technical Working Group of the 2nd International North Sea Conference

Sykes JB, Boney AD (1970) Pollution and inshore phytoplankton. Mar Poll Bull 1(3): 38–41

Tangen K (1979) Dinoflagellate blooms in Norwegian waters. In: Taylor DL, Seliger HH (eds) Proceedings of the 2nd conference on toxic dinoflagellate blooms. Elsevier, New York, Amsterdam, Oxford, pp 179–182

Tardent P (1993) Meeresbiologie. Eine Einführung. Thieme Verlag, Stuttgart, New York.

Ten Hallers-Tjabbes CC (1986) Introduction. In: Peet G (ed) Reasons for concern. Proceedings of the 2nd North Sea Seminar, Vol 2. Werkgroep Noordzee, Amsterdam, pp 10–11

Tett PB et al. (1994) Biological consequences of tidal stirring gradients in the North Sea. In: Charnock H, Dyer KR, Huthnance JM, Liss PS, Simpson JH, Tett PB (eds) Understanding the North Sea system. Chapman & Hall, London, pp 115–130

Thomann RV (1967) Part III Discussion. In: Olson TA, Burgess FJ (eds) Pollution and Marine Ecology. Interscience, New York, London, Sydney, pp 89–95

Thomas EA (1969) The Process of Eutrophication in Central European Lakes. In: Eutrophication, Causes, Consequences, Correctives. National Academy of Sciences, Washington, pp 29–49

Tibby RB, Barnard JL (1964) Some physical and biological characteristics of open coastal waters and their relationship to waste discharge. In: Pearson EA (ed) Advances in Water Pollution Research. Volume 3. Pergamon, Oxford, London, New York, Paris, pp 219–241

Timberlake L (1989) The role of scientific knowledge in drawing up the Brundtland Report. In: Andresen S, Østreng W (eds) International resource management: the role of science and politics. Belhaven Press, London, New York, pp 117–123

Tomczak G (1968) Versuch einer Klassifikation industrieller Abfallprodukte in Bezug auf die Möglichkeit einer Versenkung auf See. Helgoländer wiss Meeresunters 17: 6–24

Topping G (1976) Sewage and the sea. In: Johnston R (ed) Marine Pollution. Academic Press, London, New York, San Francisco, pp 303–351

Tougaard S, Helweg Ovesen C (eds) 1981. Environmental problems of the Wadden Sea region. Fiskeri- og Søfartsmuseet, Esbjerg

Tunberg BG, Nelson WG (1998) Do climatic oscillations influence cyclical patterns of soft bottom macrobenthic communities on the Swedish west coast? Mar Ecol Progr Ser 170: 85–94

Underdal A (1989) The role of science in international resource management: a summary. In: Andresen S, Østreng W (eds) International resource management: the role of science and politics. Belhaven Press, London, New York, pp 253–267

Underdal B, Skulbert OM, Dahl E, Aune T (1989) Disastrous bloom of *Chrysochromulina polylepis* (Prymnesiophyceae) in Norwegian coastal waters in 1988 – Mortality in marine biota. Ambio 18(5): 265–270

Ursin E, Andersen KP (1978) A model of the biological effects of eutrophication in the North Sea. Rapp P-v Reun Cons int Explor Mer 172: 366–377

Valeskaln PI, Hannerz L (1976) Recommendation. In: Åkerblom A (ed) 2nd Soviet-Swedish Symposium on the pollution of the Baltic. Ambio Special Report No. 4, p 6

Valeskaln PI, Paulsson V (1977) Recommendation. In: Åkerblom A (ed) 3rd Soviet-Swedish Symposium on the pollution of the Baltic. Ambio Special Report No. 5, pp 8–10

Vallentyne JR (1972) Panel discussion: Nutrients and eutrophication-prospects and options for the future. In: Likens GE (ed) Nutrients and Eutrophication: The limiting Nutrient Controversy. American Society of Limnology and Oceanography, Allan Press, Lawrence, Kansas, pp 306–310

Van Bennekom AJ, Gieskes WWC, Tijssen SB (1975) Eutrophication of Dutch coastal waters. Proc R Soc Lond B 189: 359–374

Van Beusekom JEE, Diel-Christiansen S (1996) Alterations of the North Sea plankton ecosystem due to changes of the regional climate, nutrient input and contaminant load: a hypothesis. In: Andersen J, Karup H, Nielsen UB (eds) Scientific Symposium on the North Sea Quality Status Report 1993. Danish Environmental Protection Agency, pp 64–73

Van Beusekom JEE, Fock H, de Jong F, Diel-Christiansen S, Christiansen B (2001) Wadden Sea specific eutrophication criteria. Wadden Sea Ecosystem Nr. 14. Common Wadden Sea Secretariat. Wilhelmshaven, Germany

Van Beusekom JEE, Brockmann U, Elbrächter M, Pätsch J, Wiltshire K (2003) Nährstoffe und Eutrophierung. In: Lozán JL, Rachor E, Reise K, Sündermann J, Von Westernhagen H (eds) Warnsignale aus Nordsee und Wattenmeer. Wissenschaftliche Auswertungen, Hamburg, pp 182–192

Van der Burgt C (1984) The birth of a Treaty. In: Van der Burgt C (ed) The Oslo and Paris Commissions. The first decade. Oslo and Paris Commissions, London, pp 1–4

Van der Voet CG (1987) Industrial developments in effluent and discharge containment. In: Peet G (ed) Proceedings of the 2nd North Sea Seminar, Werkgroep Noordzee, Amsterdam, pp 213–230

Van der Windt HJ (1992) Environmental scientists and scientific symposia in the Wadden Sea area, a changing role. In: Dankers N, Smit CJ, Scholl M (eds) Proceedings of the 7th International Wadden Sea Symposium, Ameland, 1990. Neth Inst Sea Res Publ Ser No. 20, NIOZ, Texel, pp 117–122

Van Pagee JA, Postma L (1987) North Sea Pollution. The use of modelling techniques for impact assessment of waste inputs. In: Peet G (ed) Proceedings of the 2nd North Sea Seminar. Werkgroep Noordzee, Amsterdam, pp 97–114

Van Pagee JA, Glas PCG, Hopstaken CF, Postma L (1988) Water quality modelling of the southern North Sea: a useful tool for research and management. In: Newman PJ, Agg AR (eds) Environmental Protection of the North Sea. Heinemann, Oxford, pp 597–612

Van Weering T, Kramer K (1984) Het bedreigde zeemilieu, de Noordzee. Ekologie 1984-1, No.23

Van Wijnen JF (1995) De feiten onder water. Vrij Nederland 1 July 1995

Veldhuis MJW, Colijn F, Venekamp LAH (1986a) The spring bloom of *Phaeocystis pouchetii* (Haptophyceae) in Dutch coastal waters. Neth J Sea Res 20(1): 37–48

Veldhuis MJW, Admiraal W, Colijn F (1986b) Chemical and physiological changes of phytoplankton during the spring bloom, dominated by *Phaeocystis pouchetii* (Haptophyceae): Observations in Dutch coastal waters of the North Sea. Neth J Sea Res 20(1): 49–60

Verseck K (2000) Shockwellen aus dem All. Die Woche, 6 October 2000, p 31

Vethaak AD (1992) Diseases of flounder (*Platichthys flesus*) in the Dutch Wadden Sea and their relation to stress factors. Neth J Sea Res 29: 257–277

Vidal M, Duarte CM, Sanchez MC (1999) Coastal eutrophication research in Europe: Progress and imbalances. Mar Poll Bull 38(10): 851–854

Vollenweider RA (1970) Scientific fundamentals of the eutrophication of lakes and flowing waters, with particular reference to nitrogen and phosphorus as factors in eutrophication. OECD, Paris

Von Weizäcker E, Von Moltke K, Haigh N (1988) Towards an integrated approach. In: Newman PJ, Agg AR (eds) Environmental Protection of the North Sea. Heinemann, Oxford, pp 548–562

Von Westernhagen H, Dethlefsen V (1983) North Sea oxygen deficiency and its effects on the bottom fauna. Ambio 12(5): 264–266

Wade BA, Antonio L, Mahon R (1972) Increasing organic pollution in Kingston Harbour, Jamaica. Mar Poll Bull 3(7): 106–111

Walker DL (1988) Future prospects for water quality management. In: Newman PJ, Agg AR (eds) Environmental Protection of the North Sea. Heinemann, Oxford, pp 563–575

Wass ML (1967) Biological and Physiological Basis of Indicator Organisms and Communities. Section II: Indicators of Pollution. In: Olson TA, Burgess FJ (eds) Pollution and Marine Ecology. Interscience, New York, London, Sydney, pp 271–283

Weichart G (1968) Veränderungen des Meeres durch Abfälle und die darauf basierende Einführung von fünf Gefahrenklassen für das Einbringen von Abfällen ins Meer. Helgoländer wiss Meeresunters 17: 398–410

Wetstone GS (1987) A history of the acid rain issue. In: Brooks H, Cooper CL (eds) Science for Public Policy. Pergamon Press, Oxford, New York, pp 163–195

Wettestad J, Andresen S (1990) Science and North Sea policy-making: Organization and communication. In: Freestone D, IJlstra T (eds) The North Sea: perspectives on regional environmental co-operation. Graham & Trotman/Martinus Nijhoff, London, Dordrecht, Boston, pp 111–122

Wieking G, Kröncke I (2001) Decadal changes in macrofauna communities on the Dogger Bank caused by large-scale climate variability. Senckenbergiana maritima 31(2): 125–142

Windom H (1991) GESAMP, two decades of accomplishments. IMO, London

Winsemius P (1986) Gast in eigen huis. Samson/Tjeenk Willink

Winter T (1972) Panel discussion: Nutrients and eutrophication-prospects and options for the future. In: Likens GE (ed) Nutrients and Eutrophication: The limiting Nutrient Controversy. American Society of Limnology and Oceanography, Allan Press, Lawrence, Kansas, pp 304–306

Worster D (1994) Nature's Economy. A history of ecological ideas. Cambridge University Press, Cambridge

WWF (1996a) Fundamental comments on the draft Common Procedure. OSPAR document EUT(1) 96/2/Info 2, OSPAR Commission, London

WWF (1996b) Maricult. A European research project contradicting OSPAR policy on reduction of nutrient inputs? Document OSPAR 96/5/NGO2-E, OSPAR Commission, London

Young O (1989) Science and social institutions: lessons for international resource regimes. In: Andresen S, Østreng W (eds) International resource management: the role of science and politics. Belhaven Press, London, New York, pp 7–24

Zevenboom W, de Vries I (1996) Assessment of eutrophication and prospectives. In: Andersen J, Karup H, Nielsen UB (eds) Scientific Symposium on the North Sea Quality Status Report 1993. Danish Environmental Protection Agency, pp 169–177

Zevenboom W, Bommelé M, Reuther C (2003) Signaleren, beoordelen, evalueren en reageren. Rapport NZ-2003/10. Rijkswaterstaat, Directie Noordzee

Zierhofer W (1999) Die Ruhe nach dem Sturm: Eine Evaluation der Verarbeitung des Waldsterbens in der Forschung und in den Massenmedien der Schweiz. Gaia 8(1): 8–18

Zijlstra JJ, De Wolf P (1988) Natural events. In: Salomons W, Bayne WN, Duursma EK, Förstner U (eds) Pollution of the North Sea; An assessment. Springer, Berlin, Heidelberg, pp 164–180

Zobell CE (1964) The occurrence, effects and fate of oil polluting the sea. In: Pearson EA (ed) Advances in Water Pollution Research. Volume 3. Pergamon, Oxford, London, New York, Paris, pp 85–109

Zorita E, von Storch H, Gonzales-Rouco FJ, Cubasch U, Luterbacher J, Legutke S, Fischer-Bruns I, Schlese U (2003) Simulation of the climate of the last five centuries. GKSS report 2003/12. GKSS-Forschungszentrum

Annex 1

Development of international marine pollution control regulations in the 1970s

The Oslo Convention

"The Oslo Convention for the Prevention of Marine Pollution by Dumping from Ships and Aircraft" was signed 15 February 1972. It was the first multilateral treaty in the field of environmental protection, covering substances other than oil. The Convention had been prepared and finalised within a remarkably short period of time. It was only in January 1971 that Norway had discussed problems of dumping of waste with the other Nordic countries Denmark, Finland, Iceland and Sweden. The worries of the Nordic countries had been caused by the publication of the ICES North Sea report (see 2.5.1) and press releases about dumping of chemical waste from western Europe into the North Sea and the Northeast Atlantic Ocean (Van der Burgt 1984). The Nordic countries decided to ban the dumping of persistent and harmful substances and to invite the members of the Northeast Atlantic Fisheries Convention (NEAFC) to do the same. To this end these members were invited by Norway to a meeting in Oslo in October 1971. With the aim of preparing common proposals for this meeting, France, the United Kingdom, The Netherlands, Germany and Belgium came together three times in 1971. These states sympathised with the idea of regulating ocean dumping, although they did not support a complete ban. They succeeded in coming to a common viewpoint about which substances could be dumped and under which circumstances.

Public opinion had played an important role in this achievement: in July 1971 there had been much uproar about the intended dumping of chlorinated hydrocarbons into the North Sea (Van der Burgt, loc.cit.). Also the political preparations for the Stockholm Conference (2.6.1) had a positive influence on the negotiations. The proposals of the five countries were brought in at the Oslo conference (19–22 October 1971) and in which, next to the already mentioned five west-European countries, Spain, Portugal, Norway, Sweden, Denmark, Ireland and Iceland participated. The result of this conference was a draft convention text, which, after some legal polishing, was signed on 15 February 1972. It entered into force in April 1974.

The Oslo Convention, which covers the waters of the Northeast Atlantic Ocean, including the North Sea but not including the Baltic Sea, principally applies a system of licensing of substances to be discharged or dumped. The licensing is based upon so-called black- and grey-listed substances. Black-list substances, contained in Annex 1 to the Convention, may in principle not be dumped or discharged. The black list contains organohalogen compounds, mercury, cadmium, persistent oils and plastics. Grey list substances, listed in Annex 2, are, among others, organic phosphorus and tin compounds, the heavy metals copper, chromium, zinc and lead, and non-persistent plastics, and may only be dumped or discharged with a license. In a third Annex the conditions for dumping to be specified in the license, are given. The Convention also obliges contracting parties to establish complementary or joint scientific and technical research programmes and to institute complementary or joint monitoring programmes to monitor the distribution and effects of pollutants. For the implementation of the Convention a Commission (Oslo Commission: Oscom) was set up, which held its first meeting in October 1974. To facilitate the work of the commission the Standing Advisory Committee for Scientific Advice (SACSA) and a secretariat, based in London, were established.

The London Dumping Convention

In November 1972, at a conference in London, 80 states reached agreement about a global convention on marine dumping, the London Dumping Convention (LDC). The LDC is structured in the same way as the Oslo Convention, according to a black list, containing substances which may not be dumped and a grey list for substances which may only be dumped with a license. Also the contents of the lists from both Conventions is similar to a large extent.

The Paris Convention

Upon the invitation of the French Government a diplomatic conference was held in December 1972, in which those states participated that were signatories to the Oslo Convention. Also the EEC took part in the negotiations. The aim of the conference was to come to an agreement about a convention for the prevention of pollution of the marine environment from land-based sources. The area under consideration was the marine area covered by the Oslo Convention. After three more meetings the Paris Convention was open for signatures in June 1974. The Convention entered into force in 1978.

Also the Paris Convention applies the methodology of black and grey lists. Pollution of the marine environment by substances, which are on the black list, must be eliminated, if necessary by stages. Pollution of substances from the grey list should be strictly limited. The substances on the Paris Convention lists are identical to those of the Oslo lists. In the text of the Paris Convention the criteria for placing substances on either one of the two lists is given. The black list contains substances that are not readily biodegradable, give rise to dangerous accumulation in the food chain, endanger the welfare of living organisms causing undesirable changes in the marine ecosystems, or interfere seriously with the harvesting of sea food or other legitimate uses of the sea or finally, because it is considered that pollution by these substances necessitates urgent action. Substances on the grey list have similar characteristics, but seem less noxious or more easily biodegradable.

There is an important principal difference between the two Conventions (Ospar 1984). The Oslo Convention forbids (black list) or regulates (grey list) a deliberate and specific action, namely the dumping of waste into the marine environment. The Paris Convention, on the other hand, deals with many different activities which, in the end, may cause pollution of the marine environment. The aim to eliminate, reduce or prevent pollution of the marine environment must be achieved by means of programmes and measures, such as the setting of discharge standards. The Paris Convention can, therefore, be qualified as a framework convention. In §10 and 11 of the Convention, Contracting Parties agree to establish complementary or joint research and monitoring programmes. With regard to monitoring the agreement is more specific than the Oslo Convention. According to §11 a permanent monitoring system must progressively be set up which allows "the earliest possible assessment of the existing level of marine pollution" and "the assessment of the effectiveness of measures for the reduction of marine pollution from land-based sources."

As is the case with the Oslo Convention, the implementation of the Paris Convention is a responsibility of a commission, the Paris Commission (Parcom), which is assisted by a

Technical Working Group (TWG) and by a secretariat. The latter is the same as for the Oslo Commission. One more body is shared by the two Conventions. It is the Joint Monitoring Group (JMG), providing scientific advice about the monitoring carried out in the Convention area. In 1978 the Commissions decided to establish a Joint Monitoring Programme (JMP). ICES had played an important role in the preparation of the programme, since it had, upon request of the Commissions, carried out a baseline study for various pollutants. The programme was based upon national programmes and a number of principles developed by the JMG. In the first three years of the programme, the period 1979–1981, parameters measured were mercury, cadmium and PCBs in organisms and mercury and cadmium in water. ICES generally co-ordinated intercalibration exercises (Ospar, 1984).

The European Community

The year 1972 may be regarded as the start of the environmental policies of the EEC. At a summit of the heads of state of the nine member countries of the European Economic Community (EEC) in Paris, it was decided to establish an Environmental Action Programme. The first EEC Environmental Action Programme was adopted by the Council in November 1973 and covered the period 1973–1976. The Action Programme contained common objectives and principles of the Community environment policy, such as the objective to prevent, reduce and eliminate pollution and the "polluter pays" principle.

Special measures were announced with regard to "The serious problems posed by the pollution of certain zones of common interest (marine pollution, pollution of the Rhine basin and certain frontier zones)."

According to Johnson and Corcelle (1989) the Action Plan recognised that "of all forms of pollution, marine pollution is undoubtedly one of the most dangerous due to the effects it has on the fundamental biological and ecological balances governing life on our planet, the level of pollution already reached, the diversity of pollution sources and the difficulty of ensuring that any measures adopted are complied with." The Community actions regarding marine pollution would consist "in particular of co-ordinating and harmonising the rules for implementing international conventions, and of implementing projects to combat land-based marine pollution." It should be noted here that the Commission was observer to the Oslo Convention and involved in the preparation of the Paris Convention.

In line with the first Action Plan, a proposal for a Directive on the discharge of dangerous substances to the aquatic environment of the Community was presented to the Commission in October 1974. The aquatic environment includes inland waters, internal coastal waters and regional seas. The proposal aimed at harmonising the legal requirements of the EEC member states, which would result from three Conventions that were being drafted, namely the Paris Convention, the Strasbourg Convention on the protection of international water courses against pollution and the Rhine Convention (see below). Like the Oslo, Paris and London Dumping Conventions, this Directive applies the black and grey list approach. Black-listed substances should be eliminated and the pollution caused by grey list substances be reduced. The substances on both lists are similar to those from the lists of the above described Conventions. However, the grey list of the Directive also contains inorganic phosphorus compounds and substances which have an adverse effect on the oxygen balance, particularly ammonia and nitrites. The Directive was adopted in May 1976 after intensive negotiations between the United Kingdom on the one hand and the other EEC countries on the other. The UK maintained its position of applying Ecological Quality Objectives (EQOs) for pollutants, whereas the other member states favoured the application of Uniform Emission Standards (UESs) (see also 4.2.3). The compromise reached was that both approaches would be possible (Johnson and Corcelle,1989).

In 1973 the European Council also agreed upon the first environmental research programme of the EEC. The programme, which covered the period 1973–1976, was established in order to provide scientific and technical support to the section "reduction of pollution and nuisance" of the first Action programme (Johnson and Corcelle, loc.cit.) and contained no specific marine pollution issues.

The Rhine Convention

Already in 1950 the pollution of the Rhine had led Switzerland, France, Germany, The Netherlands and Luxembourg to establish the International Commission for the Protection of the Rhine against Pollution. The Commission was formalised through the Bern Convention of 29 April 1963. The tasks of the Commission were to carry out research into pollution, to propose appropriate measures to the governments and to prepare intergovernmental regulations. The mandate of the Commission was considerably extended by the first Rhine Ministers Conference (The Hague, 1972). It was decided that the Commission would prepare regulations regarding chemical pollution, pollution by chloride and thermal pollution. In 1973 the second Rhine Ministers Conference in Bonn discussed the progress made by the Commission in preparing a regulation for chemical pollution. In the following years the negotiations became more problematic because Germany wanted to await the developments with regard to the EEC Dangerous Substances Directive (see above). After this Directive had been agreed upon, the Convention for the Protection of the Rhine against Chemical Pollution was signed in December 1976. In 1977 also the EEC Council became party to the Convention.

In the preamble of the Convention it is stated that the considerations leading to the signing of the Convention not only concern the fact that chemical pollution of the Rhine threatens its flora and fauna, but also that it has undesired effects on seawater. The Convention is in many respects comparable to the Dangerous Substances Directive. It also applies the principle of black and grey listed substances, which are similar to those contained in the Directive. There is, however, one important difference. The Rhine Convention does not apply the dual approach of UESs and EQOs, but uses UESs only.

The Helsinki Convention

On 22 March 1974, in Helsinki, the Convention on the Protection of the Marine Environment of the Baltic Sea Area was signed by the governments of Finland, Sweden, Denmark, the Federal Republic of Germany, the German Democratic Republic, Poland and the Soviet-Union.

In Article 5 of the Convention, contracting parties agree to counteract the inputs by water, air or otherwise to the Baltic Sea Area of PCBs and DDT and its derivatives. For 16 substances, including noxious heavy metals, such as mercury, cadmium, lead, zinc and copper, certain organic substances, including persistent pesticides, lignin, oil, radioactive materials and certain floating or suspended substances, it was agreed to prevent land-based pollution caused by direct and diffuse inputs (Article 6).

The Helsinki Convention was the only marine pollution convention in which eutrophication was addressed, although in rather vague terms. In Annex III of the Convention "contracting parties shall endeavour to attain the goals and apply the criteria and measures enumerated in this Annex in order to control and minimise land-based pollution." According to paragraph 1 of the Annex "Municipal sewage shall be treated in an appropriate way so that the amount of organic matter does not cause harmful changes in the oxygen content of the Baltic Sea area and the amount of nutrients does not cause harmful eutrophication of the Baltic Sea area." Paragraph 3 refers to pollution by industries, amongst which nutrients.

For the implementation of the Convention a Commission was installed, the Helsinki Commission (Helcom) and a secretariat established in Helsinki. The Helsinki Convention entered into force in 1980.

Annex 2

Common Procedure for the Identification of the Eutrophication Status of the Maritime Area of the Oslo and Paris Conventions (Source: OSPAR 1997)

Preface

This document defines a common procedure for the identification of the eutrophication status of the maritime area of the Oslo and Paris Conventions (the "Common Procedure"). The Common Procedure will be an integral part of a Strategy to Combat Eutrophication. The purpose of the Common Procedure is to characterise the maritime area in terms of problem areas, potential problem areas and non-problem areas with regard to eutrophication. Action with respect to measures required following the identification of the eutrophication status of the maritime area will be specified within a Strategy to Combat Eutrophication.

The procedures specified in this document are without prejudice to existing and future legal requirements, including European Community legislation where appropriate.

1. Introduction

The Common Procedure comprises a stepwise process. The purpose of the Common Procedure is to characterise the maritime area in terms of problem areas, potential problem areas and non-problem areas with regard to eutrophication and to enable regional comparisons of eutrophication status on a Convention-wide basis. The intention of the Common Procedure is to enable regional comparisons of eutrophication status on a common basis.

The first step in the Common Procedure comprises a screening procedure. This is a preliminary ("broad brush") process which is likely to be applied once only in any given area. The screening procedure is intended to identify those areas which in practical terms are likely to be non-problem areas with regard to eutrophication, but for which there is insufficient information to apply the comprehensive procedure.

Following the application of the screening procedure, all areas which are not identified as non-problem areas with regard to eutrophication shall be subject to the comprehensive procedure and monitoring shall be undertaken in accordance with the minimum monitoring requirements for potential problem areas with regard to eutrophication in accordance with the Nutrient Monitoring Programme[1].

The second step in the Common Procedure is the comprehensive procedure. The comprehensive procedure is an iterative procedure and may be applied as many times as necessary. The outcome of the comprehensive procedure should enable a classification of the maritime area in terms of problem areas, potential problem areas and non-problem areas with regard to eutrophication.

---

[1] The Nutrient Monitoring Programme was adopted by OSPAR 1995 (cf. OSPAR 95/15/1, Annex 12)

The screening procedure is to be applied to all areas for which there is insufficient information to apply the comprehensive procedure. The selection of the size of the area to be assessed using the screening procedure is critical. Selection of areas should take into account hydrodynamic characteristics and proximity to nutrient sources. It is for the Contracting Parties concerned to decide on the size of the areas to be assessed.

2. Aim

The purpose of the Common Procedure is to characterise the maritime area in terms of problem areas, potential problem areas and non-problem areas with regard to eutrophication in accordance with the assessment procedure specified at Section 4. These areas are defined as follows:
a. problem areas with regard to eutrophication are those areas for which there is evidence of an undesirable disturbance to the marine ecosystem due to anthropogenic enrichment by nutrients;
b. potential problem areas with regard to eutrophication are those areas for which there are reasonable grounds for concern that the anthropogenic contribution of nutrients may be causing or may lead in time to an undesirable disturbance to the marine ecosystem due to elevated levels, trends and/or fluxes in such nutrients;
c. non-problem areas with regard to eutrophication are those areas for which there are no grounds for concern that anthropogenic enrichment by nutrients has disturbed or may in the future disturb the marine ecosystem.

3. The Screening Procedure

In their assessment of eutrophication status Contracting Parties are invited to obtain information to the extent possible for the following types of information, *inter alia*:
a. demographic/hydrodynamic/physical information

- demographic data: population and waste water treatment;
- agriculture and industry;
- hydrodynamic/physical features (for example fronts, upwelling, turbidity, flushing rates, residence times, water transport and currents);

b. optical observations

- relevant optical observations made by ship, aircraft or satellite (for example the presence of, or evidence to the contrary of, algal blooms or fish kills);

c. nutrient-related information

- voluntary data held by ICES, such as nutrient concentrations from international research cruises. ICES data is useful for screening large areas, but in coastal areas, fjords and small estuaries other data may be more appropriate (although such data may not be easily available);
- input data (for example, atmospheric inputs, riverine inputs or direct discharges);
- nutrient budgets (including the total nutrient component and the anthropogenic nutrient component);
- information from monitoring carried out under European Community Directives (where applicable).

When applying the screening procedure Contracting Parties are encouraged to use the sequence of information types specified at points a–c above. Reporting procedures are specified at Section 5.1.

4. The Comprehensive Procedure

4.1 Scope of the comprehensive procedure
The comprehensive procedure should be applied to all areas except those classified as non-problem areas with regard to eutrophication following the application of the screening procedure described in Section 3. Repeated applications of the comprehensive procedure should identify any change in the eutrophication status of a particular area.

4.2 Principles of the comprehensive procedure
The comprehensive procedure consists of a set of assessment criteria that may be linked to form an holistic assessment of the eutrophication status of the maritime area. The biological, chemical and physical assessment criteria may be organised into five categories of information.
These categories comprise:
a. the causative – nutrient enrichment related – factors;
 and
b. the supporting environmental factors;
 which together produce
c. the direct effects of nutrient enrichment;
d. the indirect effects of nutrient enrichment;
 and
e. other possible effects of nutrient enrichment.
It should be noted however that some anthropogenic activities other than those leading to nutrient enrichment may result in a number of these effects. The different assessment parameters in each category are listed at Section 4.2.1, the assessment process that links the assessment parameters is described at Section 4.2.2 and the application of quantitative assessment criteria is described at Section 4.2.3.

4.2.1 checklist for an holistic assessment
The qualitative assessment parameters are as follows:
a. the causative factors
 the degree of nutrient enrichment
- with regard to inorganic/organic nitrogen
- with regard to inorganic/organic phosphorus
- with regard to silicon

 taking account of:
- sources (differentiating between anthropogenic and natural sources)
- increased/upward trends in concentration
- elevated concentrations
- increased N/P, N/Si, P/Si ratios
- fluxes and nutrient cycles (including across boundary fluxes, recycling within environmental compartments and riverine, direct and atmospheric inputs)

b. the supporting environmental factors, including:
- light availability (irradiance, turbidity, suspended load)
- hydrodynamic conditions (stratification, flushing, retention time, upwelling, salinity, gradients, deposition)
- climatic/weather conditions (wind, temperature)
- zooplankton grazing (which may be influenced by other anthropogenic activities)

c. the direct effects of nutrient enrichment
 i. phytoplankton;

- increased biomass (e.g. chlorophyll a, organic carbon and cell numbers)
- increased frequency and duration of blooms
- increased annual primary production
- shifts in species composition (e.g. from diatoms to flagellates, some of which are nuisance or toxic species)

ii. macrophytes, including macroalgae;

- increased biomass
- shifts in species composition (from long-lived species to short-lived species, some of which are nuisance species)
- reduced depth distribution

iii. microphytobenthos;

- increased biomass and primary production

d. the indirect effects of nutrient enrichment

i. organic carbon/organic matter;

- increased dissolved/particulate organic carbon concentrations
- occurrence of foam and/or slime
- increased concentration of organic carbon in sediments (due to increased sedimentation rate)

ii. oxygen;

- decreased concentrations and saturation percentage
- increased frequency of low oxygen concentrations
- increased consumption rate
- occurrence of anoxic zones at the sediment surface ("black spots")

iii. zoobenthos and fish;

- mortalities resulting from low oxygen concentrations

iv. benthic community structure;

- changes in abundance
- changes in species composition
- changes in biomass

v. ecosystem structure;

- structural changes

e. other possible effects of nutrient enrichment

i. algal toxins (still under investigation – the recent increase in toxic events may be linked to eutrophication)

4.2.2 Principles for using the qualitative assessment parameters

4.2.2.1 selection of the qualitative assessment parameters

Regional differences with respect to demographic and hydrodynamic conditions will influence the selection of assessment parameters for different areas. Since it is the intention of the Common Procedure to enable regional comparisons of eutrophication status on a common basis, Contracting Parties shall harmonise the selection of assessment parameters to the extent possible. The basic assessment parameters to be used for assessment throughout the whole maritime area are those contained in the Nutrient Monitoring Programmeme. Additional parameters (e.g. the list at appendix 1) may be applied where necessary to aid the assessment process and to increase our current understanding. Assessments can take account of information supplied from monitoring, research and modelling.

4.2.2.2 links between the assessment parameters

The overall assessment of the eutrophication status of an area will take into account the interaction of the causative – nutrient-enrichment related – factors and the supporting environmental factors (cf. 4.2.1). For example, apart from nutrients, sufficient light is required to allow phytoplankton to grow and reduced zooplankton grazing could allow increased phytoplankton biomass. Linking these categories of information will enable the cause of the direct and indirect effects of nutrient enrichment to be established and will allow appropriately targeted measures to be applied where necessary. Control measures are generally applied to the causative – nutrient-enrichment related – factors as these are the factors most directly influenced by anthropogenic activities.

4.2.3 Application of the quantitative assessment criteria

All relevant assessment parameters should be considered when applying the comprehensive procedure, although there is a need to recognise that regional differences (for example in terms of hydrography) and differences in data availability are likely to affect the assessment parameters actually used in the assessment procedure. It should also be noted that although the assessment tools (eg. background/reference concentrations) may be region-specific the methodology for applying the assessment criteria is based on a common approach.

Many areas are likely to be assessed using a stepwise approach: a preliminary investigation using the screening procedure followed by the comprehensive procedure. The stepwise approach has several advantages including *inter alia*:

a. the outcome of the screening procedure applied as a broad brush technique to a large area may, in some cases, indicate areas for which more detailed investigations using the comprehensive procedure would be appropriate;

b. the outcome of the screening procedure may help focus the selection of assessment parameters for use in the comprehensive procedure;

c. the outcome of the screening procedure may be of use in helping to refine particular assessment criteria.

Areas for which there is much existing information (for example parts of the North Sea) are likely to be subject to the comprehensive procedure at an earlier date than areas for which there is little information. Nevertheless the first iteration of the comprehensive procedure should be undertaken soon after applying the screening procedure. This is particularly important for areas which will be identified as problem areas and potential problem areas with regard to eutrophication, since it will be necessary to start rapidly appropriate monitoring activities and to initiate action programmemes in these areas.

It should be pointed out that despite large anthropogenic nutrient inputs and high nutrient concentrations an area may exhibit few if any adverse effects. However, Contracting Parties should take into account the risk that nutrients input may be transferred to adjacent areas where they can cause detrimental environmental effects and Contracting Parties shall recognise problem areas and potential problem areas with regard to eutrophication outside their national jurisdiction.

5. Reporting

5.1 Screening Procedure

When reporting on their application of the screening procedure Contracting Parties are required to explain their selection of areas and information types. For each selected area, the responsible Contracting Party should prepare a statement which summarises the relevant information available and concludes whether, on the basis of that information, the area can be classified as a non-problem area or is an area which will need to be subject to the compre-

hensive procedure. Such statements should be examined within OSPAR no later than at OSPAR 1999.
Information available by the third quarter of 1998 may be used in the preparation of the regional quality status reports, with a view to its inclusion in the QSR 2000.

5.2 Comprehensive Procedure

In principle, reporting on the implementation of the comprehensive procedure although later in time should be in accordance with that for the screening procedure. For a given area, the outcome of the screening procedure may affect the implementation of the comprehensive procedure.

6. Timing

The timing of the implementation of the screening procedure and the comprehensive procedure is likely to vary for the Contracting Parties concerned; this reflects variations in the availability of relevant information. Nevertheless Contracting Parties shall as a minimum implement the screening procedure in accordance with the following schedule:

1. A first progress report on the implementation of the screening procedure shall be prepared, with a view to ASMO 1998 examining progress.
2. A final report on the results of applying the screening procedure shall be submitted to OSPAR 1999 for consideration.

Appendix 1

Additional assessment parameters

The additional assessment parameters may include the following:

- total nitrogen
- organic nitrogen
- organic phosphorous
- dissolved organic carbon
- dissolved organic nitrogen
- dissolved organic phosphorous
- sedimentation rate
- nutrients in sediments
- microphytobenthos (biomass and primary production)
- zoobenthos mortality
- fish mortality
- ecosystem structure
- algal toxins

Annex 3

OSPAR Strategy to Combat Eutrophication (source: OSPAR 1998)

RECALLING the Convention for the Protection of the Marine Environment of the North-East Atlantic, 1992 ("OSPAR Convention"), and in particular Article 2.1(a) in which Contracting Parties agree to take all possible steps to prevent and eliminate pollution and to take the necessary measures to protect the maritime area against adverse effects of human activities so as to safeguard human health and to conserve marine ecosystems and, when practicable, restore marine areas which have been adversely affected;
RECALLING Article 2.2 of the OSPAR Convention in which Contracting Parties agree to apply the precautionary principle and the polluter pays principle;
The Contracting Parties to the Convention for the Protection of the Marine Environment of the North-East Atlantic ADOPT the following objective and strategy for the purpose of directing the work of the Commission with regard to combating Eutrophication[2]

1. Objective

1.1 In accordance with the general objective, OSPAR's objective with regard to eutrophication is to combat eutrophication in the OSPAR maritime area, in order to achieve and maintain a healthy marine environment where eutrophication does not occur.

2. Guiding Principles

2.1 The strategy will use the following principles as a guide:
a. the precautionary principle;
b. that preventive action should be taken;
c. that environmental damage should, as a priority, be rectified at source; and
d. that the polluter should pay.

3.Strategy

3.1 Areas of the maritime area, for which actions are needed, will be identified by the Common Procedure for the Identification of the Eutrophication Status of the Maritime Area (the "Common Procedure") which will be used to characterise each part of the maritime area as a problem area or a potential problem area or a non-problem area with regard to eutrophication. In implementing the Common Procedure, the Commission will:
a. develop and adopt common assessment criteria;
b. assess the results of its application by Contracting Parties.
The identification of the eutrophication status of their parts of the maritime area will be made by Contracting Parties.

---

2 A number of terms used in this strategy are defined in Appendix 1.

3.2 Actions required, within their respective functions, by the Commission, or individually or jointly, by Contracting Parties, will depend upon that classification as follows:
a. in the case of non-problem areas with regard to eutrophication, the status of the area with regard to eutrophication will be reassessed by applying the Common Procedure if there are grounds for concern that there has been a substantial increase in the anthropogenic nutrient load;
b. in the case of potential problem areas with regard to eutrophication, preventive measures should be taken in accordance with the Precautionary Principle.
Furthermore, there should be urgent implementation of monitoring and research in order to enable a full assessment of the eutrophication status of each area concerned within five years of its being characterised as a potential problem area with regard to eutrophication;
c. in the case of problem areas with regard to eutrophication:

(i) measures shall be taken to reduce or to eliminate the anthropogenic causes of eutrophication;

(ii) reports shall be provided on the implementation of such measures;

(iii) assessments shall be made of the effectiveness of the implementation of the measures on the state of the marine ecosystem.

3.3 Actions should comprise an integrated target-oriented and source-oriented approach, as described in the following paragraphs.
3.4 The main elements of the target-orientated approach are as follows:
a. an evaluation of the situation in the maritime area that is expected following the implementation of agreed measures;
b. the development, where possible, of an agreed procedure to derive ecological quality objectives and the adoption of such objectives, possibly in the form of region-specific ecological quality objectives, aimed at avoiding harm to marine ecosystems.
Such quality objectives should reflect the state of region-specific marine ecosystems in areas for which there are no grounds for concern that anthropogenic nutrient enrichment has caused eutrophication or may in future do so. The development of appropriate assessment criteria in the Common Procedure is fundamental to the development of an agreed procedure to derive ecological quality objectives. The agreed assessment criteria with regard to non-problem areas, which should be the starting point for this development, will need to be defined in the Common Procedure.
These ecological quality objectives should be reviewed, and if necessary revised, in the light of scientific developments.
In the current state of knowledge there is limited scope for deriving ecological quality objectives because of the variability and interactions of physical and biological factors;
c. the setting of intermediate targets, in order to work towards attaining such objectives. Such targets should be combined with an indication of the size of further nutrient reductions required, estimated on the basis of an evaluation of the situation that is expected following the implementation of agreed measures, and possible means to achieve these reductions, taking into account § 3.5.
3.5 The source-oriented approach has the following main elements:
a. throughout the Convention area the following basic requirements:

(i) the implementation of any national or international measures as adopted by individual Contracting Parties for the reduction of nutrients in discharges/emissions from industry, sewage treatment plants, agriculture and other diffuse sources;

(ii) the promotion of good housekeeping in industry and sewage treatment and of good agricultural practice and ecological agriculture including proper use of the approach of aiming to strike a balance between the amounts of nutrients in the fertiliser applied and the requirements of the crop, and that proper attention is given to ammonia emissions;

b. in all areas from which nutrient inputs are likely, directly or indirectly, to contribute to inputs into problem areas with regard to eutrophication the following additional requirements:

(i) the implementation by Contracting Parties concerned of:
- PARCOM Recommendation 88/2 on the Reduction in Inputs of Nutrients to the Paris Convention Area;
- PARCOM Recommendation 89/4 on a Coordinated Programme for the Reduction of Nutrients;
- PARCOM Recommendation 92/7 on the Reduction of Nutrients Inputs from Agriculture into Areas where these Inputs are likely, directly or indirectly, to cause Pollution;
- any future OSPAR instruments updating these Recommendations;

(ii) the implementation of any further national or international measures for specific areas as adopted by individual Contracting Parties for the reduction of nutrients in discharges/emissions from industry, sewage treatment plants, agriculture and other diffuse sources;

(iii)the application of further measures, in all areas from which anthropogenic nutrient inputs to the maritime area continue to affect problem areas with regard to eutrophication or to be a cause for concern (following the implementation of the measures mentioned above and/or anticipated on the basis of § 3.4), i.e. the most appropriate combination *inter alia* of:
- BAT [Best Available Techniques] specifically designed for nitrogen and phosphorus removal from urban and industrial sewage;
- BAT and/or BEP [Best Environmental Practice]for agriculture (including horticulture), forestry and aquaculture;
- other measures relating to other sectors.

Such further measures should take into account their feasibility, cost-effectiveness, region-specific factors and seasonal factors. They should be complemented, as appropriate, by steps by the competent international bodies for the reduction of atmospheric emission of nitrogen.

c. in all areas from which nutrient inputs are likely, directly or indirectly, to contribute to inputs into potential problem areas with regard to eutrophication, preventive measures have to be taken in accordance with the precautionary principle. Contracting Parties concerned should report to the Commission on proposed action in this respect and should explain their expected results.

3.6 The source-orientated component should be developed and applied without delay.

3.7 When and where it is established that problem areas and potential problem areas with regard to eutrophication have achieved the status of non-problem areas with regard to eutrophication, measures should be kept at a level that ensures that this improved status is maintained. Ecological quality objectives, as soon as they are developed and adopted by OSPAR, could serve as complementary tools for establishing whether the measures for the reduction of nutrients at source are sufficient.

3.8 The further measures mentioned under §3.5b(iii) should include more stringent measures in areas where BAT and BEP are insufficient to achieve either the ecological quality objectives or, where applicable, the intermediate targets.

4. Timeframe

4.1 The Commission will implement this strategy progressively by making every effort to combat eutrophication in the maritime area, in order to achieve, by the year 2010, a healthy marine environment where eutrophication does not occur. To this end, the Commission will take the following immediate steps, so as to achieve:

by the year 2000

a. an evaluation of the situation in the maritime area that is expected following the implementation of agreed measures including those listed in paragraph 3.5b(i);

b. the identification of non-problem areas with regard to eutrophication through the screening procedure contained in the Common Procedure;

by the year 2002

c. the identification of the eutrophication status of all parts of the maritime area;

d. the agreement on any additional programmes and measures deemed necessary to achieve by 2010 a healthy marine environment where eutrophication does not occur, including, as appropriate, further intermediate targets for specific areas and the development of ecological quality objectives.

5. Implementation

5.1 This strategy will be implemented and the details developed under the Commission's Action Plan, which will establish priorities, assign tasks, and set deadlines and targets.

5.2 In order to facilitate this work, priority shall be given to:

a. the application of the Common Procedure to the OSPAR maritime area;

b. the development of appropriate reporting procedures;

c. the identification and quantification of the various sources of nutrients (e.g. by sector, sub-catchment, catchment, region, nation and/or other relevant subdivision);

d. the development of measures to combat eutrophication; and

e. the establishment of the direct and indirect links between the various sources of nutrients and any eutrophication problems, and hence the significance of those sources.

5.3 The implementation of this strategy will take place within the framework of the obligations and commitments of the various Contracting Parties in this field, in particular:

a. the obligations of the Member States of the European Community and the European Economic Area to implement the measures adopted for the reduction of nutrient discharges and emissions, *inter alia*, Council Directive 91/271/EEC (Urban Waste Water Directive) and Council Directive 91/676/EEC (Nitrate Directive); and the IPPC Directive (96/61/EC), and the provisions of the Council Regulation 2078/92/EEC;

b. measures stipulated in the Protocol Concerning the Control of Emissions of Nitrogen Oxides or their Transboundary Fluxes adopted within the framework of the Convention on Long-Range Transboundary Air Pollution (LRTAP Convention).

c. for those Contracting Parties concerned, the commitments of the North Sea States made at the North Sea Conferences, in particular paragraph 31 of the Esbjerg Declaration.

6. Overall Evaluation and Review of Progress

6.1 The Commission will develop appropriate machinery to enable a quinquennial review of progress achieved through this strategy. Based upon this review the Commission will, if necessary, revise the strategy. The first review should take place by the next ministerial meeting of the Commission, and should take account of *inter alia*:

a. any new information (e.g. on the links between causes and effects);

b. feedback on the effectiveness of measures;

c. the experience gained with this strategy;

d. the results of the quality assessment of the whole maritime area (i.e. QSR 2000);

e. progress achieved in the development of assessment criteria and their application within the framework of the Common Procedure.

Appendix 1

Definitions

1. For the purpose of this strategy:

a. "Eutrophication" means the enrichment of water by nutrients causing an accelerated growth of algae and higher forms of plant life to produce an undesirable disturbance to the balance of organisms present in the water and to the quality of the water concerned, and therefore refers to the undesirable effects resulting from anthropogenic enrichment by nutrients as described in the Common Procedure;

b. "anthropogenic" within the context of this strategy qualifies any human activities which:

(i) can result in, or contribute to, eutrophication in the marine environment; and

(ii) can be managed and/or whose contribution to eutrophication can be prevented, reduced or eliminated;

c. "to combat" means to prevent, reduce and, to the extent possible, eliminate;

d. "problem areas with regard to eutrophication" are those areas for which there is evidence of an undesirable disturbance to the marine ecosystem due to anthropogenic enrichment by nutrients;

e. "potential problem areas with regard to eutrophication" are those areas for which there are reasonable grounds for concern that the anthropogenic contribution of nutrients may be causing or may lead in time to an undesirable disturbance to the marine ecosystem due to elevated levels, trends and/or fluxes in such nutrients;

f. "non-problem areas with regard to eutrophication" are those areas for which there are no grounds for concern that anthropogenic enrichment by nutrients has disturbed or may in the future disturb the marine ecosystem;

2. The following working definitions, which will be reviewed from time to time in the light of further developments, are proposed for the purpose of this strategy:

a. "ecological quality" is an expression of the structure and function of the ecological system taking into account natural physiographic, geographic and climatic factors as well as biological, physical and chemical conditions including those from human activities;

b. "ecological quality reference level" is the level of ecological quality where the anthropogenic influence on the ecological system is minimal;

c. "ecological quality objective" is the desired level of ecological quality relative to the reference level.

Index

Printing: Krips bv, Meppel
Binding: Stürtz, Würzburg